T0214072

Communications
in Computer and Information Science   1086

Commenced Publication in 2007
Founding and Former Series Editors:
Phoebe Chen, Alfredo Cuzzocrea, Xiaoyong Du, Orhun Kara, Ting Liu,
Krishna M. Sivalingam, Dominik Ślęzak, Takashi Washio, Xiaokang Yang,
and Junsong Yuan

Editorial Board Members

Simone Diniz Junqueira Barbosa ⓘ
   Pontifical Catholic University of Rio de Janeiro (PUC-Rio),
   Rio de Janeiro, Brazil
Joaquim Filipe ⓘ
   Polytechnic Institute of Setúbal, Setúbal, Portugal
Ashish Ghosh
   Indian Statistical Institute, Kolkata, India
Igor Kotenko ⓘ
   St. Petersburg Institute for Informatics and Automation of the Russian
   Academy of Sciences, St. Petersburg, Russia
Lizhu Zhou
   Tsinghua University, Beijing, China

More information about this series at http://www.springer.com/series/7899

Wil M. P. van der Aalst · Vladimir Batagelj ·
Dmitry I. Ignatov · Michael Khachay ·
Valentina Kuskova · Andrey Kutuzov ·
Sergei O. Kuznetsov · Irina A. Lomazova ·
Natalia Loukachevitch · Amedeo Napoli ·
Panos M. Pardalos · Marcello Pelillo ·
Andrey V. Savchenko · Elena Tutubalina (Eds.)

Analysis of Images, Social Networks and Texts

8th International Conference, AIST 2019
Kazan, Russia, July 17–19, 2019
Revised Selected Papers

 Springer

Editors
Wil M. P. van der Aalst ⓘD
RWTH Aachen University
Aachen, Germany

Dmitry I. Ignatov ⓘD
National Research University
Higher School of Economics
Moscow, Russia

Valentina Kuskova ⓘD
National Research University
Higher School of Economics
Moscow, Russia

Sergei O. Kuznetsov ⓘD
National Research University
Higher School of Economics
Moscow, Russia

Natalia Loukachevitch ⓘD
Moscow State University
Moscow, Russia

Panos M. Pardalos ⓘD
University of Florida
Gainesville, USA

Andrey V. Savchenko ⓘD
National Research University
Higher School of Economics
Nizhny Novgorod, Russia

Vladimir Batagelj ⓘD
University of Ljubljana
Ljubljana, Slovenia

Michael Khachay ⓘD
Institute of Mathematics and Mechanics
Yekaterinburg
Yekaterinburg, Russia

Andrey Kutuzov ⓘD
University of Oslo
Oslo, Norway

Irina A. Lomazova ⓘD
National Research University
Higher School of Economics
Moscow, Russia

Amedeo Napoli ⓘD
Loria
Vandoeuvre lès Nancy, France

Marcello Pelillo ⓘD
Ca' Foscari University of Venice
Venezia Mestre, Italy

Elena Tutubalina ⓘD
Kazan Federal University
Kazan, Russia

ISSN 1865-0929   ISSN 1865-0937 (electronic)
Communications in Computer and Information Science
ISBN 978-3-030-39574-2   ISBN 978-3-030-39575-9 (eBook)
https://doi.org/10.1007/978-3-030-39575-9

© Springer Nature Switzerland AG 2020
This work is subject to copyright. All rights are reserved by the Publisher, whether the whole or part of the material is concerned, specifically the rights of translation, reprinting, reuse of illustrations, recitation, broadcasting, reproduction on microfilms or in any other physical way, and transmission or information storage and retrieval, electronic adaptation, computer software, or by similar or dissimilar methodology now known or hereafter developed.
The use of general descriptive names, registered names, trademarks, service marks, etc. in this publication does not imply, even in the absence of a specific statement, that such names are exempt from the relevant protective laws and regulations and therefore free for general use.
The publisher, the authors and the editors are safe to assume that the advice and information in this book are believed to be true and accurate at the date of publication. Neither the publisher nor the authors or the editors give a warranty, expressed or implied, with respect to the material contained herein or for any errors or omissions that may have been made. The publisher remains neutral with regard to jurisdictional claims in published maps and institutional affiliations.

This Springer imprint is published by the registered company Springer Nature Switzerland AG
The registered company address is: Gewerbestrasse 11, 6330 Cham, Switzerland

Preface

This volume contains the refereed proceedings of the 8th International Conference on Analysis of Images, Social Networks, and Texts (AIST 2019)[1]. The previous conferences during 2012–2018 attracted a significant number of data scientists – students, researchers, academics, and engineers working on interdisciplinary data analysis of images, texts, and social networks.

The broad scope of AIST made it an event where researchers from different domains, such as image and text processing, exploiting various data analysis techniques, could meet and exchange ideas. We strongly believe that this may lead to the cross-fertilisation of ideas between researchers relying on modern data analysis machinery.

Therefore, AIST brought together all kinds of applications of data mining and machine learning techniques. The conference allowed specialists from different fields to meet each other, present their work, and discuss both theoretical and practical aspects of their data analysis problems. Another important aim of the conference was to stimulate scientists and people from industry to benefit from the knowledge exchange and identify possible grounds for fruitful collaboration.

The conference was held during July 17–19, 2019. The conference was organised in Kazan, the capital of the Republic of Tatarstan, Russia, on the campus of Kazan (Volga region) Federal University[2].

This year, the key topics of AIST were grouped into six tracks:

1. General Topics of Data Analysis chaired by Sergei Kuznetsov (Higher School of Economics, Russia) and Amedeo Napoli (Loria, France)
2. Natural Language Processing chaired by Natalia Loukachevitch (Lomonosov Moscow State University, Russia), Andrey Kutuzov (University of Oslo, Norway), and Elena Tutubalina (Kazan Federal University, Russia)
3. Social Network Analysis chaired by Vladimir Batagelj (University of Ljubljana, Slovenia) and Valentina Kuskova (Higher School of Economics, Russia)
4. Analysis of Images and Video chaired by Marcello Pelillo (University of Venice, Italy) and Andrey Savchenko (Higher School of Economics, Russia)
5. Optimisation Problems on Graphs and Network Structures chaired by Panos Pardalos (University of Florida, USA) and Michael Khachay (IMM UB RAS and Ural Federal University, Russia)
6. Analysis of Dynamic Behaviour Through Event Data chaired by Wil van der Aalst (RWTH Aachen University, Germany) and Irina Lomazova (Higher School of Economics, Russia)

---

[1] http://aistconf.org.

[2] https://kpfu.ru/eng.

The Programme Committee and the reviewers of the conference included 160 well-known experts in data mining and machine learning, natural language processing (NLP), image processing, social network analysis, and related areas from leading institutions of 24 countries including Argentina, Australia, Austria, Canada, Czech Republic, Denmark, France, Germany, Greece, India, Iran, Italy, Japan, Lithuania, the Netherlands, Norway, Qatar, Romania, Russia, Slovenia, Spain, Taiwan, Ukraine, and the USA. This year, we received 134 submissions, mostly from Russia but also from Australia, Belarus, Finland, Germany, India, Italy, Norway, Pakistan, Russia, Spain, Sweden, and Vietnam.

Out of 134 submissions, only 27 full papers and 8 short papers were accepted as regular oral papers. Thus, the acceptance rate was around 24% (not taking into account 21 automatically rejected papers). An invited opinion talk and a tutorial paper are also included in LNCS volume 11832. In order to encourage young practitioners and researchers, we included 36 papers in this companion volume after their poster presentation at the conference. Each submission was reviewed by at least three reviewers, experts in their fields, in order to supply detailed and helpful comments.

The conference featured several invited talks and an industry session dedicated to current trends and challenges.

The invited talks from academia were on Computer Vision and NLP, respectively:

- Ivan Laptev (Inria and VisionLabs, France), "Towards Embodied Action Understanding"
- Alexander Panchenko (Skolkovo Institute of Science and Technology, Russia), "Representing Symbolic Linguistic Structures for Neural NLP: Methods and Applications"

The invited industry speakers gave the following talks:

- Elena Voita (Yandex, Russia), "Machine Translation: Analysing Multi-Head Self-Attention"
- Yuri Malkov (Samsung AI Center, Russia), "Learnable Triangulation of Human Pose"
- Oleg Tishutin and Ekaterina Safonova (Iponweb, Russia), "Fraud Detection in Real-Time Bidding"

The programme also included a tutorial on high-performance tools for deep models:

- Evgenii Vasilyev (Lobachevski State University of Nizhni Novgorod, Russia) and Gleb Gladilov (Intel Corporation, Russia), "Intel® Distribution of OpenVINO™ Toolkit: A Case Study of Semantic Segmentation"

An invited opinion talk on comparison of academic communities formed by the authors of Russian-speaking NLP-oriented conferences was presented by Andrey Kutuzov and Irina Nikishina under the title "Double-Blind Peer-Reviewing and Inclusiveness in Russian NLP Conferences."

We would like to thank the authors for submitting their papers and the members of the Programme Committee for their efforts in providing exhaustive reviews.

According to the programme chairs, and taking into account the reviews and presentation quality, the best paper awards were granted to the following papers:

- Track 1. General Topics of Data Analysis: "Histogram-Based Algorithm for Building Gradient Boosting Ensembles of Piece-Wise Linear Decision Trees" by Alexey Gurianov
- Track 2. Natural Language Processing: "Authorship Attribution in Russian with New High-Performing and Fully Interpretable Morpho-Syntactic Features" by Elena Pimonova, Oleg Durandin, and Alexey Malafeev
- Track 3. Social Network Analysis: "Analysis of Students Educational Interests Using Social Networks Data" by Evgeny Komotskiy, Tatiana Oreshkina, Liubov Zabokritskaya, Marina Medvedeva, Andrey Sozykin, and Nikolai Khlebnikov
- Track 4. Analysis of Images and Video: "Data Augmentation with GAN: Improving Chest X-rays Pathologies Prediction on Class-Imbalanced Cases" by Tatiana Malygina, Elena Ericheva, and Ivan Drokin
- Track 5. Optimisation Problems on Graphs and Network Structures: "Efficient PTAS for the Euclidean Capacitated Vehicle Routing Problem with Non-Uniform Non-Splittable Demand" by Michael Khachay and Yuri Ogorodnikov
- Track 6. Analysis of Dynamic Behaviour Through Event Data: "Method to Improve Workflow Net Decomposition for Process Model Repair" by Semyon Tikhonov and Alexey Mitsyuk

We would also like to express our special gratitude to all the invited speakers and industry representatives.

We deeply thank all the partners and sponsors. Besides for the hosting university, our main sponsor and the co-organiser this year was the National Research University Higher School of Economics, while Springer sponsored the best paper awards.

Our special thanks go to Springer for their help, starting from the first conference call to the final version of the proceedings. Last but not least, we are grateful to Airat Khasianov and Valery Solovyev from the Higher Institute of Information Technology and Intelligent Systems of KFU, and all the organisers, especially to Yuri Dedenev, and the volunteers, whose endless energy saved us at the most critical stages of the conference preparation.

Here, we would like to mention that the Russian word "aist" is more than just a simple abbreviation (in Cyrillic) – it means "a stork". Since it is a wonderful free bird, a symbol of happiness and peace, this stork gave us the inspiration to organise the AIST conference. So we believe that this young and rapidly growing conference will likewise bring inspiration to data scientists around the world!

October 2019

Wil van der Aalst
Vladimir Batagelj
Dmitry Ignatov
Michael Khachay
Valentina Kuskova
Andrey Kutuzov
Sergei Kuznetsov
Irina Lomazova
Natalia Loukachevitch
Amedeo Napoli
Panos Pardalos
Marcello Pelillo
Andrey Savchenko
Elena Tutubalina

Organisation

Programme Committee Chairs

Wil van der Aalst	RWTH Aachen University, Germany
Vladimir Batagelj	University of Ljubljana, Slovenia
Michael Khachay	Krasovskii Institute of Mathematics and Mechanics of Russian Academy of Sciences and Ural Federal University, Russia
Valentina Kuskova	National Research University Higher School of Economics, Moscow, Russia
Andrey Kutuzov	University of Oslo, Norway
Sergei Kuznetsov	National Research University Higher School of Economics, Moscow, Russia
Amedeo Napoli	Loria, CNRS, Inria, and University of Lorraine, France
Irina Lomazova	National Research University Higher School of Economics, Moscow, Russia
Natalia Loukachevitch	Computing Centre of Lomonosov Moscow State University, Russia
Panos Pardalos	University of Florida, USA
Marcello Pelillo	University of Venice, Italy
Andrey Savchenko	National Research University Higher School of Economics, Nizhny Novgorod, Russia
Elena Tutubalina	Kazan Federal University, Russia

Proceedings Chair

Dmitry I. Ignatov	National Research University Higher School of Economics, Moscow, Russia

Steering Committee

Dmitry I. Ignatov	National Research University Higher School of Economics, Moscow, Russia
Michael Khachay	Krasovskii Institute of Mathematics and Mechanics of Russian Academy of Sciences and Ural Federal University, Russia
Alexander Panchenko	University of Hamburg, Germany, and Université Catholique de Louvain, Belgium
Andrey Savchenko	National Research University Higher School of Economics, Nizhny Novgorod, Russia
Rostislav Yavorskiy	Surgut State University and National Research University Higher School of Economics, Russia

Programme Committee

Anton Alekseev	St. Petersburg Department of Steklov Institute of Mathematics of the Russian Academy of Sciences, Russia
Ilseyar Alimova	Kazan Federal University, Russia
Vladimir Arlazarov	Smart Engines Ltd. and Federal Research Centre "Computer Science and Control" of the Russian Academy of Sciences, Russia
Aleksey Artamonov	Neuromation, Russia
Ekaterina Artemova	National Research University Higher School of Economics, Moscow, Russia
Jaume Baixeries	Universitat Politècnica de Catalunya, Spain
Amir Bakarov	National Research University Higher School of Economics, Moscow, Russia
Vladimir Batagelj	University of Ljubljana, Slovenia
Laurent Beaudou	National Research University Higher School of Economics, Moscow, Russia
Malay Bhattacharyya	Indian Statistical Institute, India
Chris Biemann	University of Hamburg, Germany
Elena Bolshakova	Moscow State Lomonosov University, Russia
Andrea Burattin	Technical University of Denmark, Denmark
Evgeny Burnaev	Skolkovo Institute of Science and Technology, Russia
Aleksey Buzmakov	National Research University Higher School of Economics, Perm, Russia
Ignacio Cassol	Universidad Austral, Argentina
Artem Chernodub	Ukrainian Catholic University and Grammarly, Ukraine
Mikhail Chernoskutov	Krasovskii Institute of Mathematics and Mechanics of the Ural Branch of the Russian Academy of Sciences and Ural Federal University, Russia
Alexey Chernyavskiy	Samsung R&D Institute, Russia
Massimiliano de Leoni	University of Padua, Italy
Ambra Demontis	University of Cagliari, Italy
Boris Dobrov	Moscow State Lomonosov University, Russia
Sofia Dokuka	National Research University Higher School of Economics, Moscow, Russia
Alexander Drozd	Tokyo Institure of Technology, Japan
Shiv Ram Dubey	Indian Institute of Information Technology, India
Svyatoslav Elizarov	Alterra AI, USA
Victor Fedoseev	Samara National Research University, Russia
Elena Filatova	City University of New York, USA
Goran Glavaš	University of Mannheim, Germany
Ivan Gostev	National Research University Higher School of Economics, Moscow, Russia
Natalia Grabar	Université de Lille, France

Artem Grachev	Samsung R&D Institute and National Research University Higher School of Economics, Russia
Dmitry Granovsky	Yandex, Russia
Dmitry Gubanov	Trapeznikov Institute of Control Sciences of the Russian Academy of Sciences, Russia
Dmitry Ignatov	National Research University Higher School of Economics, Moscow, Russia
Dmitry Ilvovsky	National Research University Higher School of Economics, Moscow, Russia
Max Ionov	Goethe University Frankfurt, Germany, and Lomonosov Moscow State University, Russia
Vladimir Ivanov	Innopolis University, Russia
Anna Kalenkova	National Research University Higher School of Economics, Moscow, Russia
Ilia Karpov	National Research University Higher School of Economics, Moscow, Russia
Nikolay Karpov	Sberbank and National Research University Higher School of Economics, Nizhny Novgorod, Russia
Egor Kashkin	Vinogradov Russian Language Institute of the Russian Academy of Sciences, Russia
Yury Kashnitsky	Koninklijke KPN N.V., The Netherlands
Mehdi Kaytoue	Infologic, Liris, and CNRS, France
Alexander Kazakov	Matrosov Institute for System Dynamics and Control Theory, Siberian Branch of the Russian Academy of Sciences, Russia
Alexander Kelmanov	Sobolev Institute of Mathematics, Siberian Branch of the Russian Academy of Sciences, Russia
Attila Kertesz-Farkas	National Research University Higher School of Economics, Moscow, Russia
Mikhail Khachay	Krasovsky Institute of Mathematics and Mechanics, Russia
Alexander Kharlamov	Institute of Higher Nervous Activity and Neurophysiology of the Russian Academy of Sciences, Russia
Javad Khodadoust	Payame Noor University, Iran
Donghyun Kim	Kennesaw State University, USA
Denis Kirjanov	National Research University Higher School of Economics, Moscow, Russia
Sergei Koltcov	National Research University Higher School of Economics, St.Petersburg, Russia
Jan Konecny	Palacký University Olomouc, Czech Republic
Anton Konushin	Lomonosov Moscow State University and National Research University Higher School of Economics, Moscow, Russia
Andrey Kopylov	Tula State University, Russia
Mikhail Korobov	ScrapingHub Inc., Ireland

Evgeny Kotelnikov	Vyatka State University, Russia
Ilias Kotsireas	Maplesoft and Wilfrid Laurier University, Canada
Boris Kovalenko	National Research University Higher School of Economics, Moscow, Russia
Fedor Krasnov	Gazprom Neft, Russia
Ekaterina Krekhovets	National Research University Higher School of Economics, Nizhny Novgorod, Russia
Tomas Krilavicius	Vytautas Magnus University, Lithuania
Anvar Kurmukov	Kharkevich Institute for Information Transmission Problems of the Russian Academy of Sciences, Russia
Valentina Kuskova	National Research University Higher School of Economics, Moscow, Russia
Valentina Kustikova	Lobachevsky State University of Nizhny Novgorod, Russia
Andrey Kutuzov	University of Oslo, Norway
Andrey Kuznetsov	Samara National Research University, Russia
Sergei O. Kuznetsov	National Research University Higher School of Economics, Moscow, Russia
Florence Le Ber	Université de Strasbourg, France
Alexander Lepskiy	National Research University Higher School of Economics, Moscow, Russia
Bertrand M. T. Lin	National Chiao Tung University, Taiwan
Benjamin Lind	Anglo-American School of St. Petersburg, Russia
Irina Lomazova	National Research University Higher School of Economics, Moscow, Russia
Konstantin Lopukhin	Scrapinghub Inc., Ireland
Anastasiya Lopukhina	National Research University Higher School of Economics, Moscow, Russia
Natalia Loukachevitch	Lomonosov Moscow State University, Russia
Ilya Makarov	National Research University Higher School of Economics, Moscow, Russia
Tatiana Makhalova	National Research University Higher School of Economics, Moscow, Russia, and Loria, Inria, France
Olga Maksimenkova	National Research University Higher School of Economics, Moscow, Russia
Alexey Malafeev	National Research University Higher School of Economics, Nizhny Novgorod, Russia
Yury Malkov	Samsung AI Center, Russia
Valentin Malykh	Vk.ru and Moscow Institute of Physics and Technology, Russia
Nizar Messai	Université François Rabelais Tours, France
Tristan Miller	Austrian Research Institute for Artificial Intelligence, Austria

Olga Mitrofanova	Saint Petersburg State University, Russia
Alexey A. Mitsyuk	National Research University Higher School of Economics, Moscow, Russia
Evgeny Myasnikov	Samara State University, Russia
Amedeo Napoli	Loria, CNRS, Inria, and Université de Lorraine, France
Long Nguyen The	Irkutsk State Technical University, Russia
Huong Nguyen Thu	Irkutsk State Technical University, Russia
Kirill Nikolaev	National Research University Higher School of Economics, Moscow, Russia
Damien Nouvel	Inalco University, France
Dimitri Nowicki	Glushkov Institute of Cybernetics of the National Academy of Sciences, Ukraine
Evgeniy M. Ozhegov	National Research University Higher School of Economics, Perm, Russia
Alina Ozhegova	National Research University Higher School of Economics, Perm, Russia
Alexander Panchenko	University of Hamburg, Germany, and Skolkovo Institute of Science and Technology, Russia
Panos Pardalos	University of Florida, USA
Marcello Pelillo	University of Venice, Italy
Georgios Petasis	National Centre for Scientific Research "Demokritos", Greece
Anna Petrovicheva	Xperience AI, Russia
Alex Petunin	Ural Federal University, Russia
Stefan Pickl	Bundeswehr University Munich, Germany
Vladimir Pleshko	RCO LLC, Russia
Aleksandr I. Panov	Institute for Systems Analysis of the Russian Academy of Sciences, Russia
Maxim Panov	Skolkovo Institute of Science and Technology, Russia
Mikhail Posypkin	Dorodnicyn Computing Centre of the Russian Academy of Sciences, Russia
Anna Potapenko	National Research University Higher School of Economics, Russia
Surya Prasath	Cincinnati Children's Hospital Medical Center, USA
Andrea Prati	University of Parma, Italy
Artem Pyatkin	Novosibirsk State University and Sobolev Institute of Mathematics of the Siberian Branch of the Russian Academy of Sciences, Russia
Irina Radchenko	ITMO University, Russia
Delhibabu Radhakrishnan	Kazan Federal University and Innopolis University, Russia
Vinit Ravishankar	University of Oslo, Norway
Evgeniy Riabenko	Facebook, UK
Anna Rogers	University of Massachusetts Lowell, USA
Alexey Romanov	University of Massachusetts Lowell, USA

Yuliya Rubtsova	Ershov Institute of Informatics Systems, Siberian Branch of the Russian Academy of Sciences, Russia
Alexey Ruchay	Chelyabinsk State University, Russia
Eugen Ruppert	TU Darmstadt, Germany
Christian Sacarea	Babes-Bolyai University, Romania
Aleksei Samarin	Saint Petersburg State University, Russia
Grigory Sapunov	Intento, USA
Andrey Savchenko	National Research University Higher School of Economics, Nizhny Novgorod, Russia
Friedhelm Schwenker	Ulm University, Germany
Oleg Seredin	Tula State University, Russia
Tatiana Shavrina	National Research University Higher School of Economics, Moscow, Russia
Andrey Shcherbakov	Intel, Australia
Sergey Shershakov	National Research University Higher School of Economics, Moscow, Russia
Oleg Slavin	Institute for Systems Analysis of Russian Academy of Sciences, Russia
Henry Soldano	Laboratoire d'Informatique de Paris Nord, France
Alexey Sorokin	Lomonosov Moscow State University, Russia
Dmitry Stepanov	Program System Institute of Russian Academy of Sciences, Russia
Vadim Strijov	Dorodnicyn Computing Centre of Russian Academy of Sciences, Russia
Irina Temnikova	Qatar Computing Research Institute, Qatar
Yonatan Tariku Tesfaye	University of Central Florida, USA
Martin Trnecka	Palacký University Olomouc, Czech Republic
Christos Tryfonopoulos	University of Peloponnese, Greece
Elena Tutubalina	Kazan Federal University, Russia
Ella Tyuryumina	National Research University Higher School of Economics, Moscow, Russia
Sergey Usilin	Institute for Systems Analysis of the Russian Academy of Sciences, Russia
Sebastiano Vascon	University Ca' Foscari of Venice, Italy
Evgeniy Vasilev	Lobachevsky State University of Nizhny Novgorod, Russia
Kirill Vlasov	Moscow Institute of Physics and Technology, Russia
Ekaterina Vylomova	The University of Melbourne, Australia
Dmitry Yashunin	HARMAN International, USA
Dmitry Zaytsev	National Research University Higher School of Economics, Moscow, Russia
Alexey Zobnin	National Research University Higher School of Economics, Moscow, Russia
Nikolai Zolotykh	University of Nizhni Novgorod, Russia
Olga Zvereva	Ural Federal University, Russia

Additional Reviewers

Saba Anwar
Vladimir Bashkin
Anna Berger
Vladimir Berikov
Rémi Cardon
Sofia Dokuka
Dmitrii Egurnov
Elizaveta Goncharova
Ivan Gostev
Artem Grachev

Natalia Korepanova
Anna Muratova
Victor Nedel'Ko
Alexander Plyasunov
Vitaly Romanov
Alexander Semenov
Grigory Serebryakov
Andrei Shcherbakov
Oleg Slavin
Kirill Struminskiy

Organising Committee

Andrey Novikov (Head of Organisation)	National Research University Higher School of Economics, Moscow, Russia
Elena Tutubalina (Local Organising Chair)	Kazan Federal University, Russia
Yury Dedenev (Venue Organisation)	Kazan Federal University, Russia
Anna Ukhanaeva (Communications and Website Support)	National Research University Higher School of Economics, Moscow, Russia
Valeria Andrianova (Travel Support)	National Research University Higher School of Economics, Moscow, Russia
Ilseyar Alimova	Kazan Federal University, Russia
Anna Kalenkova	National Research University Higher School of Economics, Moscow, Russia
Ayrat Khasyanov	Kazan Federal University, Russia
Zulfat Miftahutdinov	Kazan Federal University, Russia
Valery Solovyev	Kazan Federal University, Russia

Volunteer

Anita Kurova   Kazan Federal University, Russia

Sponsors

National Research University Higher School of Economics, Russia
Springer

Contents

Analysis of Images and Video

Optimization Problems on Graphs and Network Structures

Analysis of Dynamic Behavior Through Event Data

General Topics of Data Analysis

Optimizing Q-Learning with K-FAC Algorithm

Roman Beltiukov$^{(\boxtimes)}$ iD

Peter the Great St. Petersburg Polytechnic University, Saint-Petersburg, Russia
`maybe.hello.world@gmail.com`

Abstract. In this work, we present intermediate results of the application of Kronecker-factored Approximate curvature (K-FAC) algorithm to Q-learning problem. Being more expensive to compute than plain stochastic gradient descent, K-FAC allows the agent to converge a bit faster in terms of epochs compared to Adam on simple reinforcement learning tasks and tend to be more stable and less strict to hyperparameters selection. Considering the latest results we show that DDQN with K-FAC learns more quickly than with other optimizers and improves constantly in contradiction to similar with Adam or RMSProp.

Keywords: Q-learning · K-FAC · Reinforcement learning · Natural gradient

1 Introduction

During the last years, many successes in the field of reinforcement learning were achieved by using Deep Q-network (DQN) algorithms. Starting with [1] where authors combined Q-learning with latest convolutional neural networks, rather big amount of articles were published and proposed both architecture and algorithm changes to achieve higher scores and stability of Q-learning [2–4].

Being a Q-function optimizing algorithm compared to policy-optimization algorithms Q-learning is subject to instability, leading to unpredictable degradation of an agent. Majority of articles mentioned above stabilized the learning process and allowed to reach higher results but the problem is still relevant.

According to [5], using natural gradient descent can improve the speed of convergence process measured by a number of steps by increasing the time of calculation of each step. There are only few articles about applying second-order optimization methods to Deep Q-networks algorithms which either use heavy computational realizations of Fisher matrix calculation leading to significantly increased training time [6] or suppose serious neural network architecture changes for natural gradient descent realization [7]. Meanwhile, in the area of policy-gradient algorithms, second-order optimization techniques become more and more popular and widely used [8,9]. These reasons together with a request from OpenAI to research this topic led to ideas of further work.

© Springer Nature Switzerland AG 2020
W. M. P. van der Aalst et al. (Eds.): AIST 2019, CCIS 1086, pp. 3–8, 2020.
https://doi.org/10.1007/978-3-030-39575-9_1

In this article, we attempted to apply second-order optimization techniques like natural gradient descent to the Q-learning in order to improve the stability of learning and decrease training time.

2 Background

2.1 Reinforcement Learning

We consider the standard RL framework [10] where an agent is involved in Markov Decision Process (MDP) with infinite horizon. An MDP is defined by the tuple (S, A, R, P, γ), which consists of a set of states S, a set of actions A, a reward function $R(s, a)$, transition probability function $P(s, a)$ and discount factor γ. At step t the agent observes state $s_t \in S$, selects some action $a_t \in A$ and then receives scalar reward r and next state s_{t+1} given by transition function. The agent aims to maximize γ-discounted cumulative reward by finding optimal policy π_θ.

Value-based methods optimize policies by optimizing the corresponding value function $Q^\pi(s, a)$ which estimates expected future reward that can be obtained by the agent that takes action a in given state s and then acts according to the policy π. The optimal value function $Q^*(s, a)$ in this case provides the most correct prediction of reward and is determined by solving the Bellman equation:

$$Q^*(s, a) = E\left[R(s, a) + \gamma \max_{a'} E\left[Q^*(s', a')\right]\right] \qquad (1)$$

The optimal policy π^* then is defined as $\pi^*(s) = argmax_{a \in A} Q^*(s, a)$.

2.2 Deep Q-Networks

Presented in [1], DQN approach approximates the Q-function with a deep neural network that outputs values of all possible actions for a given observation. Usually, during training DQN agents use such techniques as replay buffer to store states, actions, rewards and transitions to learn and separate target networks for stabilizing learning process.

The double deep Q-learning (DDQN) method [2] proposes using the current network for calculating argmax over next state values with different loss function:

$$J = R(s, a) + \gamma Q(s_{t+1}, argmax Q^t(s_{t+1}, a_{t+1})) \qquad (2)$$

Using such an approach leads to better performance and reducing overestimations of action values.

2.3 Kronecker-Factored Approximate Curvature

Natural gradient descent [5] (NGD) is a second-order optimization method that use Fisher information matrix to take steepest descent direction in model distribution space instead of parameter space. Using NGD usually allows to converge in a fewer number of iterations than with stochastic gradient descent methods.

A recently proposed Kronecker-factored approximate curvature algorithm [11] (K-FAC) uses a computationally efficient approximation of the Fisher matrix through Kronecker factorization to perform natural gradient updates. By increasing computing time of single step approximately by 30% for full-connected layers K-FAC allows to converge faster in term of number of iterations then SGD with momentum.

3   Method

For this research Tensorflow open-source implementation of K-FAC algorithm was used [12]. For optimizer testing, OpenAI Baselines' [13] DDQN implementation was adopted and necessary changes for K-FAC optimizer stabilization were also made. Two network architectures were used for evaluating optimizer:

1. For discrete environments from "classic control environments list" of Open-AI Gym [14], and also for LunarLander-v2, 3-layer fully connected neural network was used. The first and second layers have 64 and 16 neurons with hyperbolic tangent activation, the last layer has as many neurons as a number of available actions in the corresponding environment.
2. For Atari environments from OpenAI Gym a network, similar to [1], was used.

List of current changes mainly includes merging K-FAC implementation code with both networks, changing some hyperparameters of networks, etc. The current implementation is a work-in-progress and subject to change in the future.

Experiments were held on GPU partition with Nvidia Tesla K40. Full code and preprocessing settings for each environment and also current results are available in projects' repository[1].

4   Current Status and Intermediate Results

By now the implementation of DDQN with K-FAC optimizator has been completed and intermediate results were obtained for some classic control environments of OpenAI Gym framework. For each optimizer, optimal hyperparameters were selected by a grid search with averaging across 20 different seeds. In particular, in *CartPole-v1* K-FAC-DDQN realization showed the faster speed of convergence on optimal hyperparameters than similar DDQN realization with *Adam* [15] or *RMSProp* [16]. Corresponding results are shown in Fig. 1. Spending about 30% more time for each iteration, K-FAC-DDQN implementation requires a fewer number of iterations for reaching a similar reward. Additional results are shown in Table 1 and Fig. 2. For some environments (like *MountainCar-v0*) memory ability of network is rather low and plateau of reward plot is not close to state-of-the-art examples.

During the training process, another interesting observation was made – K-FAC implementation requires less attention to hyperparameters of optimizator (except dumping) than *Adam* or *RMSProp*. On average, the K-FAC-DDQN

---

[1] https://github.com/maybe-hello-world/qfac.

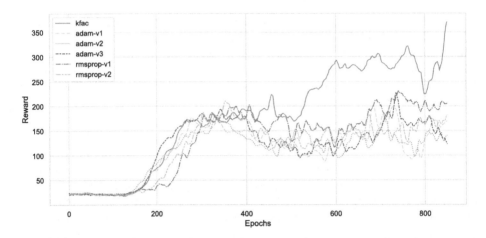

Fig. 1. Mean performance of optimizers with optimal hyperparameters across 20 random seeds on *CartPole-v1*.

Table 1. The average total reward among the last 100 games for various optimizers by running an ϵ-greedy policy with $\epsilon = 0.1$ for a fixed number of steps.

Optimizer	MountainCar-v0	CartPole-v1	Acrobot-v1	LunarLander-v2
Adam	-122	183	-108	-46
RMSProp	-119	194	-97	-136
K-FAC	-124	**321**	-92	**38**

implementation obtained more stable and higher results during the grid search process over the hyperspace of hyperparameters than other with optimizers.

Also sometimes [17,18] while working with graphical input data, researchers use byte interpretation with integer range [0..255] without scaling to [0..1] float range. Unfortunately together with K-FAC optimizer, this leads to instability of Cholesky transformation and fail to compute, resulting in obligatory scaling of input data for the network.

Summarizing, K-FAC optimizer tends to be more smoothly growing and avoiding sudden loss of performance during training. Possibly for very simple tasks, other optimizers could be much more efficient hopping up to optimum, but for more hard environments K-FAC should be considered as an option. Although, it's important to remember that using second order optimization algorithms requires more memory for Fisher matrix computation and additional computational time for the algorithm. Amount of memory and computational time depends on network layers (from about 30% for fully-connected layers and up to 2–3 times for convolution layers) and input data so in case of training heavy convolutional networks using K-FAC could be impossible or less effective, if use certain K-FAC options.

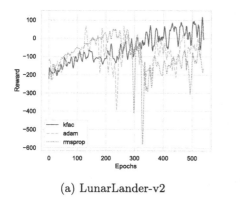

(a) LunarLander-v2

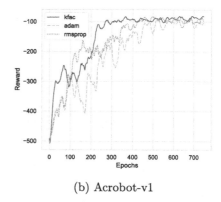

(b) Acrobot-v1

Fig. 2. Mean performance of optimizers with optimal hyperparameters in different environments. On average K-FAC mainly do not allow burst improvements but also tends to avoid performance degradation.

5   Discussion

At present, the hypothesis that using K-FAC RL agent can achieve a more stable and predictable learning process is neither proven nor disproved – this requires the searching for optimal hyperparameters and numerous experiments in several Gym environments. Nevertheless, intermediate results allow us to speak about the possibility of using K-FAC algorithm for Q-learning agents. In case of stabilizing learning process by K-FAC, it could be possible to achieve higher total rewards or train with less number of iterations.

At this moment, only DQN and DDQN algorithms were implemented due to the difficulty of adapting the K-FAC method to newer algorithms such as Rainbow DQN [19]. However, using a less stable implementation of DQN agent may lead to a more noticeable impact of K-FAC on the result.

Further, we plan to adapt K-FAC method to later DQN improved implementations, like Rainbow DQN, as well as conducting numerous experiments with the existing implementation in various OpenAI Gym environments.

Acknowledgments. We would like to thank Olga Tushkanova for helpful comments, constructive criticism and useful feedback. The results of the work were obtained using the computational resources of Peter the Great Saint-Petersburg Polytechnic University Supercomputing Center (www.scc.spbstu.ru). The research was partially funded by 5-100-2020 program and SPbPU university.

References

1. Mnih, V., et al.: Playing atari with deep reinforcement learning, December 2013. http://arxiv.org/abs/1312.5602
2. van Hasselt, H., Guez, A., Silver, D.: Deep reinforcement learning with double Q-learning, September 2015. http://arxiv.org/abs/1509.06461

3. Bellemare, M.G., Dabney, W., Munos, R.: A distributional perspective on reinforcement learning, July 2017. http://arxiv.org/abs/1707.06887
4. Fortunato, M., et al.: Noisy networks for exploration, June 2017. http://arxiv.org/abs/1706.10295
5. Amari, S.I.: Natural gradient works efficiently in learning. Neural Comput. **10**(2), 251–276 (1998). https://doi.org/10.1162/089976698300017746. http://www.mitpressjournls.org/10.1162/089976698300017746
6. Knight, E., Lerner, O.: Natural gradient deep Q-learning, March 2018. http://arxiv.org/abs/1803.07482
7. Desjardins, G., Simonyan, K., Pascanu, R., Kavukcuoglu, K.: Natural neural networks, July 2015. http://arxiv.org/abs/1507.00210
8. Schulman, J., Levine, S., Moritz, P., Jordan, M.I., Abbeel, P.: Trust region policy optimization, February 2015. https://arxiv.org/abs/1502.05477
9. Wu, Y., Mansimov, E., Liao, S., Grosse, R., Ba, J.: Scalable trust-region method for deep reinforcement learning using Kronecker-factored approximation, August 2017. http://arxiv.org/abs/1708.05144
10. Sutton, R.S., Barto, A.G., et al.: Introduction to Reinforcement Learning, vol. 135. MIT Press, Cambridge (1998)
11. Martens, J., Grosse, R.: Optimizing neural networks with Kronecker-factored approximate curvature, March 2015. http://arxiv.org/abs/1503.05671
12. Martens, J., Tankasali, V., Duckworth, D., Johnson, M., Zhang, G., Koonce, B.: tensorflow\kfac. https://github.com/tensorflow/kfac. Accessed 25 Mar 2019
13. Dhariwal, P., et al.: OpenAI baselines (2017). https://github.com/openai/baselines
14. Brockman, G., et al.: OpenAI Gym, June 2016. http://arxiv.org/abs/1606.01540
15. Kingma, D.P., Ba, J.: Adam: a method for stochastic optimization, December 2014. http://arxiv.org/abs/1412.6980
16. Hinton, G.: RMSProp. https://www.cs.toronto.edu/~tijmen/csc321/slides/lecture_slides_lec6.pdf. Accessed 25 Mar 2019
17. Ecoffet, A., Huizinga, J., Lehman, J., Stanley, K.O., Clune, J.: Go-Explore: a new approach for hard-exploration problems, January 2019. http://arxiv.org/abs/1901.10995
18. Resnick, C., Raileanu, R., Kapoor, S., Peysakhovich, A., Cho, K., Bruna, J.: Backplay: "Man muss immer umkehren", July 2018. http://arxiv.org/abs/1807.06919
19. Hessel, M., et al.: Rainbow: combining improvements in deep reinforcement learning, October 2017. http://arxiv.org/abs/1710.02298

Evolutionary Algorithms for Constructing an Ensemble of Decision Trees

Evgeny Dolotov[1,2]([✉]) and Nikolai Zolotykh[1,2]

[1] Yandex, Moscow, Russia
{dolotov-e,nikzolotykh}@yandex-team.ru
[2] National Research University Higher School of Economics, Moscow, Russia

Abstract. Most decision tree induction algorithms are based on a greedy top-down recursive partitioning strategy for tree growth. In this paper, we propose several methods for induction of decision trees and their ensembles based on evolutionary algorithms. The main difference of our approach is using real-valued vector representation of decision tree that allows to use a large number of different optimization algorithms, as well as optimize the whole tree or ensemble for avoiding local optima. Differential evolution and evolution strategies were chosen as optimization algorithms, as they have good results in reinforcement learning problems. We test the predictive performance of this methods using several public UCI data sets, and the proposed methods show better quality than classical methods.

Keywords: Classification · Decision tree induction · Evolutionary algorithm · Differential evolution

1 Introduction

Decision trees are a popular method of machine learning for solving classification and regression problems. Because of their popularity many algorithms exists to build decision trees [1,2]. However, the task of constructing optimal or near-optimal decision tree is very complex. Most decision tree induction algorithms are based on a greedy top-down recursive partitioning strategy for tree growth. They use different variants of impurity measures, such as information gain [2], gain ratio [3], gini-index [4] and distance-based measures [5], to select an input attribute to be associated with an internal node. One major drawback of the greedy search is that it usually leads to sub-optimal solutions. The underlying reason is that local decisions at each nodes are in fact interdependent and cannot be found in this way.

A popular approach that can partially solve these problems is the induction of decision trees through evolutionary algorithms (EAs) [15]. In this approach, each "individual" in evolutionary algorithms represents a solution to the classification problem. Each solution is evaluated by a fitness function, which measures the quality of it. At each new generation, the best solutions have a higher probability

© Springer Nature Switzerland AG 2020
W. M. P. van der Aalst et al. (Eds.): AIST 2019, CCIS 1086, pp. 9–15, 2020.
https://doi.org/10.1007/978-3-030-39575-9_2

of being selected for reproduction. The selected solutions undergo operations inspired by genetics, such as crossover and mutation, producing new solutions which will replace the parents, creating a new population of solutions. This process is repeated until a stopping criterion is satisfied. Instead of a local search, EAs perform a robust global search in the space of candidate solutions. As a result, EAs tend to cope better with attribute interactions than greedy methods and avoid local optima.

In this paper we propose an approach that encodes a decision tree as a real-valued homogeneous vector, since we also encode indices of features by real numbers and decode them using the operation of finding the minimum. This approach allows to use a large number of different optimization algorithms, such as differential evolution [6] and evolution strategies [7].

2 Related Work

The number of proposed evolutionary algorithms for decision tree induction has grown in the past few years, mainly because they report good predictive accuracy whilst keeping the comprehensibility of decision trees. There are two the most common approaches to encoding decision trees for evolutionary algorithms: tree-based encoding and fixed-length vector encoding. They all use different methods to encode indices of features, threshold values, leaves, and operators in nodes. The main differences in tree-based approaches are the presence of pointers to nodes and the ability to encode trees of various sizes. Axis-parallel decision trees are the most common type found in the literature, mainly because this type of tree is usually much easier to interpret than an oblique tree. A node in axis-parallel decision tree can be described by two parameters: index of tested feature and threshold value. A popular approach [8] to encoding such trees is to encode each node with one integer and one real number, but in this case, we get heterogeneous and more complex representation of decision tree, than in approach proposed in this article, which makes the process of finding the optimal solution more complex. Authors of article [9] describe a very similar approach to encoding oblique decision trees with real-valued vectors and optimizing them with differential evolution algorithms, but in this article we propose a more compact representation specifically for axis-parallel decision trees. A more detailed overview of an evolutionary methods for constructing decision trees can be found here [10].

3 Proposed Approach

In this paper we propose a new approach to construct axis-parallel decision tree for classification problems using evolutionary algorithms.

3.1   Real-Valued Vector Representation

In axis-parallel trees each node splits dataset according to the following rule:

$$f(x) = \begin{cases} 1, & \text{if } a_i \leq t \\ 0, & \text{otherwise} \end{cases} \tag{1}$$

Thus, each node of the tree is described by two parameters: the index of a feature and a threshold value.

Suppose we have a fixed-length real-valued vector with values in the segment $[0, 1]$. This vector consists of two parts of equal length – the first part encodes feature indices, and the second part encodes threshold values. Also suppose that all features of objects belong to the segment $[0, 1]$. If this is not the case, then we normalize features using the maximum and minimum values from the training dataset. To restore the index of a feature from the vector, we should find the position of the minimum value in the first part of this vector and find its remainder of integer division by the number of features. The value in the second part of the vector in this position is used as threshold value in the corresponding node. After that, the next minimal value in the vector, the corresponding index of the feature in the node and a threshold value should be found. This operation is repeated until the entire vector is used. Using the indices of features and their threshold values for all nodes, the decision tree without leaves can be built by sequentially adding the nodes. After that, the leaves are added to the decision tree by using training dataset and the majority rule. Thus, we can construct a decision tree from a real-valued vector and evaluate its characteristics.

3.2   Differential Evolution

The differential evolution (DE) [6] is an effective evolutionary algorithm designed to solve optimization problems with real-valued parameters. A population in DE consists of N individuals:

$$P = \{x_1, x_2, ..., x_N\} \tag{2}$$

The j-th value of the individual x_i in the initial population is calculated as follows:

$$x_j^i = x_j^{\min} + r(x_j^{\max} - x_j^{\min}), \tag{3}$$

where $r \in [0, 1]$ is a uniformly distributed random number.

The evolutionary process implements an iterative scheme to evolve the initial population. At each iteration of this process, known as the generation, a new population of individuals is generated from the previous one. Each individual is used to build a new vector by applying the mutation and crossover operators:

– Mutation. Three randomly chosen individuals are linearly combined as follows:

$$v^i = x^{j_1} + \alpha(x^{j_2} - x^{j_3}), \tag{4}$$

where α is a user-specified constant.

– Crossover. The mutated vector is recombined with the target vector to build the trial vector:

$$u_j^i = \begin{cases} v_j^i, & \text{if } r \leq CR \text{ or } j = l \\ x_j^i, & \text{otherwise} \end{cases} \tag{5}$$

where $r \in [0, 1]$ is uniformly distributed random number and CR is the crossover rate.
– Selection: A one-to-one tournament is applied to determine which individual is selected as a member of the new population.

In the final step, when a stop condition is fulfilled, DE returns the best individual in the current population.

3.3 Evolution Strategies

Unlike the method of differential evolution, the population in the method of evolution strategies [7] consists of only one individual:

$$P \sim x \tag{6}$$

The initial individual is calculated as follows:

$$x_j = x_j^{\min} + r(x_j^{\max} - x_j^{\min}), \tag{7}$$

where $r \in [0, 1]$ is a uniformly distributed random number. We sample several off-sets which are represented as a normal distributed random vector $e_1, e_2, ..., e_n \sim \mathcal{N}(0, I)$. Then we shift the individual in the direction of the weighted sum of the offsets, which approximate the gradient:

$$x \leftarrow x + \alpha \frac{1}{n\sigma} \sum_{i=1}^{n} f(x + \sigma e_i) e_i, \tag{8}$$

where α and σ are user-specified constants.

3.4 Construction of Ensembles

Two of the most popular approaches for constructing the ensembles of decision trees is bagging and boosting. Example of method that use bagging approach is random forest and example of method that use boosting approach is AdaBoost. In this part of the paper we propose to replace classical algorithms to induction of decision trees in these methods by evolutionary algorithms described earlier. Thus, we obtain two new methods: evolutionary random forest (EvoRF) as the analogue of random forest and EvoBoost as the analogue of AdaBoost. In addition to this, we consider the method (EvoEnsemble) in which each individual in the population is the whole ensemble, and representation of the ensemble is a large real-valued vector obtained by concatenation with a vector for each tree from the ensemble. Thus, in this method – evolutionary ensemble, we optimize the whole ensemble at once, which theoretically should lead to a better result.

4   Experiment

For experiments we use several popular datasets from UCI repository. Experiments are divided into two parts.

First, we evaluate classification accuracy of the methods based on evolutionary algorithms and compare their results with the classical methods for solve classification problems. Experiments show that using of the proposed methods does not allow to exceed the results of classical algorithms for constructing decision trees on some datasets, but on the vast majority of datasets using evolution strategies allows to achieve the significant improvement in the accuracy of prediction by several percent (Table 1). Therefore, we decide to use this algorithm to build ensembles in subsequent experiments.

Table 1. Comparison of popular classification algorithms such as CART [11] and multilayer perceptron (MLP) [12] with the proposed approaches: differential evolution (DE) and evolution strategies (ES).

Dataset	CART	MLP	DE	ES	Dataset	CART	MLP	DE	ES
car	96.74	**98.32**	90.59	91.18	molecular-p	75.85	**86.54**	85.57	86.01
tic-tac-toe	**93.65**	93.38	87.96	86.39	diabets	74.49	73.89	75.03	**75.07**
glass	71.42	71.36	73.02	**73.45**	balance-scale	78.07	79.68	**80.05**	80.04
iris	94.45	97.13	96.97	**97.24**	ionosphere	88.23	89.78	**91.32**	91.17
australian	85.67	86.12	**86.42**	86.05	cmc	54.83	**56.05**	55.89	56.01
wine	92.43	93.75	94.58	**94.65**	vehicle	69.75	**72.31**	71.96	72.18
liver-disoder	67.73	66.96	**68.36**	68.25	lympth	77.97	78.13	**78.42**	78.36
haberman	73.25	74.89	75.43	**75.76**	dermatology	94.32	93.56	95.67	**95.75**
heart-statlog	78.75	75.43	79.34	**80.20**	sonar	75.33	77.35	76.49	**79.43**
page-blocks	96.98	95.76	**97.35**	97.03	credit-g	72.25	**75.43**	74.32	73.85

Second, we evaluate classification accuracy of several approaches to constructing ensembles of decision trees. For these experiments we use datasets which have only two different labels, in other words, we are solving the problem of binary classification. Various hyperparameter of the random forest algorithm and AdaBoost such as, the depth of trees and maximum number of trees was selected using the method of grid search, and then these same parameters were used for their evolutionary analogues. As well as in the case of using evolutionary algorithms for constructing a single decision tree, experiments show that using of the proposed methods does not allow to exceed the results of classical algorithms for constructing ensemble of decision trees on some datasets, but the method that represent whole ensemble as one real-valued vector are showing best accuracy on most datasets (Table 2).

Table 2. Comparison of random forest (RF) [13] and AdaBoost [14] with the proposed approaches for constructing the ensembles of decision trees: evolutionary version of random forest (EvoRF), AdaBoost, EvoBoost and evolution ensemble (EvoEnsemble)

Dataset	RF	EvoRF	EvoEnsemble	AdaBoost	EvoBoost
tic-tac-toe	97.48	97.76	**97.84**	96.31	96.91
australian	92.03	91.59	**92.73**	91.36	90.93
liver-disoder	**77.32**	75.27	76.73	76.31	76.45
molecular-p	89.24	90.64	**91.03**	90.21	90.40
diabetes	82.31	83.74	83.67	82.23	**85.07**
ionosphere	92.35	92.89	**93.11**	91.76	92.17
haberman	79.45	80.12	80.79	79.21	**80.69**
heart-statlog	83.43	84.24	83.78	83.09	**83.85**
sonar	85.14	86.38	**86.19**	85.02	86.03
credit-g	77.31	77.24	79.15	79.07	**79.63**

5 Conclusion and Future Work

In this paper, we have proposed several methods that use different evolutionary algorithms to construct decision trees and their ensembles. The main contribution of this paper is method to construct real-valued vector representation of decision tree that allows to use different evolutionary algorithms for constructing decision trees and their ensembles. The proposed algorithms show better quality than classical methods such as CART, random forest and AdaBoost on popular datasets from UCI repository, but in order to achieve such high results, it takes more time than using classical algorithms. This is due to the fact that the methods using evolutionary algorithms during training several times build trees and evaluate their quality, while classical algorithms do it only once.

A detailed analysis of the computational performance of the proposed methods, parallel computations in evolutionary algorithms, initialization of initial population by results of the classical decision tree inductions algorithms and evolutionary analogue of gradient boosting are possible areas for further research.

References

1. Duda, R.O., Hart, P.E., Stork, D.G.: Pattern Classification, 2nd edn. Wiley, Hoboken (2001)
2. Quinlan, J.R.: Induction of decision trees. Mach. Learn. **1**(1), 81–106 (1986)
3. Quinlan, J.R.: C4.5: Programs for Machine Learning. Morgan Kaufmann Publishers Inc., Massachusetts (1993)
4. Breiman, L., Friedman, J.H., Olshen, R.A., Stone, C.J.: Classification and Regression Trees. Chapman and Hall/CRC, Boca Raton (1984)
5. De Mantaras, R.L.: A distance-based attribute selection measure for decision tree induction. Mach. Learn. **6**(1), 81–92 (1991)

6. Tasgetiren, M., Liang, Y., Sevkli, M., Gencyilmaz, G.: Differential evolution algorithm for permutation flowshop sequencing problem with makespan criterion. In: Proceedings of the 4th International Symposium on Intelligent Manufacturing Systems (IMS 2004), pp. 442–452 (2004)
7. Rechenberg, I., Eigen, M.: Evolutionsstrategie: Optimierung Technischer Systeme nach Prinzipien der Biologischen Evolution. Frommann-Holzboog, Stuttgart (1973)
8. Jankowski, D., Jackowski, K.: Evolutionary algorithm for decision tree induction. In: Saeed, K., Snášel, V. (eds.) CISIM 2014. LNCS, vol. 8838, pp. 23–32. Springer, Heidelberg (2014). https://doi.org/10.1007/978-3-662-45237-0_4
9. Rivera-Lopez, R., Canul-Reich, J.: Differential evolution algorithm in the construction of interpretable classification models. In: Artificial Intelligence-Emerging Trends and Applications, pp. 49–73 (2018)
10. Basgalupp, M.P., Carvalho, A., Barros, R.C., Freitas, A.: A survey of evolutionary algorithms for decision-tree induction. IEEE Trans. Syst. Man Cybern. Part C Appl. Rev. **42**(3), 291–312 (2012)
11. Breiman, L., Friedman, J.H., Olshen, R.A., Stone, C.J.: Classification and Regression Trees, Wadsworth, Belmont. Republished by CRC Press, CA (1984)
12. Haykin, S.: Neural Networks: A Comprehensive Foundation. Prentice Hall PTR, New Jersey (1994)
13. Breiman, L.: Random forests. Mach. Learn. **45**(1), 5–32 (2001)
14. Freund, Y., Schapire, R.E.: A decision-theoretic generalization of on-line learning and an application to boosting. J. Comput. Syst. Sci. **55**(1), 119–139 (2001)
15. Back, T.: Evolutionary Algorithms in Theory and Practice: Evolution Strategies, Evolutionary Programming Genetic Algorithms. Oxford University Press, Oxford (1996)

An Algorithm for Constructing a Topological Skeleton for Semi-structured Spatial Data Based on Persistent Homology

Sergey Eremeev$^{(\boxtimes)}$ and Semyon Romanov

Vladimir State University, Vladimir, Russia
`sv-eremeev@yandex.ru, cwwc@bk.ru`

Abstract. The construction and use of a topological skeleton for processing semi-structured information based on persistent homology methods is considered in the article. In the work, the main topological feature for analysis of object is a hole. The application of the developed algorithm to solve the actual problem of geoinformatics in the matching of spatial objects at different scales of map is shown. Comparison of topological skeletons at different tree depths is demonstrated.

Keywords: Persistent homology · Topological skeleton · Geoinformatics · Multi-scale maps

1 Introduction

To calculate the characteristics of an object in images, it is necessary to use different models to represent them. The resulting set of features is the basis for solving various problems. These are searching for objects by pattern, comparing images, identifying objects. The results are widely used in all spheres of human activity. For example, in chemistry when searching for an object with a certain three-dimensional structure, in criminology when comparing fingerprints, in computer vision systems when tracking an object, and others.

A particular challenge is the analysis of objects that have a deformation. And this is especially observed in the problems of geoinformatics, because we often deal with objects on different scales [1]. In addition, objects on a map of the same scale may not even be related to each other (see Fig. 1(a)), and after generalization they represent a single object (see Fig. 1(b)). That is, disparate and semi-structured data are perceived by a person as a whole, and the computer perceives each object separately among unrelated data. Matching, searching and identifying such objects is an actual problem, because it is not clear what characteristics we should use to analyze the object.

On the other hand, objects at different scales can be compared in a rough estimate to reveal only the general structure or in a detailed form, where high

© Springer Nature Switzerland AG 2020
W. M. P. van der Aalst et al. (Eds.): AIST 2019, CCIS 1086, pp. 16–26, 2020.
https://doi.org/10.1007/978-3-030-39575-9_3

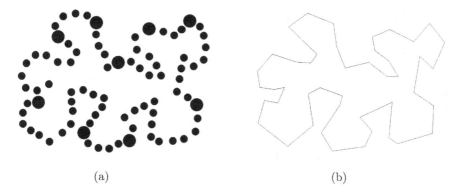

(a)   (b)

Fig. 1. (a) A set of features on a map of the same scale in the form of unrelated data, (b) representation of a group of features after generalization in the form of a single polygon feature.

accuracy of details is important. That is, it is necessary to use a model that will allow us to compare objects depending on the accuracy of detail.

The purpose of the article is to represent the initial semi-structured information in the form of a set of characteristics that would reveal the overall structure of the object and compare with other objects at the required level of detail.

2 Related Work

Consider the existing methods for calculating the characteristics of objects that are deformed. Currently, methods of continuous representation of binary images in the form of borders, skeletons and circulars are actively used [2, 3]. The basis of these methods are Jordan curves, which are the mapping of a circle into a plane $\mathbb{R}^2$. That is, a binary image is a union of continuous shapes. To compare the images, the characteristics of these figures are analyzed. These methods are well suited for comparing single-linked binary images. However, there are difficulties if the task is to compare semi-structured data, for example, such as in Fig. 1(a). A separate group are the methods to extract sketched features of the object. These methods are used to search for $3d$ objects based on Gabor filters [4], allow us to analyze deformable objects with further comparison of their graph models [5, 6], and also use convolutional neural networks to extract objects [7]. Important studies are carried out in the analysis of the skeleton of contour deformations of the object [8]. There are also developments in which a dictionary of deformable forms of the object is created [9].

Promising new research is aimed at the analysis of unorganized data and the creation of a persistent skeleton [10]. The basis of these studies are methods of topological analysis, namely persistent homology [11–14]. Data at different scales are also poorly organized. For the analysis and comparison of such structures the methods of topological analysis are best suited.

3   Methodology

3.1   Notation and Topological Features

This work is a continuation of studies to compare objects at different scales, which showed good results [15]. In order to compare two objects at different scales with the deformation for each object a barcode is constructed in [15], which describes the topological features of the objects and allows us to find their general structure. However, this approach still has serious problems in the generalization of the object. In the process of changing the object, as a rule, its key points also change, which leads to a change in topological features.

To solve this problem, we propose to extend the analysis of topological characteristics of the object by introducing relations between topological features and constructing a topological skeleton. The main topological characteristic in the proposed approach is a hole. Holes are formed by the iterative connection of the contour points of the object in ascending order of their distance from each other.

The initial data for the topological analysis are two spatial objects X and Y, consisting of a sparse space of points which are a set of key points:

$$X = \{x_1, x_2, ..., x_n\}, Y = \{y_1, y_2, ..., y_m\},$$

where x_i are key points of the object $X (i = 1, 2, ..., n)$,
y_j are key points of the object $Y (j = 1, 2, ..., m)$,
n and m denote number of key points of objects X and Y, respectively.

An example of a sparse point space is shown in Fig. 2.
Topological features are formed on the basis of a set of key points. The main topological characteristic in the proposed approach is a hole.

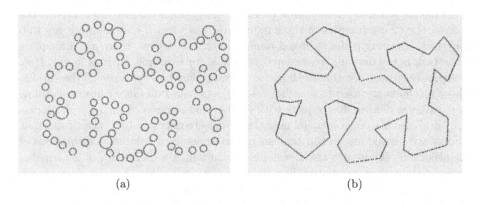

(a)   (b)

Fig. 2. A sparse point space for objects in Fig. 1.

Definition 1. *A hole H is a subset of key points that forms a closed contour of four or more edges.*

To compute holes at each iteration step, we use an approach based on alpha complex [12] in which the contour points connect to each other according to the Delaunay triangulation rules. Thus, the object is divided into a set of triangles.

Figure 3 shows the iterative process of hole separation. First, straight lines connect the points that are closest to each other (see Fig. 3(a)). Next, a closed loop is formed as a hole at number 1. Incrementally, the initial point space forms a set of triangles with the creation of intermediate holes (see Fig. 3(b–h)). The finite set of triangles is shown in Fig. 3(i).

This iterative appearance of holes looks like a tree whose root indicates to a hole denoting the boundaries of the original object, and its branches are further dividing the hole when connecting the key points of the object.

Let l be the number of holes formed as a result of the iterative hole separation process.

Definition 2. *An inheritance tree T is a connected acyclic graph whose vertices are holes, and edges show a parent-child relationship when separating holes:*

$$T = \{V, E\},$$

where $V = \{H_1, H_2, ..., H_l\}$ is a set of tree vertices that correspond to holes, E is a set of edges formed between elements $v_i, v_j \in V$ $(i, j = 1, 2, ..., l; i \neq j)$ in the process of hole separation.

Obviously, the root of the tree is the parent hole, and each of the leaves is a child hole. For example, a hole with index 1 is the parent, and holes with indexes 2, 3, ..., 10 are its child holes (see Fig. 3).

Thus, the hole H_1 is separated and two holes are formed. The smallest of the two holes is denoted by H_2, and the name H_1 of the another hole is preserved. Gradually, the hole H_1 is reduced and new child holes are formed. Similarly, the hole H_2 has a similar separation process. The scheme of such separation of holes is shown in Fig. 4.

The hole inheritance tree contains information about the topological appearance of holes, but does not contain information about the spatial relationships between holes.

Definition 3. *A topological skeleton is a connected acyclic graph whose vertices are holes with spatial relations between them:*

$$T' = \{V, E'\},$$

where $V = \{H_1, H_2, ..., H_l\}$ is a set of tree vertices that correspond to holes, E' is a set of edges that describe the spatial relationships between elements $v_i, v_j \in V$ $(i, j = 1, 2, ..., l; i \neq j)$.

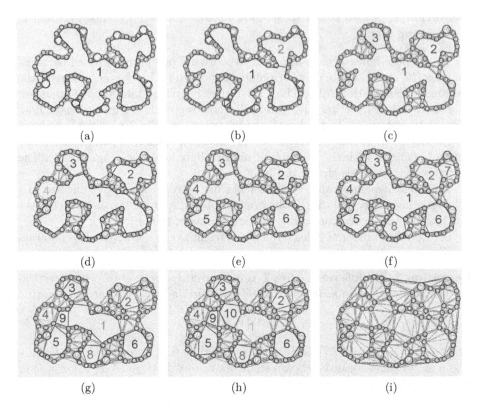

Fig. 3. An iterative process of splitting the parent hole (a) to child holes (b–h), (i) finite set of triangles.

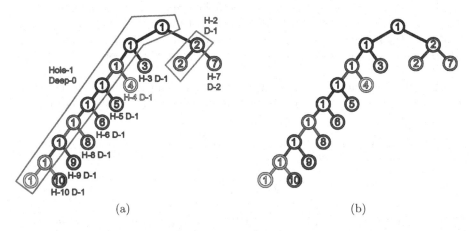

Fig. 4. (a) A scheme of separation of holes according to Fig. 3(a–h), (b) detailed diagram showing depth information for each hole. A detailed diagram in a more convenient form as a tree inheritance holes is shown in Fig. 5

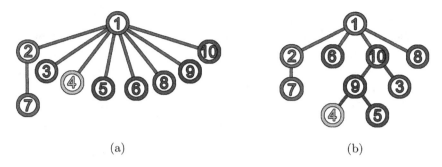

(a)   (b)

Fig. 5. (a) A tree of inheritance of holes according to Fig. 3(a–h), (b) topological skeleton tree that describes the spatial relationships between holes in Fig. 3(h).

Creating a topological skeleton will allow us to establish relationships between holes according to their spatial relationships. An example of a topological skeleton consisting of the vertices of the inheritance tree in Fig. 5(a) is shown in Fig. 5(b).

3.2 Creating a Topological Skeleton

To obtain information about the size of the hole and their further use it is necessary to describe the hole, which is a set of points, by the single object. To do this, we propose to describe the points of the hole in a circle (see Fig. 6(a)) whose center is the center of the hole and its radius is the size.

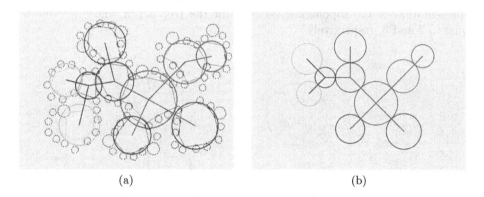

(a)   (b)

Fig. 6. (a) A topological skeleton according to Fig. 5(b) with the level of depth of the tree of 4, (b) representation of a topological skeleton based only on information about the relationships between holes and their sizes.

Let H_1, H_2 be any two holes. Let's consider that $r(H_1, H_2)$ is the distance between holes H_1 and H_2 which is calculated as the distance between the two closest points of these holes.

To create a topological skeleton, we will set all vertices of the source tree as direct child vertices of the parent. Next, for each hole H_k we look for the smallest distances $\min\limits_{k \neq i}[r(H_k, H_i)](k, i = 1, 2, .., l)$ with other $l - 1$ holes. Having data on the distances between holes, we begin to iterate through all the child holes of the parent hole.

Consider three holes such as parent H_p and two child holes H_1 and H_2. For hole H_1, we will set a new parent H_2 if the two conditions in Eq. 1 are satisfied:

$$r(H_1, H_2) < r(H_1, H_p) \ and \ r(H_2, H_p) < r(H_1, H_p) \tag{1}$$

Thus, for each child it is necessary to find an adjacent hole the distance to which will be less than the distance between the parent hole and the child. Also note that the distance between the adjacent hole and the parent hole must also be less than the distance between the child and the parent hole. If several adjacent holes satisfy Eq. 1 then we choose the adjacent hole that is closest to the child. In this case, the relationship between the child and the parent hole is broken, and the adjacent hole becomes a parent for the child. After iterating through all the child holes of the parent hole, we obtain the following structure of relationships shown in Fig. 6(a).

To store topological information about an object, it is sufficient to have a structure of connections between holes and their sizes. Based on this data, the representation of the object skeleton looks as shown in Fig. 6(b). This approach allows us to store the overall structure of the object with less computer memory.

By redefining the order of inheritance of holes according to the topological skeleton, we can choose different levels of depth. We can use this to choose the accuracy when comparing skeletons. For example, Fig. 7 shows the following representation of the topological skeleton for the tree depth whose a value is equal to 2 and 3, respectively.

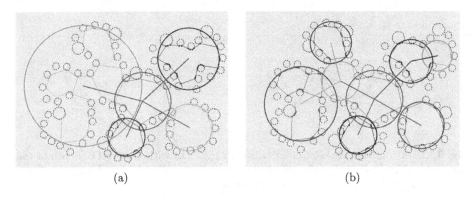

(a)   (b)

Fig. 7. A topological skeleton (a) with the level of tree depth of 2, (b) with the level of tree depth of 3.

3.3   Comparison of Topological Skeletons

To compare objects we use an algorithm that counts the number of child holes for each node. We assume that vertices are identical if they have the same number of child holes and are inherited from vertices previously recognized as identical. The depth value is used as the accuracy of the comparison.

We can see in Fig. 8 a semi-structured object and deformed after generalization which consist of holes having a similar number of child holes at each level of detail. Therefore, we identify these two objects as the same.

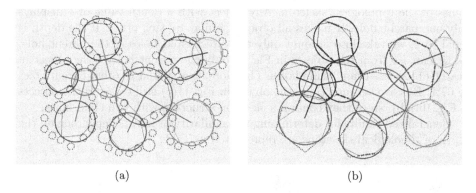

(a)	(b)

Fig. 8. Comparison of topological skeletons in which depth level of each tree is equal to 4.

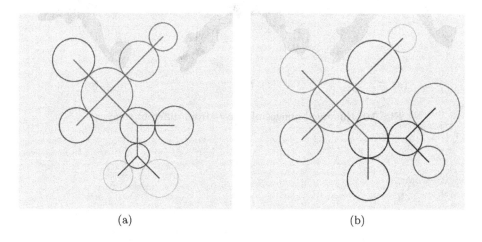

(a)	(b)

Fig. 9. Representation of topological skeletons according to Fig. 8.

In addition, if we recreate the skeleton from existing trees (see Fig. 9) then we will notice that Fig. 9(a) obtained from Fig. 8(a) differs in size from Fig. 9(b) obtained from Fig. 8(b). But since we assume that the object was deformed, we do not take into account the size of the holes, and compare only the topological features.

4   Results

Consider the sequence of objects for experimental studies. We take parts of the forest maps from different sources and binarize them (see Fig. 10). We construct their topological skeletons from binarized images (see Fig. 11).

Now let's compare the objects shown in Fig. 11 using different tree depth values of the topological skeleton. Any object with a depth value of 1 displays only one parent hole. It makes all objects similar to each other. If the depth is equal to 2, we take into account only the direct child holes of the parent hole. As we can see from the objects in Fig. 11 only the parent hole of the object in Fig. 11(c) has less than 3 child holes. Therefore, it is similar to other parent holes by 67%. But at the same time, the object in Fig. 11(c) still exists in the objects in Fig. 10(a,b), since they contain a skeleton structure with two child holes.

Generalized results for determining the similarity of forests depending on the depth of topological skeletons are represented in Table 1.

(a)   (b)   (c)

Fig. 10. Binarized images of forests from different sources.

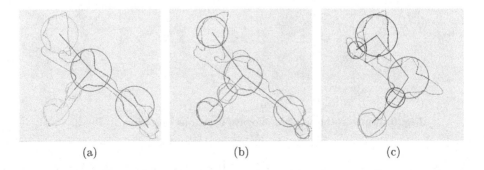

(a)   (b)   (c)

Fig. 11. Topological skeletons for the initial objects in Fig. 10.

Table 1. Results showing the similarity in percent between objects in Fig. 10 based on their topological skeletons taking into account the depth of the tree.

Skeleton tree depth	Object	a	b	c
1	a	100	100	100
	b	100	100	100
	a	100	100	100
2	a	100	100	67
	b	100	100	67
	a	100	100	67
3	a	100	100	67
	b	100	100	67
	a	100	100	67

5   Conclusion

The problem for processing semi-structured data is considered in the article. It is shown that currently the most effective models for the representation and processing of graphic objects are continuous models that approximate the analyzed object in the form of geometric objects and skeletons. It is required to use the methodology for semi-structured data at different scales, which allows us to identify the overall structure of such objects. To solve this problem, we propose an algorithm for calculating the topological skeleton of semi-structured data based on persistent homology. The iterative process of creating a topological skeleton tree is shown. Experimental studies are presented to compare topological skeletons with different depth levels. The topological skeleton shows the spatial relationships between the holes, which are formed as a result of the algorithm using persistent homology.

As a further research, the problem of analysis of spatial relations between holes is promising. The developed algorithm can be used to compare different-scale maps in geoinformatics, when searching for images of a similar structure and for processing semi-structured information.

Acknowledgment. The reported study was funded by RFBR and Vladimir region according to the research project №17-47-330387.

References

1. Eremeev, S.V., Andrianov, D.E., Komkov, V.A.: Comparison of urban areas based on database of topological relationships in geoinformational systems. Pattern Recogn. Image Anal. **25**(2), 314–320 (2015)
2. Lomov, N.A., Mestetskiy, L.M.: Area of the disk cover as an image shape descriptor. Comput. Opt. **40**(4), 516–525 (2016)

3. Vizilter, Yu.V., Sidyakin, S.V.: Comparison of shapes of two-dimensional figures with the use of morphological spectra and EMD metrics. Pattern Recogn. Image Anal. **25**(3), 365–372 (2015)
4. Eitz, M., Richter, R., Boubekeur, T., Hildebrand, K., Alexa, M.: Sketch-based shape retrieval. ACM Trans. Graph. **31**(4), 1–10 (2012)
5. Li, Y., Hospedales, T.M., Song, Y.Z., Gong, S.: Fine-grained sketch-based image retrieval by matching deformable part models. In: BMVC (2014)
6. Dugat, V., Gambarotto, P., Larvor, Y.: Qualitative geometry for shape recognition. Appl. Intell. **17**, 253–263 (2002)
7. Wang, F., Kang, L., Li, Y.: Sketch-based 3D shape retrieval using convolutional neural networks. In: CVPR, pp. 1875–1883 (2015)
8. Bai, X., Latecki, L.J.: Path similarity skeleton graph matching. IEEE Trans. Pattern Anal. Mach. Intell. **30**(7), 1282–1292 (2008)
9. Hariharan, B., Zitnick, C.L., Dollar, P.: Detecting objects using deformation dictionaries. In: CVPR, pp. 1995–2002 (2014)
10. Kalisnik, S., Kurlin, V., Lesnik, D.: A higher-dimensional homologically persistent skeleton. Adv. Appl. Math. **102**, 113–142 (2019)
11. Zomorodian, A., Carlsson, G.: Computing persistent homology. Discret. Comput. Geom. **33**(2), 249–274 (2005)
12. Edelsbrunner, H.: Computational Topology: An Introduction. American Mathematical Society, Providence (2009)
13. Eremeev, S., Kuptsov, K., Seltsova, E.: Algorithm for selecting homogeneous regions from a set of spatial objects. In: SCM, pp. 109–113 (2017)
14. Eremeev, S., Kuptsov, K.: Spatial objects classification algorithm on the basis of topological features of a form. In: RPC, pp. 44–48 (2017)
15. Eremeev, S., Kuptsov, K., Romanov, S.: An approach to establishing the correspondence of spatial objects on heterogeneous maps based on methods of computational topology. In: van der Aalst, W., et al. (eds.) AIST 2017. LNCS, vol. 10716, pp. 172–182. Springer, Cham (2018). https://doi.org/10.1007/978-3-319-73013-4_16

Comparison of Machine Learning Algorithms in Restaurant Revenue Prediction

Stepan Gogolev$^{(\boxtimes)}$ (ID) and Evgeniy M. Ozhegov (ID)

Research Group for Applied Markets and Enterprises Studies,
National Research University Higher School of Economics, Perm, Russia
s.l.gogolev@gmail.com, tos600@gmail.com

Abstract. In this paper, we address several aspects of applying classical machine learning algorithms to a regression problem. We compare the predictive power to validate our approach on a data about revenue of a large Russian restaurant chain. We pay special attention to solve two problems: data heterogeneity and a high number of correlated features. We describe methods for considering heterogeneity—observations weighting and estimating models on subsamples. We define a weighting function *via* Mahalanobis distance in the space of features and show its predictive properties on following methods: ordinary least squares regression, elastic net, support vector regression, and random forest.

Keywords: Weighted regression · Machine learning · Revenue prediction

1 Introduction

A global trend of collecting and storing information creates the demand for methods of analyzing and exploiting it. For instance, firms are interested in quantifying some qualitative features, explaining and creating predictions of consumers behavior.

Nowadays, a large number of machine learning methods provides algorithms of model creation to optimize almost any business process. However, some models and methods should be applied to some specific kinds of data only [5]. There is never a unique answer to the question about the choice of models and methods. Therefore, researchers benefit as much from the use of a methodology the most suitable to their dataset and a specific problem.

In the paper, we discuss the advantages of common machine learning methods applied to the problem of restaurant revenue prediction in Russian cities. A

The article was prepared within the framework of the Academic Fund Program at the National Research University Higher School of Economics (HSE University) in 2019–2020 (grant #19-04-048) and within the framework of the Russian Academic Excellence Project "5–100".

© Springer Nature Switzerland AG 2020
W. M. P. van der Aalst et al. (Eds.): AIST 2019, CCIS 1086, pp. 27–36, 2020.
https://doi.org/10.1007/978-3-030-39575-9_4

similar problem is studied in paper [6] where decision trees are employed to heterogeneous objects analysis. We follow this paper by applying other methods and show its suitability to a problem of franchise restaurant revenue prediction and choice of the best location for a new restaurant.

In the paper, we use information collected from operating restaurants to predict revenue for potential restaurants in other cities. According to the industrial organization literature (See, for ex., [2,5], and [12]), revenue of the restaurant depends on a number of characteristics including number of people in the local market, the average wage, a size of different target age groups, the number of direct and indirect competitors, characteristics of nearest competitors, etc. These features are closely related to each other, so the problem of partial feature collinearity occurs.

Heterogeneity of cities is another typical challenge of the location choice problem [13]. The problem is to study cities with a population from 10 thousand up to 12 million people in Russia. The proposed methodology is able to predict revenue for restaurants in cities of various size, taking into account their qualitative differences and variation in population and average wage.

2   Problem Statement

In this paper, we compare different methods for a problem of revenue prediction for franchise restaurants in cities where are no such restaurants. Restaurants belong to a franchise fast-food industry where each restaurant within a franchise is very standardized between cities. There are major distinctions in possible revenue due to the difference in cities, a location of a restaurant within a city and the degree of competition within a city.

Revenue is a relevant factor of success in a franchise as it reflects the number of clients visiting a restaurant and it does not depend on the quality of management, costs at the period and other in a franchise. Franchise system guarantees equal costs and profitability of restaurants in different cities. The reason is in a common technology of production and similar pricing on raw materials. It allows to concentrate efforts on comparison of cities suggesting other factors being equal.

The chosen franchise has about 300 restaurants in 184 cities. We analyze monthly revenue for the last 3 years[1]. From the starting point the data more than half of the restaurants were opened. Some restaurants were opened less than three years ago, so we collected 5889 observations as an unbalanced panel. Maintaining a panel structure instead of aggregating revenue is necessary to avoid seasonal bias.

According to the main goal of the study, we focus on objects' features (characteristics of cities) and creating a prediction for an average restaurant in the city given that there are no restaurants of the chosen franchise. We have three major groups of predictors: seasonal factors, specific restaurant characteristics

---

[1] Data is available at goo-gl.ru/5vIE.

(operating period and part of the revenue from delivery), and market environment features.

The market environment features consist of demographic[2] and competitors' characteristics[3]. The first group of variables includes detailed information about consumers: market size, its specific segments, consumer income. The second one describes firms' behavior on the market: a number of direct and indirect competitors, average restaurant bill, average estate price, and wage. Fast-food restaurants compete simultaneously on several markets: some types of cafes, restaurants, food delivery, etc. These markets are closely linked to each other and have common features. However, employing market and city characteristics raise the challenge of partial features collinearity that may provide a prediction bias.

Heterogeneity is another feature of the data. We analyze heterogeneity through heteroscedasticity of the errors and the presence of outliers. Common White test [14] proves the presence of heteroscedasticity at the 1% significance level. It can be interpreted as follows. Estimating model with linear regression gives different variation of error (degree of model accuracy) for different values of predictors. As for the outliers, the dataset contains some non-representative objects (cities). For instance, Moscow with a population equal 12 million inhabitants is almost 100 times than the average population of cities under consideration. Using the coefficient of variation [3] we check homogeneity of cities reveal the heterogeneity by following features: population, average wage, number of opened cafes, pizzerias and restaurants.

3 Methodology

3.1 Model Comparison Algorithm

In order to overcome the issue with the presence of heteroscedasticity and outliers described above, we follow [15] and use MAPE (mean absolute percentage error) instead of MSE (mean squared error) as a prediction quality criterion. We compare the predictive power of models by MAPE in order to give lower weight to predictions with non-representatively large errors. It does not have the property of underestimation the largest errors like mean squared error or other metrics using squared errors [15]. As we calculate model errors for all objects, MAPE shows the average absolute error of the model in percent of the average value of the target variable in our case. The lower MAPE is, the more predictive power of the model has. It will be useful for further interpretation of the metric.

The next important step is choosing the technique for assessing the prediction power of the models. We compare out-of-sample predictions due to possible overfitting problem by the procedure of leave-one-out cross-validation for model parameters and 10-fold cross-validation for hyperparameters. Now we move to describe the steps of cross-validation in detail.

---

[2] Data are taken from gks.ru.
[3] Data taken from 2GIS.

The main idea is averaging the error of prediction among all available objects through predicting new data that was not used in estimating. For this, we choose one city in leave-one-out and $\frac{1}{10}$ of all cities in 10-fold cross-validation as a test sample and exclude appropriate observations from the dataset, remaining observations from the training sample. Then we train a model on different training objects and choose optimal hyperparameters according to mean absolute error on 10 test samples. Finally, we train the model hyperparameters optimally selected at the previous step on different training samples and create predictions for the test observations related to one object. We repeat these steps for other test cities. As a result, we obtain a vector of out-of-sample predictions for all cities in the dataset. Then we compare mean absolute error between predictions and actual values and test what model gives the best results. To validate our results and estimate the possible overfitting issue, we also calculate the in-sample coefficient of determination (R^2) that shows the proportion of the explained variance in target variable and reflects the goodness of fit of a model on a training data.

3.2   Prediction Models

We follow [5,6] and use four different methods of regression estimation: linear regression (OLS), elastic net (ELNET), support vector regression (SVR) and random forest regression (RF). The last three methods allow to overcome a feature collinearity problem.

Firstly, we make revenue predictions using naive model and linear regression model to compare other results with these baselines. Naive model is OLS with a constant only. We calculate confidence intervals for models fit *via* bootstrapping with 200 replications test the statistical difference in a fit.

In the linear regression model we minimize the sum of squared errors to obtain optimal parameters values β in a linear index:

$$\hat{\beta}_{OLS} = \underset{\beta}{\operatorname{argmin}}(y - x\beta)'(y - x\beta). \tag{1}$$

where y is monthly revenue of restaurant in the city in the month, x is a vector of features including characteristics of the city and competitors in city, specific characteristics of the restaurant in the period and seasonal factors that depends on the period, and β are parameters to be estimated.

This method may suffer from a high degree of partial multicollinearity. We check the variance inflation factors (VIF) [9] for the group of competitors characteristics. In this group, all 12 factors have VIF more than 20 that signals on the problem of correlated factors.

An elastic net regularization method is one of the solutions to the multicollinearity issue [7]. This method minimizes the sum of squared errors penalized on the absolute and squared values of estimated parameters:

$$\hat{\beta}_{ELNET} = \underset{\beta}{\operatorname{argmin}}(y - x\beta)'(y - x\beta) + \lambda_1||\beta||_1 + \lambda_2||\beta||_2^2. \tag{2}$$

where λ_1 and λ_2 are parameters of regularization to be estimated.

As a result of the optimization problem solving, some predictors can be excluded from target variable prediction when its parameter value β shrunk to 0. Therefore, the model does not take into account additional information that excluded variables contain.

SVR provides another way of estimating model parameters. Unlike least squares methods, SVR avoids explicit specification of the regression equation [11]. SVR training process depends on the kernel function that defines the relationship between the target variable and predictors. Hence, it is crucial to concentrate on the choice of a kernel function. We check common kernel functions including Gaussian, linear, and polynomial. Generally, type of the kernel function can be chosen based on the type of relation between features if it is known. We use 10-fold cross-validation to select the best kernel function and calibrate its hyperparameters (regularization parameter C, tolerance ε) and degree for polynomial kernel function.

The last method we apply is a RF regression—an ensemble of regression trees. Training of a tree is an iterative process where the input data is split by predictors into smaller groups with different predicted value in each partition group. Combination of such trees is an ensemble that allows to reduce prediction variance and improve out-of-sample prediction power. Another advantage of using regression trees is the revelation of a nonlinear relation between the target variable and predictors [8].

The quality of a RF model mainly depends on the following parameters: the number of trees in an ensemble and the number of predictors randomly sampled in each split. The former should be large enough to reduce the variance of prediction, raised as a result of correlated variables in input data. The last parameter corresponds to the quality of the model. The higher the number of variables used, the better the quality of the model and the higher probability of overfitting. We tune both parameters using the out-of-bag estimation of the model. It is based on the sampling of test observations and calculating prediction error for observations which were not used in the training process of the model. It is proved that out-of-bag error estimations tend to leave-one-out cross-validation estimation what makes them a reliable method for selecting parameters of RF [4].

3.3 Accounting for Heterogeneity

Accounting for a heterogeneity requires the use of specific methods. We use two common ways: weighting of observations and training of model on subsamples through data partition [1]. The first method consists of giving various weights to different observations in the process of model training, while the second way assumes reducing objects in training dataset to the most relevant ones.

Both approaches use implicitly a function that assigns to all objects in the dataset (cities in our case) a value that reflects the proximity of objects. We can define this function as a distance function between two points in the space of objects characteristics. Let us describe steps on implementation of methods accounting for heterogeneity to a problem of revenue prediction.

To create a model that provides a prediction of revenue for the restaurant in a test city i, we should train the model on the remaining dataset $-i = \{j \in N, j \neq i\}$ (observations related to training cities). After that, we calculate distances from each training city in $-i$ to the test city i. In the case of weighting observations, the next step is transforming distances into weights and estimating the model. Naturally, we put higher weight to observation with a lower distance to the test city. Therefore, it is possible to use the inverse function to transform distance into weight. In this work, we use inverse power function. The definition of weight is follows:

$$w_{ij} = d_{ij}^{-\gamma} \tag{3}$$

where γ is a parameter to be estimated.

Now we turn to another case of training model on a subsample. We introduce the rule that defines an interval of values of distance that indicates whether to include the city in training dataset or not. We define bounds of the interval so that there are 75% of observations the most similar to the test city. The percent of observations that will be included in the training dataset is chosen according to the size of the overall dataset. That is to say, we include observations related to object j in training dataset if $d_{ij} \leq Q_{0.75}$, where $Q_{0.75}$ is a 75%-th quantile of the distance distribution among all j.

After that, we define a space of characteristics and a distance function between objects. As we can distinguish the most heterogeneous variables, a possible solution is to consider all of them in a distance function. We can construct overall distance as a sum of distances in all dimensions only if dimensions are orthogonal. Otherwise, distances in dimensions responding for correlated predictors would be overfitted. Mahalanobis distance function allows to include values from different dimensions with different weights [10]. It measures the difference between the object and the distribution of other objects in terms of standard deviations. The distance between each training observations and test observation i with the covariance matrix of predictors Ω is follows:

$$d_{ij} = \sqrt{(x_i - x_j)'\Omega^{-1}(x_i - x_j)}. \tag{4}$$

Mahalanobis distance is applicable to correlated variables, hence researcher can choose any combination of variables that forms a space of objects characteristics. In this work, we include three the most heterogeneous variables in weighting function: population, average wage and the number of restaurants competitors in the city.

After describing two procedures of heterogeneity eliminating it is necessary to discuss the compatibility of these procedures with 4 ML methods, starting with the simplest OLS and ELNET methods. The addition of weighting function to them modifies objective functions presented in Eqs. (1 and 2) into weighted errors minimization problems:

$$\hat{\beta}_{OLSW}^i = \underset{\beta}{\mathrm{argmin}}(y_{-i} - x_{-i}\beta)'\,\mathrm{diag}(w_i)(y_{-i} - x_{-i}\beta). \tag{5}$$

$$\hat{\beta}^i_{ELNETW} = \underset{\beta}{\mathrm{argmin}}(y_{-i} - x_{-i}\beta)'\,\mathrm{diag}(w_i)(y_{-i} - x_{-i}\beta) + \lambda_1||\beta||_1 + \lambda_2||\beta||_2^2. \quad (6)$$

Estimation on subsamples for these methods is acceptable but has a significant drawback. Subsampling reduces the size of the dataset and estimation efficiency. Strict selecting of observations in training dataset may results in the poor model due to insufficient information in selected data. At the same time, soft selection can keep training dataset unchanged.

Turning to a SVR model, the use of weighting there is not recommended. The algorithm of SVR assumes estimation of the model, based on the training data points nearest to the hyperplane. This means that the model is automatically trained on observations closest to the "average" observation, while outliers are ignored. The most suitable way to train the SVR model for some test object is estimating on a subsample where test object represents average observation. Such a model shows better predictive power despite the small training sample size.

The problem of heterogeneity in RF regression is eliminated automatically due to splitting input data into smaller groups. In this model quality of prediction mostly depends on the number of training objects similar to test ones. If it is large enough, regression trees are able to divide observations into groups better than other methods. However, in the lack of similar objects and observations, RF regression often does not show good fir due to low ability to extrapolate relations.

In the next section, we show the comparison of these algorithms and provide results for an ensemble of simple predictors. We find optimal weights for models in an ensemble using constrained linear regression and explain the resulting weights.

4   Results

Out-of-sample prediction for a city assumes creating a training model on the sample that does not contain any information about the city for what we make a prediction for. Table 1 shows measures of accuracy—MAPE and R^2—for out-of-sample prediction as an error percentage of mean overall monthly revenue and in-sample coefficient of determination.

The naive model gives the baseline out-of-sample prediction for comparison with other methods as we assumed. 95% confidence interval for MAPE in this model is from 46.3% to 72.6%. The highest MAPE in other methods is 41.3% (in OLS estimated on a subsample), so we come to the conclusion that all described methods are statistically significant and provide a better fit than the prediction by mean y. With the improvement model from OLS to ELNET method, MAPE decreases from 38 to 34.5%. It proves the benefits of regularization methods usage in the case of the high number of correlated variables. Modifying the least squares method with weighting function (optimal value of parameter γ is equal to 0.8) also improves the predictive power of the model. It decreases the variability of predictions (SD falls) with error level. Estimating model on a subsample does not

Table 1. Prediction power of models

	Mean	SD	Out-of-sample MAPE	In-sample R^2
y	2 643 306	1 571 221	–	–
Model for $\hat{y}$				
Naive model	2 643 306	0	59.44%	0.00
OLS	2 367 107	1 476 862	38.17%	0.58
OLS on a subsample	2 332 341	1 622 088	41.26%	0.37
OLS with weighting	2 365 270	1 461 773	36.64%	0.56
ELNET	2 324 616	1 123 046	34.54%	0.45
ELNET on a subsample	2 453 445	1 518 215	37.65%	0.53
ELNET with weighting	2 296 467	1 096 400	33.29%	0.42
SVR	2 387 787	1 271 851	32.76%	0.65
SVR on a subsample	2 353 508	1 301 179	33.37%	0.70
RF	2 379 884	959 417	30.70%	0.93
Ensemble	2 372 320	1 307 328	23.59%	0.97
Number of observations	5 889			
Number of objects	184			
Number of predictors	43			

improve any model due to the decrease of efficiency of models trained on smaller samples. Similar conclusions are associated with estimating SVR on a subsample. Overall, combining several methods (elastic net method and weighting function) allows to achieve the best quality of out-of-sample prediction for the least squared method.

SVR and RF regressions outperform results of linear regressions: MAPE is 32.8 and 30.7% respectively. The random forest model works better than other models at the regression problem with heterogeneity by construction. RF does not reveal averaged relations and does not extrapolate relations between variables to uncommon values of them. That is the reason why it is useless for predictions revenue in atypical cities. However, it is the best among the considered method for predicting. A higher value of in-sample R^2 indicates on a possible overfitting problem there. Using the ensemble of models improves results in terms of the coefficient of determination and MAPE because of overcoming overfitting problem and combining advantages of considered methods. Out-of-sample error in the ensemble is 2.5 times fewer that error in naive model and the lowest among all methods.

Although we show a statistically significant difference in accuracy by comparison errors of prediction with errors of the naive model, there is a lot of ways to improve results. As we solve problem of cities comparison and not the problem of optimal location within a city, we cannot include in the model some specific features (for instance, spatial characteristics about competitors inside the city).

To sum up, we compare the predictive power of some models on the dataset with heterogeneity and correlated predictors. Results show that the ensemble has properties to overcome both problems and has the lowest mean absolute error and the highest coefficient of determination. Moreover, we show the advantages of weighting observations in the estimating process and possible drawbacks of estimating models on subsamples.

5 Conclusion

In this paper, we summarize the methodology of constructing a model with the best predictive power to forecast revenue in the restaurant in the out-of-sample city. We describe methods of heterogeneity elimination in the model: observations weighting and data partition with the following estimation on subsamples. Additionally, we suggest some ways of dealing with collinearity problem: an elastic net method, support vector and random forest regressions. We show advantages of those methods under different assumptions and validate these statements at the problem of revenue prediction.

Basically, the paper can be extended in two ways. First of all, it is possible to consider other methods of solving problems: for instance, principal component analysis for reducing the number of correlated predictors or more detailed analysis of ensemble trees algorithms (bagging, boosting, etc.). The second way is to use a more accurate approach to compare model prediction power. For each model we can calculate a MAPE confidence interval using the bootstrap. Computing confidence intervals allows to compare the predictive power of models with more certainty.

References

1. Athey, S., Imbens, G.: Recursive partitioning for heterogeneous causal effects. Proc. Nat. Acad. Sci. **113**(27), 7353–7360 (2016)
2. Berthon, P., Holbrook, M., Hulbert, J.: Beyond market orientation. A conceptualization of market evolution. J. Interact. Mark. **14**(3), 50–66 (2000)
3. Bennett, B.: On an approximate test for homogeneity of coefficients of variation. In: Ziegler, W.J. (ed.) Contribution to Applied Statistics, pp. 169–171. Birkhäuser, Basel (1976). https://doi.org/10.1007/978-3-0348-5513-6_16
4. Breiman, L.: Heuristics of instability and stabilization in model selection. Ann. Stat. **24**(6), 2350–2383 (1996)
5. Chiang, W., Chen, J., Xu, X.: An overview of research on revenue management: current issues and future research. Int. J. Revenue Manag. **1**(1), 97–128 (2007)
6. Kim, S., Upneja, A.: Predicting restaurant financial distress using decision tree and AdaBoosted decision tree models. Econ. Model. **36**, 354–362 (2014)
7. Lee, D., Lee, W., Lee, Y., Pawitan, Y.: Sparse partial least-squares regression and its applications to high-throughput data analysis. Chemometr. Intell. Lab. Syst. **109**(1), 1–8 (2011)

8. Liu, S., et al.: Learning accurate and interpretable models based on regularized random forests regression. BMC Syst. Biol. **8**(3), S5 (2014)
9. Mansfield, E., Helms, B.: Detecting multicollinearity. Am. Stat. **36**(3a), 158–160 (1982)
10. Neale, M.C.: Individual fit, heterogeneity, and missing data in multigroup SEM. In: Modeling Longitudinal and Multiple-Group Data: Practical Issues, Applied Approaches, and Specific Examples. Lawrence Erlbaum Associates, Hillsdale (2000)
11. Cristianini, N., Shawe-Taylor, J.: An Introduction to Support Vector Machines and Other Kernel-Based Learning Methods. Cambridge University Press, Cambridge (2000)
12. Walzer, N., Blanke, A., Evans, M.: Factors affecting retail sales in small and mid-size cities. Community Dev. **49**(4), 69–484 (2018)
13. Wang, K., Wai, K., Liang, L., Fue, X.: Entry patterns of low-cost carriers in Hong Kong and implications to the regional market. J. Air Transp. Manag. **64B**, 101–112 (2017)
14. White, H.: A heteroskedasticity-consistent covariance matrix estimator and a direct test for heteroskedasticity. Econometrica **48**(4), 817–838 (1980)
15. Willmott, C., Matsuura, K.: Advantages of the mean absolute error (MAE) over the root mean square error (RMSE) in assessing average model performance. Clim. Res. **30**(1), 79–82 (2005)

Online Augmentation for Quality Improvement of Neural Networks for Classification of Single-Channel Electrocardiograms

Valeriia Guryanova[✉][iD]

Lomonosov Moscow State University, Leninskie Gory 1, Moscow 119991, Russia
valeriiaguryanova@gmail.com

Abstract. Currently, on the market, there are mobile devices that are capable of reading a person's single-lead electrocardiogram (ECG). These ECGs can be used to solve problems of determining various diseases. Neural networks are onearameters of augmentations of the approaches to solving such problems. In this paper, the usage of online augmentation during the training of neural networks was proposed to improve the quality of the ECGs classification. The possibility of using various types of online augmentations was explored. The most promising methods were highlighted. Experimental studies showed that the quality of the classification was improved for various tasks and various neural network architectures.

Keywords: Deep learning · Neural networks · Single-lead ECG classification · Online augmentation

1 Introduction

ECG is a signal that displays the electronic activity of the heart. Each ECG recording shows the potential difference between two electrodes located on the surface of the body. Each of the measured potential differences is called lead. In medical institutions, 12-lead ECGs are commonly used. There are plenty of works that show that many diseases can be identified using 12-lead ECG [1,2].

Currently, on the market, there are mobile devices that are capable of reading an electrocardiogram (ECG) of a person, for example, AliveCor [3], CardioQvark [4]. Such devices read only one of the 12 leads. The question is what diseases can be detected using such ECGs and what quality of the detection can be achieved. Some researches use deep learning approaches to analyze such ECGs [5,6].

Usually, when using neural networks, for example, in the field of image analysis, augmentation of data is often employed to improve the quality of classification. Augmentation is the enrichment of data by the addition of synthetic samples. Augmentation can be applied in two ways: either a one-time addition of synthetic samples at the data preparation stage (offline augmentation) or

© Springer Nature Switzerland AG 2020
W. M. P. van der Aalst et al. (Eds.): AIST 2019, CCIS 1086, pp. 37–49, 2020.
https://doi.org/10.1007/978-3-030-39575-9_5

the replacement of a part of data with a synthetic one during training with some probability (online augmentation). Online augmentation has the following advantages in comparison with offline: it reduces the time and storage costs for generating synthetic data, which simplifies the selection of the optimal combination of parameters and augmentations. It should be noted that augmentation can be viewed as some regularization to prevent over-fitting.

There are libraries, for example, [7], which provide implemented augmentation methods for images, allow you to create a combination of various methods and also simplify the process of training neural networks. There are no such libraries for ECG signals. The task of studying the possibility of improving the quality of classification due to online augmentation for single-channel ECGs is relevant today. It is the main goal of this article.

The paper is organized as follows. Initially, the general structure of the library prototype and the methods that have been implemented in it are described. Then, the used representations of ECG signals and architectures of neural networks that were used for the experiments are given. Then datasets on which experimental researches were carried out are reported. Finally, the structure of the experiments and the results obtained are provided.

2   The Library for Online Augmentation and Implemented Methods

There are following general principles of the implemented library: synthetic objects are generated based on a predetermined signal using some transformations; each transformation of the signal is performed by the transformer function, which is applied with a specific probability; the transformer can be chosen randomly from a certain group with a particular probability.

There are two groups of transformers for online augmentations in the library: one-signal transformers that use only one given signal from the training set; two-signal transformers that besides the main signal additionally use a signal from a specific group of the training set.

2.1   One-Signal Transformers

SetRandomZeros (RZ). This method was used in [8]. It sets some signal points to zero. The specific number of signal points for each transformation is set at random each time from a specified interval. The boundaries of this interval are the transformer parameters.

SetRandomZeroWindows (RZW). This method was deployed in [6]. It is similar to the previous method, but it sets to zero some windows. The specific window size is set at random each time from a specific interval. The boundaries of this interval are the transformer parameters. Other parameters of the transformer are the same as in the previous method.

AddNoise (N). This method was described in [8]. This method adds Gaussian noise to the signal. The average and standard deviation of noise are set as parameters of the transformer.

SignalSlice (SS). This method was used in [9]. This method takes a signal slice. The specific size of a slice is set at random each time from a certain interval. The boundaries of this interval are the transformer parameters.

Resample (R). This method was utilized in [9]. This method performs signal resampling. The specific resampling coefficient is set at random each time from a certain interval. The boundaries of this interval are the transformer parameters.

ResamplePart (RP). This method was proposed for this paper. It is similar to the previous method, but it resamples only signal parts, so the signal diversity is increasing. The specific size of the signal part is set at random each time from a certain interval. The boundaries of this interval are the transformer parameters. Other parameters of the transformer are the same as in the previous method.

RandomFilter (RF). This method was proposed for this paper. The main motivation was to make a neural network more robust to changes in very small and high frequencies. This method applies a Butterworth filter to the signal. Highpass or lowpass filter can be used as a filter type. Only second-order filters are used. The critical frequency is set at random each time from a certain interval. The boundaries of this interval are the transformer parameters.

Combine3Signals (3S). This method was described in [10]. It consists of the following steps: the signal is normalized; a sinusoid is generated with an initial phase from $[-180; -90]$ and an end phase from $[90; 180]$, the amplitude is randomly selected in the interval from $[-2; 2]$; two signals are combined with each other and with a random Gaussian noise with a mean of 0 and a standard deviation of 0.05. For the proper training and evaluation, all signals (even without augmentation and from test and validation sets) are normalized for this method.

2.2   Two-Signal Transformers

ReplaceRPeak (RR). This method was proposed for this paper. A random R-peak in a given signal is replaced with a random R-peak from a signal of the same class of the training sample. The primary motivation for this and the next approach was to try to create a new patient with the same disease. For some heart problems, single R-peak could determine the whole problem. The parameters of the transformer set the number of points before and after the R-peak, which participate in the replacement.

CombineRPeaks (CR). This method was proposed for this paper. This method is similar to the previous one, but the R-peak is not replaced but averaged with the R-peak of another signal.

All previous methods were applied directly to the signal. The following methods apply to the representation of a signal for neural network input.

MixUp (M). This method was proposed by [11]. Let (x_1, y_1) and (x_2, y_2) be two random objects of the training sample. Then, a new training sample object (x_3, y_3) will be constructed as ($\lambda \in [0, 1]$ – parameter of method, but in this paper only $\lambda = 0.2$ was used):

$$x_3 = \lambda x_1 + (1 - \lambda) x_2,$$
$$y_3 = \lambda y_1 + (1 - \lambda) y_2. \tag{1}$$

Autoencoder (AE). The main idea of this method was described in [12] and this method uses the autoencoder architecture for the construction of new examples. It is as follows:

- each example of the training set is projected into feature space by feeding it through the encoder, extracting the resulting context vector;
- for each sample in the training set, it is found K nearest neighbours in the space of context vectors, in this paper $K = 10$ and for each pair of neighbouring context vectors, a new context vector can then be generated using interpolation:

$$c' = (c_k - c_j)\lambda + c_j, \tag{2}$$

where c' is the synthetic context vector, c_i and c_j are neighbouring context vectors, and λ is a variable in the range $(0, 1)$ that controls the degree of interpolation. In this paper only $\lambda = 0.5$ was used. This new context vector can be used to generate new signal representation

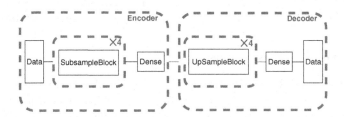

Fig. 1. Autoencoder architecture for online augmentation

In this paper, the autoencoder architecture that is shown in Fig. 1 is used. The encoder consists of four subsample blocks and dense layer. The general

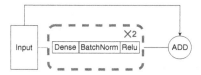

Fig. 2. The general architecture of upsample and subsample blocks

structure of the subsample block is shown in Fig. 2. There are a dense layer, batch normalization layer, and relu as the activation function. The first dense layer in a subsample block produces output two times less than the input vector. The second dense layer in the subsample block leaves the size of the output vector the same as the size of the input vector. The input vector and the output vector of two dense layers are summed up. The decoder consists of four upsample blocks and dense layer. The general structure of upsample block is the same as subsample block. The only difference is that it produces the output vector twice the input vector.

All methods can be divided into methods that are behind the motivation to create some new patients: ReplaceRPeak, CombineRPeaks, Autoencoder, 3Signals, and methods that are trying to make the neural network resistant to standard signal transformations that do not change the target: SetRandomZeroWindows, AddNoise, SignalSlice, Resample, ResamplePart, RandomFilter. A separate method is a MixUp whose authors tried to make neural network stable to adversarial examples and corrupted labels.

3   Signal Representation

Various signal representations can be used as the input vector for the neural network. Only representations with constant shape were deployed for the simplicity of training a neural network. The following representations were used in this paper.

- The R-peak represents ECG beat and is one of the most important components of a signal. The average R-peak, which is calculated over the entire signal can be used as a signal representation for the neural network (AverageRPeak). The method parameters are the number of points before and after the peak.
- Zero padding up to maximal signal length (ZeroPadding). All of the following methods use this method so that representations of signals with different length have the same size.
- The approximation coefficients from the signal wavelet transformation (Wavelets) can be used as a signal representation. In this paper, Daubechies wavelets [13] were used as wavelet functions.
- Signal spectrogram with the logarithmic transformation (LogSpectrogram) can be used as a signal representation. For the calculation of spectrogram

length of each segment was set to 3 of signal rate sample rate. Number of overlapped points was set to signal sample rate.

4   Neural Network Architectures

For this paper, three different neural network architectures were used.

The general structure of the first neural network(CardioResNet1D) is shown in Fig. 3. The architecture of the convolutional block (CBlock) is shown in Fig. 4. This architecture is similar to the one described in [14]. However, due to high computational complexity, it has only six convolution blocks instead of 15 and increments the number of filters every second convolution block instead of fourth respectively. ZeroPadding, AverageRPeak, and Wavelets can be used as input for this network.

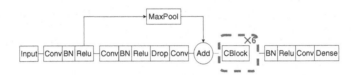

Fig. 3. The architecture of CardioResNet1D network

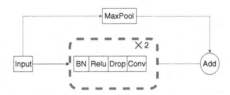

Fig. 4. The architecture of CardioResNet1D convolution block

The general structure of the second neural network(CardioResNet1D) is shown in Fig. 5. The architecture of the convolutional block (ConvBlock) is shown in Fig. 6. This architecture is similar to the one described in [6] as CNN architecture. To improve quality of classification and increase the stability of training it was proposed to use residual connections [15] in ConvBlocks as shown in Fig. 6. LogSpectrogram can be used as input for this network.

The third architecture is the simple architecture of two stacked LSTM layers with hidden size equals to 64 and dense layer with softmax. ZeroPadding, AverageRPeak, and Wavelets can be used as input for this network.

Fig. 5. The architecture of CardioResNet1D network

Fig. 6. The architecture of CardioResNet2D convolution block

5   Data for the Experiments

Three different datasets were used for experiments evaluation of online augmentation.

The first dataset is the data for classification of single-lead ECG according to the rhythm. This data was provided from the AliveCor device for the PhysioNet/CinC Challenge 2017 [16]. There are four classes of ECG: with a normal rhythm, with atrial fibrillation rhythm (AF), other rhythm and noisy recordings with the length of signal from 9 to 60 s. For this paper, only the first three classes were considered. There are 5154 recordings of normal rhythm, 771 recordings of AF rhythm and 2557 recordings of other rhythms. There is no information in the dataset description about the number of patients in the sample, and what cardiogram belongs to which patient. Therefore, the overall classification quality can be overstated. ECG recordings were sampled at 300 Hz, and they have been filtered by the AliveCor device. For subtracting trend, the median filter with a kernel size of 107 was used. It should be noted that in this paper the test set that was used for evaluation in a lot of articles, for example, in [6] was not publicly available.

The following datasets were collected using CardioQvark API [4]. There is dataset for binary classification of coronary heart disease (CHD) and tuberculosis (TB). ECG recordings from both datasets were sampled as 1000 Hz, and they have been filtered by the CardioQvark device. The length of ECG recordings varies from 30 s to 300 s. For subtracting trend, the median filter with a kernel size of 187 was used. The TB dataset consists of 4045 zero-labeled ECGs from 368 patients and 1232 positive-labeled ECGs from 136 patients. The CHD dataset includes of 3213 zero-labeled ECGs from 244 patients and 1378 positive-labeled ECGs from 236 patients.

6   Experiments

Due to the high computational complexity of experiments for each data set, only two different network architectures were used. The main goal of these experiments was not to achieve the best possible quality on a single dataset, but to show how augmentations can or can not improve the quality on a fixed neural network. The used architectures are shown in Table 1. This table shows which signal representation, network architecture, and signal sampling frequency were used, and the code name for reference to the model. For AverageRPeak representation, it is indicated in parenthesis the number of points before and after.

Table 1. Models description for different datasets

Dataset	Signal representation	Sample rate	Network architecture	Code Name
AliveCor	Wavelet	300	CardioResNet1D	A1
AliveCor	LogSpectogram	300	CardioResNet2D	A2
TB	LogSpectrogram	66	CardioResNet2D	T2
TB	AverageRPeak (500, 1000)	1000	CardioResNet1D	T1
CHD	AverageRPeak (500, 1000)	1000	CardioResNet1D	C1
CHD	AverageRPeak (1000, 1000)	1000	LSTM	CL

For all network architectures, the cross-entropy loss was used as a training objective, depending on the target, a binary or multiclass version was deployed. For all experiments, the data samples were divided into three parts: training, validation, and test sample. If for each cardiogram was indicated to which patient it belongs, then the splitting was done in a way that patients do not overlap between three samples. The training sample was used to train the model. The validation sample was used to select the parameters of the neural network and augmentations. Also, validation sampling was used for an early stop at the deterioration in quality during ten epochs. A test sample was used to evaluate the method. Parameters for training each neural network are listed in Table 2. Parameters that are not listed in the table were used by default.

Table 2. Parameters of neural networks

Model	Optimizer	Learning rate	Batch size
A1	Adam	0.00018	8
A2	SGD	0.013	16
C1	Adam	0.0001	4
CL	Adam	0.01	8
T1	SGD	0.01	4
T2	Adam	0.0001	8

Augmentations parameters that were different in experiments for different models are shown in Tables 3 and 4. For N method, the mean was set to 0 for all models. For the R method the resampling factor was in one of the intervals: (0.54, 1) or (1, 1.5). Each time the certain interval was chosen at random with equal probability from those intervals. In the table, the case when one of two values for a parameter is chosen randomly from two values is shown as |. In the RP method the resampling factor was in one of the intervals: (0.4, 1) or (1, 6). In the RF method, hp means high-pass filter and lp means low-pass filter. Due to high computational complexity, the AE method was used only for three models.

Table 3. Parameters of augmentations

Model	RZ	RZW		N	SS	AE
	Zeros amount	Zeros amount	Window length	Std	Slice size	Context size
C1	(1, 800)	(1, 800)	(2, 30)	0.005	(200000, 299990)	—
CL	(1, 800)	(1, 800)	(2, 30)	0.005	(200000, 299990)	100
T1	(1, 800)	(1, 800)	(2, 30)	0.005	(200000, 299990)	100
T2	(1, 100)	(1, 100)	(2, 30)	0.001	(15000, 19998)	—
A1	(3, 90)	(3, 90)	(2, 30)	0.001	(17000, 18285)	300
A2	(3, 90)	(3, 90)	(2, 30)	0.001	(17000, 18285)	—

Table 4. Parameters of augmentations

Model	RP	RR/CR		RF
	Part length	Points before	Points after	Frequency and type
C1	(600, 3200)	1000	1000	(2.5, 5)hp| (20, 30)lp
CL	(600, 3200)	1000	1000	(2.5, 5)hp| (20, 30)lp
T1	(600, 3200)	1000	1000	(1.5, 4)hp| (18, 25)lp
T2	(40, 250)	60	60	(2.5, 5)hp| (20, 30)lp
A1	(30, 1000)	180	180	(1.34, 3)hp| (9, 13)lp
A2	(30, 1000)	180	180	(1.34, 3)hp| (9, 13)lp

Methods were compared using various quality metrics (QM). For binary classification problems, the following quality metrics were utilized: the binary cross-entropy loss (Loss), accuracy (Acc), f-score (F1), roc-AUC score (AUC). For multiclass classification problems, the following quality metrics were considered: the cross-entropy loss(Loss), accuracy (Acc). Because of the high computational complexity, only the binary cross-entropy loss and the cross-entropy loss were used as metrics for an early stop.

The best model out of ten launches was used for models comparison. The results are shown in Table 5. The None column represents the case when no augmentations were used. Pr means the probability of augmentation with which

Table 5. Experimental results

QM(M)	None	RZ	RZW	N	SS	R	RP	RF	3S	RR	CR	M	AE
Loss (A1)	0.56	0.548	0.538	0.540	0.538	0.537	**0.521**	0.525	0.544	0.525	0.536	0.531	0.540
Pr		0.1	0.5	0.1	0.4	0.2	0.2	0.1	0.3	0.1	0.3	0.2	0.4
Acc (A1)	0.79	0.799	0.806	0.791	0.792	0.804	**0.809**	0.80	0.797	0.804	0.796	0.798	0.801
Pr		0.1	0.1	0.1	0.5	0.1	0.2	0.3	0.1	0.5	0.3	0.2	0.2
Loss (A2)	0.603	0.577	0.57	0.567	0.58	0.563	**0.561**	0.581	0.569	0.584	0.578	0.565	—
Pr		0.1	0.1	0.3	0.5	0.5	0.4	0.5	0.5	0.4	0.5	0.4	—
Acc (A2)	0.756	0.762	0.763	0.761	0.759	0.77	**0.772**	0.757	0.767	0.764	0.76	0.765	—
Pr		0.4	0.1	0.2	0.3	0.5	0.5	0.1	0.5	0.4	0.5	0.2	—
Loss (T1)	0.381	0.366	0.36	0.365	0.347	0.365	0.3635	0.349	0.356	0.365	0.351	**0.344**	0.352
Pr		0.4	0.5	0.3	0.5	0.4	0.1	0.2	0.3	0.1	0.4	0.1	0.2
Acc (T1)	0.843	0.859	0.863	.850	0.859	0.864	**0.866**	0.842	0.831	0.851	0.858	0.844	0.859
Pr		0.4	0.5	0.1	0.4	0.4	0.1	0.5	0.1	0.2	0.4	0.2	0.5
F1 (T1)	0.681	0.694	0.69	0.713	0.712	0.711	0.706	0.691	0.693	0.708	**0.72**	**0.72**	0.704
Pr		0.5	0.5	0.1	0.5	0.2	0.1	0.4	0.3	0.2	0.4	0.3	0.1
AUC (T1)	0.877	0.889	0.885	0.882	0.883	0.886	**0.891**	0.883	0.883	0.878	0.891	0.887	0.882
Pr		0.5	0.5	0.2	0.1	0.4	0.1	0.2	0.4	0.2	0.5	0.1	0.3
Loss (T2)	0.4062	0.380	0.379	0.354	0.361	0.365	0.361	0.374	0.379	0.368	0.35	**0.3489**	—
Pr		0.2	0.3	0.1	0.5	0.4	0.1	0.3	0.4	0.3	0.3	0.3	—
Acc (T2)	0.834	0.845	0.858	0.854	0.838	0.852	0.858	0.849	0.845	0.8547	0.844	**0.8638**	—
Pr		0.3	0.3	0.1	0.5	0.3	0.1	0.2	0.2	0.1	0.4	0.2	—
F1 (T2)	0.6256	0.687	0.679	0.687	0.636	0.68	**0.709**	0.662	0.686	0.6644	0.67	0.6865	—
Pr		0.2	0.3	0.2	0.4	0.3	0.1	0.5	0.5	0.3	0.5	0.3	—
AUC (T2)	0.862	0.869	0.879	0.873	0.883	0.883	**0.894**	0.866	0.865	0.8690	0.876	0.879	—
Pr		0.2	0.5	0.2	0.5	0.3	0.1	0.2	0.1	0.3	0.2	0.3	—
Loss (C1)	0.4785	0.459	0.466	0.479	0.462	0.472	0.456	0.462	0.464	0.451	**0.443**	0.4531	0.457
Pr		0.3	0.3	0.2	0.1	0.2	0.2	0.2	0.3	0.3	0.2	0.2	0.2
Acc (C1)	0.789	0.803	0.802	0.792	0.802	0.801	0.801	**0.809**	0.799	0.796	0.798	0.803	0.8
Pr		0.3	0.2	0.1	0.1	0.3	0.3	0.3	0.4	0.4	0.2	0.1	0.2
F1 (C1)	0.677	0.688	0.685	0.675	0.689	0.681	0.688	0.687	0.692	**0.7**	0.682	0.672	0.681
Pr		0.3	0.2	0.2	0.1	0.1	0.2	0.3	0.2	0.4	0.4	0.1	0.3
AUC (C1)	0.865	0.87	0.859	0.861	0.863	0.872	**0.874**	0.872	0.869	0.87	0.864	0.854	0.872
Pr		0.2	0.2	0.3	0.1	0.5	0.4	0.1	0.1	0.2	0.2	0.2	0.3
Loss (CL)	0.4965	0.481	0.479	0.489	**0.463**	0.475	0.477	0.472	0.467	0.4877	0.469	0.468	—
Pr		0.5	0.4	0.3	0.4	0.2	0.1	0.5	0.2	0.2	0.3	0.3	—
Acc (CL)	0.770	0.774	0.775	0.784	0.779	0.78	0.773	**0.79**	0.774	0.7682	0.784	0.774	—
Pr		0.5	0.3	0.2	0.4	0.3	0.1	0.1	0.3	0.2	0.4	0.2	—
F1 (CL)	0.606	0.624	0.621	0.623	0.64	0.621	0.635	0.621	0.6319	0.636	0.638	**0.645**	—
Pr		0.5	0.3	0.4	0.4	0.5	0.3	0.2	0.4	0.3	0.2	0.2	—
AUC (CL)	0.799	0.810	0.811	0.819	0.828	0.826	0.816	0.819	0.814	0.8259	0.824	**0.829**	—
Pr		0.3	0.1	0.3	0.2	0.1	0.1	0.2	0.3	0.3	0.5	0.4	—

this quality was achieved. The following probabilities were considered: 0.1, 0.2, 0.3, 0.4, 0.5.

The best quality from all online augmentations is highlighted in bold. As follows from the above table, it was possible with the help of online augmentations to improve the quality of various criteria for various models and data. The methods that showed the best improvement: ResamplePart, Mixup, CombinerRPeaks, ReplaceRPeaks, Resample, SignalSlice, and RandomFilter. For each method that gives the best improvement for the combination of model and qual-

Table 6. Statistical significance

Model	Loss		Acc		F1		AUC	
	Method	P-value	Method	P-value	Method	P-value	Method	P-value
A1	RP	0.00085	RP	0.02	—	—	—	—
A2	RP	9.08e-05	RP	0.00084	—	—	—	—
T1	M	0.003	RP	0.005	CR	0.0002	RP	0.04
T2	M	0.001	M	0.04	RP	0.0001	RP	0.04
C1	CR	0.001	RP	0.008	3S	0.007	RP	0.065
CL	SS	0.0004	RF	0.02	M	0.0001	M	0.003

ity metric statistical tests were conducted to check for the significance of changes in the quality obtained. The Mann – Whitney U test was used for this purpose. The results are shown in Table 6. The lowest p-value observed on loss(binary cross-entropy or cross-entropy) quality metric. The reason for this is probably covered in that the training procedure used this metric for early stopping on the validation set, so the best model were chosen based on this metric. The main goal of this article was to explore the improvement in quality on at least one quality metric but a moderate improvement in other quality metrics can be observed especially in the f-score metric. It varies from 0.023 to 0.083. This is the important metric for signal classification as a sample is often imbalanced.

ResamplePart can be considered the best augmentation method since it gives the best quality improvement on several datasets and is one of the simplest to implement and it does not take a second signal or R-peak calculation. It can be assumed that this method achieves good quality as it preserves the general shape of the signal, but it changes the number of points in some of its parts. Thus, the network is learning not to focus on specific points, but to extract general information, which in general has remained intact. MixUp can be considered as the next best method, the justification why it can work well is in the article [11].

RandomZeros and RandomZeroWindows do not give a big increase in quality, it can be assumed that the reason for this is their similarity to the dropout method that is present in the used architectures. While autoencoder gives above-average quality improvement, it requires great development effort and significant additional computational power, which makes it not the most optimal augmentation to use in an average deep learning pipeline.

It is difficult to draw a conclusion about the probabilities of augmentation on which the best quality is achieved based. The best approach at the moment seems to look at various probabilities to maximize the quality.

7   Conclusion

In this paper online-augmentation for single-lead ECGs was presented. Some new methods were introduced. Besides them: ResampleParts, RandomFilter,

ReplaceRPeaks and CombineRPeaks. ResampleParts shows the best quality improvement on different datasets and neural network architectures. Different methods for online-augmentation were compared on three different data sets and various neural network architectures. Experimental studies showed that the quality of the classification by using for comparison different quality criteria was improved for various tasks and various neural network architectures.

In future works, various combinations of the studied methods may be investigated. Also, the possibility of using some libraries for the selections of optimal hyperparameters combinations can be explored.

Acknowledgements. This article contains the results of a project carried out within the implementation of the Program of the Center for Competence of the National Technology Initiative "Center for Storage and Analysis of Big Data", supported by the Ministry of Science and Higher Education of the Russian Federation under the Lomonosov Moscow State University Project Support Fund 13/1251/2018 from 11.12.2018.

References

1. Ribeiro, A.H., et al.: Automatic Diagnosis of Short-Duration 12-Lead ECG using a Deep Convolutional Network. arXiv preprint arXiv:1811.12194 (2018)
2. Houssein, E.H., Kilany, M., Hassanien, A.E.: ECG signals classification: a review. Int. J. Intell. Eng. Inform. **5**, 376 (2017). https://doi.org/10.1504/ijiei.2017.087944
3. AliveCor. https://www.alivecor.com
4. CardioQvark. http://www.cardioqvark.ru
5. Yao, Z., Zhu, Z., Chen, Y.: Atrial fibrillation detection by multi-scale convolutional neural networks. In: 2017 20th International Conference on Information Fusion (Fusion). IEEE (2017). https://doi.org/10.23919/ICIF.2017.8009782
6. Zihlmann, M., Perekrestenko, D., Tschannen, M.: Convolutional recurrent neural networks for electrocardiogram classification. In: 2017 Computing in Cardiology (CinC). IEEE (2017)
7. Buslaev, A., et al.: Albumentations: fast and flexible image augmentations. arXiv preprint arXiv:1809.06839 (2018)
8. Schlüter, J., Grill, T.: Exploring data augmentation for improved singing voice detection with neural networks. In: ISMIR (2015)
9. Le Guennec, A., Malinowski, S., Tavenard, R.: Data augmentation for time series classification using convolutional neural networks. In: ECML/PKDD Workshop on Advanced Analytics and Learning on Temporal Data (2016)
10. Tan, J.H., et al.: Application of stacked convolutional and long short-term memory network for accurate identification of CAD ECG signals. Comput. Biol. Med. **94**, 19–26 (2018)
11. Zhang, H., et al.: mixup: Beyond empirical risk minimization. arXiv preprint arXiv:1710.09412 (2017)
12. DeVries, T., Taylor, G.W.: Dataset augmentation in feature space. arXiv preprint arXiv:1702.05538 (2017)
13. Liu, C.L.: A Tutorial of the Wavelet Transform, p. 1–72. National Taiwan University, Department of Electrical Engineering (NTUEE), Taiwan (2010)
14. Rajpurkar, P., et al.: Cardiologist-level arrhythmia detection with convolutional neural networks. arXiv preprint arXiv:1707.01836 (2017)

15. He, K., et al.: Deep residual learning for image recognition. In: Proceedings of the IEEE Conference on Computer Vision and Pattern Recognition (2016)
16. Clifford, G.D., et al.: AF classification from a short single lead ECG recording: the PhysioNet/Computing in Cardiology Challenge 2017. In: 2017 Computing in Cardiology (CinC). IEEE (2017)

Guided Layer-Wise Learning for Deep Models Using Side Information

Pavel Sulimov, Elena Sukmanova, Roman Chereshnev,
and Attila Kertész-Farkas$^{(\boxtimes)}$ (iD)

Department of Data Analysis and Artificial Intelligence, Faculty of Computer
Science, National Research University Higher School of Economics (HSE),
3 Kochnovsky Proezd, Moscow, Russian Federation
akerteszfarkas@hse.ru

Abstract. Training of deep models for classification tasks is hindered by local minima problems and vanishing gradients, while unsupervised layer-wise pretraining does not exploit information from class labels. Here, we propose a new regularization technique, called diversifying regularization (DR), which applies a penalty on hidden units at any layer if they obtain similar features for different types of data. For generative models, DR is defined as divergence over the variational posteriori distributions and included in the maximum likelihood estimation as a prior. Thus, DR includes class label information for greedy pretraining of deep belief networks which result in a better weight initialization for fine-tuning methods. On the other hand, for discriminative training of deep neural networks, DR is defined as a distance over the features and included in the learning objective. With our experimental tests, we show that DR can help the backpropagation to cope with vanishing gradient problems and to provide faster convergence and smaller generalization errors.

Keywords: Deep learning · Variational methods · Abstract representation

1 Introduction

Deep models [4], especially deep neural networks (DNNs), iteratively process data through various abstraction levels as $\mathbf{x} = \mathbf{h}_0 \to \mathbf{h}_1 \to \mathbf{h}_2 \to \cdots \to \mathbf{h}_L = y$. The first layer $\mathbf{x}$ is the raw data layer, and higher layers $\mathbf{h}_l$ aim to give a higher abstraction of the data, often referred to as features. The last layer can correspond either to class labels y in classification tasks or some other high-level cause in generative tasks. Each layer utilizes a monotone, non-linear so-called activation function $g_l(\mathbf{h}_l^T \theta_l) \to \mathbf{h}_{l+1}$ to transform features $\mathbf{h}_l$ to $\mathbf{h}_{l+1}$, where θ_l denotes the parameterization of the feature transformation at the given layer.

Currently, there are two main approaches for training deep models: convolutional and non-convolutional. In the non-convolutional approach, the data distribution is modeled by generative probabilistic distributions via maximizing the likelihood. One of the standard approaches is an unsupervised layer-wise

© Springer Nature Switzerland AG 2020
W. M. P. van der Aalst et al. (Eds.): AIST 2019, CCIS 1086, pp. 50–61, 2020.
https://doi.org/10.1007/978-3-030-39575-9_6

pretraining [2,3,20,24,25]. This is a bottom-up approach, in which a deep model is built via iteratively stacking single-layer generative models until the desired depth is achieved, and the procedure is finalized by using either backpropagation for discriminative fine-tuning or, e.g., a wake-sleep algorithm for generative fine-tuning [10]. In each step, a generative model is often trained to model either a joint distribution $P(\mathbf{h}_l, \mathbf{h}_{l+1}; \theta_l)$ using restricted Boltzmann machines (RBMs) or $P(\mathbf{h}_{l+1}|\mathbf{h}_l)$ using sigmoid belief networks (SBNs), where $\mathbf{h}_l$ represents the input data obtained from the previous layer, and $\mathbf{h}_{l+1}$ is treated as latent variable. Then, the subsequent layer is to learn a distribution of data obtained by sampling the hidden units of the previous layer. For the sake of simplicity, let us denote the input data by $\mathbf{x}$ and the latent variables by $\mathbf{h}$ at any layers. For a given set of data $\mathcal{D} = \{\mathbf{x}_i\}$, the model parameters are estimated by maximizing the data log likelihood, which is given as:

$$l(\theta; \mathcal{D}) = \log P(\mathcal{D}; \theta) = \sum_{\mathbf{x}_i \in \mathcal{D}} \log \sum_{\mathbf{h}} P(\mathbf{x}_i, \mathbf{h}; \theta). \tag{1}$$

The optimization of Eq. 1 w.r.t. θ is often done via the Markov chain Monte Carlo (MCMC) technique or via variational methods. The former tends to suffer from slow mixing rates and is too computationally expensive to be practical [1,23]. Variational methods [16] utilize an auxiliary so-called variational distribution $Q_{\mathbf{x}}(H; \phi)$ over the hidden variables for every given data $\mathbf{x}$ in order to approximate $P(H|\mathbf{x}; \theta)$ wherein the approximation is solved via optimization. Here, we introduce $Q_{\mathbf{x}}^{\phi}$ and $P_{\mathbf{x}}^{\theta}$ to shorten $Q(H|x; \phi)$ and $P(H|x; \theta)$, respectively. Then the log likelihood can be bounded from below as:

$$\begin{aligned} l(\theta; \mathcal{D}) &\geq \sum_{\mathbf{x} \in \mathcal{D}} \sum_{\mathbf{h}} Q_{\mathbf{x}}(\mathbf{h}; \phi) \log \frac{P(\mathbf{x}, \mathbf{h}; \theta)}{Q_{\mathbf{x}}(\mathbf{h}; \phi)} \\ &= \sum_{\mathbf{x} \in \mathcal{D}} \left(\mathbb{E}_{Q_{\mathbf{x}}}[\log P(\mathbf{x}, \mathbf{h}; \theta)] - \mathbb{E}_{Q_{\mathbf{x}}}[\log Q_{\mathbf{x}}^{\phi}] \right) \quad = \mathcal{L}(\theta, \phi; \mathcal{D}). \end{aligned} \tag{2}$$

The last term $\mathcal{L}$ is a lower bound on the original log likelihood and can be decomposed into two parts as:

$$\mathcal{L}(\theta, \phi; \mathcal{D}) = \sum_{\mathbf{x} \in \mathcal{D}} \log P(\mathbf{x}; \theta) - \sum_{\mathbf{x} \in \mathcal{D}} KL\left(Q_{\mathbf{x}}^{\phi}, P_{\mathbf{x}}^{\theta}\right). \tag{3}$$

This means that a tight lower bound can be achieved by minimizing the Kullback-Leibler (KL) divergence between the variational distribution Q and the exact posterior distribution P. Following [6,16] in variational perspectives, the parameters maximizing the lower bound of the log likelihood $\mathcal{L}$ can be obtained by iteratively alternating between the following two optimization steps:

E-step: $\phi^{(t+1)} \leftarrow \mathrm{argmax}_{\phi \in M} \mathcal{L}(\theta^{(t)}, \phi; \mathcal{D})$
M-step: $\theta^{(t+1)} \leftarrow \mathrm{argmax}_{\theta \in \Omega} \mathcal{L}(\theta, \phi^{(t+1)}; \mathcal{D})$

until convergence, where M and Ω are the corresponding parameter spaces.

In the convolutional approach, deep neural network models are made of convolutional layers, rectified linear units, and pooling layers [10,21,22] inspired by the mammal's visual cortex [9,14], and they are referred to as convolutional neural networks (CNNs). CNNs are said to yield better gradients for the backpropagation algorithm in contrast to fully connected, deep, sigmoid neural networks. As a result, CNNs usually provide improved performance on visual-related or speech recognition-related tasks.

In both approaches, convolutional and non-convolutional, the training of θ_l parameters in deep models is notoriously hard, and it is often viewed as an art rather than a science. There are four main problems with training deep models for classification tasks: (i) Training of deep generative models via an unsupervised layer-wise manner does not utilize class labels, therefore essential information might be neglected. (ii) When a generative model is learned, it is difficult to track the training, especially at higher levels [10]. For DNNs, the backpropagation method suffers from a problem known as vanishing gradients [8]. (iii) In principle, a generative model can be fitted to data arbitrarily well [12,13,26], in practice, the optimization procedure with latent variables can stuck in a poor local minima. (iv) The structure of the model is often specified in advance, and the designed model might not fit the data well. In particular, the number of hidden units or layers is often defined by the experimenter's intuition or habits; however, it is hard to give a bone fide estimation on the numbers of the latent components.

Here, we introduce a new regularization method, called diversifying regularization (DR), on the hidden units for training deep models for classification tasks. In principle, the proposed regularizer favors different abstract representation for two data samples belonging to different classes. This regularization is denoted by $D(\mathbf{h}_p^{(l)}, \mathbf{h}_q^{(l)})$, where $\mathbf{h}_p^{(l)}$ and $\mathbf{h}_q^{(l)}$ are abstract representations of data $\mathbf{x}_p$ and $\mathbf{x}_q$ (resp.) at layer l, and the data are of different types ($y_p \neq y_q$). For maximum likelihood estimation of generative models, we define DR in terms of divergence function $D(Q_{x_p}, Q_{x_q})$ and include it in Eq. 3 as an additive term. For discriminative learning of DNNs using backpropagation, we introduce DR as a distance function and include it in the learning objective as an additive cost.

We anticipate that DR helps cope with the aforementioned four problems. (i) DR is constructed based on class labels, and it can be employed on any abstract representation pairs at any hidden layer. Therefore, it implicitly includes information about the data classes in the unsupervised layer-wise pretraining. (ii) As a consequence of (i), DR can guide the pretraining at higher levels toward solution which obtain good classification performance. (iii) DR can help direct the gradient ascent optimizer toward an optimum where the internal representation of different types of data is more different. This could be particularly significant in the early steps in the optimization procedure when the gradients could be directed toward regions containing such solutions. (iv) Good regularization techniques can mitigate the impact of model structure construction problems, and they should help cope with overfitting problems and provide smaller generalization errors over a wider range of structures.

It could also be possible to reward hidden units if they learn similar abstract representations for data of the same type. In our opinion, this hampers the model's ability to discover fine subgroups in the data. For instance, considering a model that distinguishes between, e.g., cats and dogs, we might not want to force the model to learn similar representations for tigers and Persian cats, especially at lower abstraction levels.

There have been several attempts to incorporate class labels into RBMs. For instance, Larochelle et al. [19] coupled a data vector with the one-hot-encoded class labels as $(\mathbf{x}, \mathbf{y})$ and trained a RBM. The drawback of this representation is that it requires significantly more weights, which linearly depend on the number of the classes.

This article is organized as follows. In the next section, we introduce DR for probabilistic generative models, where particular emphasis is placed on RBMs and Variational Autoencoders (VAEs). In Sect. 3, we introduce DR for discriminative DNN. In Sect. 4, we present and discuss experimental results, and finally, we summarize our conclusions in the last section.

2 Diversifying Regularization Based on Side Information for Generative Models

Let us introduce a regularization method for latent variable generative models using side information. In general, side information refers to knowledge that is neither in the input data nor in the output label space but includes useful information for learning. The idea of side information was introduced in [28] and applied recently in [15]. Let $\mathcal{D} = \{(\mathbf{x}_i, y_i)\}$ be a given labeled training set. Using class labels from the training data, we construct side information by defining a set of data pairs, where the members of each pair belong to different classes, that is, if $(\mathbf{x}_p, \mathbf{x}_q) \in \mathcal{N}$ then $y_p \neq y_q$. We note that the size of $\mathcal{N}$ can be very large in the case of big data; however, we think that it is not necessary to include all possible pairs but just a smaller subset or only pairs of data that are close to each other in terms of some distance measure.

We define the regularization term, denoted by D, by divergence functions over the distribution functions over the latent variables. For instance, for two given data $\mathbf{x}_p$ and $\mathbf{x}_q$ the regularization is defined as $D\left(Q^\phi_{\mathbf{x}_p},\ Q^\phi_{\mathbf{x}_q}\right)$. This term can be included to the MLE as follows:

$$r(\theta, \phi; \mathcal{D}) = \sum_{\mathbf{x} \in \mathcal{D}} \log \sum_{\mathbf{h}} P(\mathbf{x}, \mathbf{h}; \theta) + \alpha \sum_{(\mathbf{x}_p, \mathbf{x}_q) \in \mathcal{N}} D\left(Q^\phi_{\mathbf{x}_p},\ Q^\phi_{\mathbf{x}_q}\right), \qquad (4)$$

where r stands for regularized likelihood. The second term introduces a penalty on two distributions if they are similar, but they should not be, and α is a trade-off parameter. D denotes a divergence function for probability distributions. A good source of divergence functions is provided by, e.g., Csiszár's f-divergence class [6]. It is defined as follows:

Definition 1. *Given any continuous, convex function* $f : [0, +\infty) \to R \cup \{+\infty\}$, *such that* $f(1) = 0$, *then the f-divergence between distributions* P *and* Q *is measured as*

$$D_f(P, Q) = \sum_z q_z f\left(\frac{p_z}{q_z}\right). \tag{5}$$

The Kullback-Leibler divergence is an f-divergence, if $f(t) = t \log t$. We were interested in using a symmetric divergence measure; therefore, we decided to chose the Hellinger divergence. It is defined as $D_H(P, Q) = 1 - \sum_z \sqrt{p_z q_z}$ generated by $f(t) = 1 - \sqrt{t}$. We note that the log of the Hellinger divergence is closely related to the Bhattacharyya distance (BD) (also known as a Rényi divergence at $\alpha = 0.5$), which is defined as $D_B(P \parallel Q) = -\ln \sum_z \sqrt{p_z q_z}$. BD has been successfully used in a wide range of applications, including feature extraction and selection [5], image processing [11], and speaker recognition [29].

The lower bound of r can be obtained again as

$$\mathcal{R}(\phi, \theta; \mathcal{D}) = \sum_{\mathbf{x} \in \mathcal{D}} \log P(\mathbf{x}; \theta) + \alpha \sum_{(\mathbf{x}_p, \mathbf{x}_q) \in \mathcal{N}} D_H(Q_{\mathbf{x}_p}^\phi, Q_{\mathbf{x}_q}^\phi) - \sum_{\mathbf{x} \in \mathcal{D}} KL(Q_{\mathbf{x}}^\phi, P_{\mathbf{x}}^\theta). \tag{6}$$

The optimization of Eq. 6 can be carried out with, for instance, iterative gradient-based methods. By using the EM algorithm, we get:

E-step: $\phi^{(t+1)} \leftarrow \text{argmax}_{\phi \in M} \left\{ \alpha \sum_{(\mathbf{x}_p, \mathbf{x}_q) \in \mathcal{N}} D_H(Q_{\mathbf{x}_p}^\phi, Q_{\mathbf{x}_q}^\phi) - \sum_{\mathbf{x}} KL(Q_{\mathbf{x}}^\phi, P_{\mathbf{x}}^{\theta^{(t)}}) \right\}$

M-step: $\theta^{(t+1)} \leftarrow \text{argmax}_{\theta \in \Omega} \sum_i \log P(\mathbf{x}_i; \theta)$.

The derivatives of the regularization D can be calculated analytically; however, they can be easily calculated automatically using recent toolboxes, e.g., with Theano's grad function.

2.1 Restricted Boltzmann Machines (RBMs) Using Side Information

Restricted Boltzmann machines are particular type of Markov random fields that have two types of random binary variables, visible and hidden, denoted by $\mathbf{x}$ and $\mathbf{h}$ respectively. The visible and hidden units are connected, but there are no connections among the visible units and among the hidden units. The probability of a given input in this model parameterized by θ is:

$$P(\mathbf{x}; \theta) = \frac{1}{Z(\theta)} \sum_{\mathbf{h}} \exp(-E_\theta(\mathbf{x}, \mathbf{h})), \tag{7}$$

where $E_\theta(\mathbf{x}, \mathbf{h}) = -\sum_{ij} x_i \theta_{ij} h_j$ is the energy, θ_{ij}s are the weights between visible unit x_i and hidden unit h_j, and $Z(\theta)$ denotes the partition function. Here, we omit the feature weights for $\mathbf{x}$ and $\mathbf{h}$ for the sake of simplicity; however, they can be included in the model and in our diversifying regularization. The conditional distributions over hidden units $\mathbf{h}$ and visible units $\mathbf{x}$ can be formulated by logistic functions:

$$P(\mathbf{h}|\mathbf{x};\theta) = \prod_j P(h_j|\mathbf{x};\theta), \quad \text{where } p(h_j = 1|\mathbf{x};\theta) = \sigma(\sum_i \theta_{ij}x_i)$$

$$P(\mathbf{x}|\mathbf{h};\theta) = \prod_i P(x_i|\mathbf{h};\theta), \quad \text{where } p(x_j = 1|\mathbf{h};\theta) = \sigma(\sum_j \theta_{ij}h_j),$$

(8)

where $\sigma(x) = (1 + \exp(-x))^{-1}$ is the logistic function. The derivatives of the log-likelihood $l(\theta; \mathcal{D})$ w.r.t. the model parameters using a set of data are:

$$\frac{\partial l(\theta; \mathcal{D})}{\theta_{ij}} = \sum_{\mathbf{x}_k} \sum_{\mathbf{h}} P(\mathbf{h}|\mathbf{x}_k; \theta)x_{ki}h_j - \sum_{\mathbf{x},\mathbf{h}} P(x,h)x_i h_j,$$

(9)

where $\mathbf{x}, \mathbf{h}$ runs over all possible states. The first term is called the data-dependent term, and Salakhutdinov et al. [25] proposed a mean-field algorithm to approximate the posterior distribution $Q(\mathbf{h}|\mathbf{x}; \phi)$. The posterior distribution fully factorizes and can be written in the form $Q_{\mathbf{x}}(H) = \prod_j Q_{\mathbf{x}}^j(H_j)$, where $Q_{\mathbf{x}}^j(H_j) = (\sigma(z_x^j))^{H_j}(1 - \sigma(z_x^j))^{1-H_j}$ and $z_x^j = \sum_i \theta_{ij}x_i$. Because of the factorization property, the mean filed algorithm can provide a fixed-point equations $\mu_x^j = \sigma(z_x^j + \sum_{i \neq j} \mu_x^i)$ $(\forall j)$ and, therefore, the posteriors are approximated by $Q_x^{j\phi}(H = 1) = \mu_x^j$. It has been shown that mean field approximation provides a fast estimation for the posterior in practice. For more details, we refer the reader to [25]. The second term of Eq. 9 is called the data-independent term, and it is typically approximated by stochastic MCMC sampling methods [23], such as the k-step contrastive divergence (CD) or persistent CD algorithms [27,30].

Now, we introduce the diversifying regularization for the variational optimization. The regularization is introduced on the individual factors Q_x^j in the following way:

$$D_H(Q_{\mathbf{x}_p}^\phi, Q_{\mathbf{x}_q}^\phi) = 1 - \sum_{h=\{0,1\}} \sqrt{Q_{\mathbf{x}_p}^{j\phi}(h)Q_{\mathbf{x}_q}^{j\phi}(h)} \quad = 1 - \sqrt{\overline{\sigma}(z_p^j)\overline{\sigma}(z_q^j)} - \sqrt{\sigma(z_p^j)\sigma(z_q^j)},$$

(10)

where $\overline{\sigma}(z) = 1 - \sigma(z)$. The derivation of the regularizer's gradient can be obtained as follows:

$$\frac{\partial}{\phi_{ij}} D_H(Q_{\mathbf{x}_p}^\phi, Q_{\mathbf{x}_q}^\phi) = \sum_{h=\{0,1\}} (-1)^h \frac{1}{2} \sqrt{Q_{\mathbf{x}_p}^{j\phi}(h)Q_{\mathbf{x}_q}^{j\phi}(h)} [Q_{\mathbf{x}_p}^{j\phi}(\overline{h})x_{p_i} + Q_{\mathbf{x}_q}^{j\phi}(\overline{h})x_{q_i}],$$

(11)

where $\overline{h} = 1 - h$.

We introduce the total diversifying regularization to the data log likelihood as follows:

$$r(\theta; \mathcal{D}) = \sum_{x \in \mathcal{D}} \log P(\mathbf{x}) + \alpha \sum_{(\mathbf{x}_p, \mathbf{x}_q) \in \mathcal{N}} D_H(Q_{\mathbf{x}_p}^\phi, Q_{\mathbf{x}_q}^\phi).$$

(12)

Now, the weight update rule for the data-dependent term using mean filed approximation, and for the data-independent term using stochastic approximation, we get:

$$\frac{\partial r(\theta; \mathcal{D})}{\theta_{ij}} = \sum_{\mathbf{x}_l} x_{li} \mu_{x_l}^j - x_i^{(k)} h_j^{(k)}$$

$$+ \alpha \sum_{(\mathbf{x}_p, \mathbf{x}_q) \in \mathcal{N}} \left(\sqrt{\overline{\mu}_{x_p}^j \overline{\mu}_{x_q}^j} [\mu_{x_p}^j x_{pi} + \mu_{x_q}^j x_{qi}] - \sqrt{\mu_{x_p}^j \mu_{x_q}^j} [\overline{\mu}_{x_p}^j x_{pi} + \overline{\mu}_{x_q}^j x_{qi}] \right) \tag{13}$$

where $\overline{\mu} = 1 - \mu$, the $1/2$ is absorbed in the regularization trade-off parameter α, and $x^{(k)}$ and $h^{(k)}$ are obtained with k-step Gibbs sampling.

The DR does not require any constraints on the visible units, and it can be used in combination with real-valued Gaussian-Bernoulli RBMs as well.

2.2 Variational Autoencoders (VAEs) Using Side Information

Variational Autoencoders (VAEs) [7,17] represent data generation process by an unobserved continuous latent random variable H which is decoded with neural networks. The parameters of VAEs are trained jointly by backpropagation algorithm via optimizing the following cost function:

$$\mathcal{L}(\theta, \phi; \mathcal{D}) = \sum_{\mathbf{x} \in \mathcal{D}} \mathbb{E}_{Q_{\mathbf{x}}}[\log P(\mathbf{x}|H; \theta)] - \sum_{\mathbf{x} \in \mathcal{D}} KL\left(Q_{\mathbf{x}}^{\phi}, P(H)\right). \tag{14}$$

The first term is the expected reconstruction error, while the second term, the KL-term, pushes the posterior Q toward the prior $P(H)$. We note that Q tends to "be covered" by P in this case.

The DR can be introduced over the reconstruction level and we can have several options. For instance, DR could be defined via cross-entropy as follows:

$$D_{CE}(\mathbf{x}_p, \mathbf{x}_q) = \mathbb{E}_{Q_{\mathbf{x}_q}}[\log P(\mathbf{x}_p|H; \theta)], \tag{15}$$

However, when one does not require a probabilistic interpretation over the last layer, simply an Euclidean distance could be used as: $D_2(h_{\mathbf{x}_p}, h_{\mathbf{x}_q}) = \|h_{\mathbf{x}_p} - h_{\mathbf{x}_q}\|_2^2$, where $h_{\mathbf{x}_a}$ denotes the reconstruction of data $\mathbf{x}_a$ obtained with VAE and 2 in D_2 indicates its relation to the squared L_2-norm.

3   Diversifying Regularization Based on Side Information for Discriminative Deep Models

In the case of DNNs, let $\mathbf{h}_{\mathbf{x}_p}^l$ denote the features obtained at hidden layer l by forward propagation from data $\mathbf{x}_p$. Because $\mathbf{h}_{\mathbf{x}_p}^l$ does not have any probabilistic interpretation, we can define DR for two different types of data $\mathbf{x}_p$ and $\mathbf{x}_q$ by any distance function. Here, we define it as $D_2(\mathbf{h}_{\mathbf{x}_p}^l, \mathbf{h}_{\mathbf{x}_q}^l) = \|\mathbf{h}_{\mathbf{x}_p}^l - \mathbf{h}_{\mathbf{x}_q}^l\|_2^2$. Therefore, the learning objective including D_2 can be formulated as follows:

$$J(\theta; \mathcal{D}) = \sum_{\mathbf{x} \in D} \ell(G(\mathbf{x}; \theta), y) + \alpha \Omega(\theta) - \sum_{(\mathbf{x}_p, \mathbf{x}_q) \in \mathcal{N}} \sum_{l=1}^{L-1} \alpha_l D_2(\mathbf{h}_{\mathbf{x}_p}^l, \mathbf{h}_{\mathbf{x}_q}^l), \tag{16}$$

where ℓ is a loss function such as cross entropy, $G(\mathbf{x}; \theta)$ denotes the output prediction by DNN, Ω is a parameter regularization, and αs are trade-off values. The derivatives of this learning objective can be calculated analytically; however, it can also be done automatically via, for instance, Theano's grad function.

One can apply different trade-off coefficients α_l at different layers. Perhaps, at lower level feature learning, it is not expected to obtain different representations for different types of data; however, at higher levels, closer to the prediction level, stronger regularization α_l would be desired.

4   Experimental Results and Discussion

The aim of our experiments is to demonstrate that training algorithms achieve better generalization performance with using DR than without using it under exactly the same parameter settings. We did not use any sophisticated methods such as adaptive learning rate, momentum methods, second order optimization methods, data augmentation, image preprocessing methods, etc, which would make it more difficult to reveal the contribution of DR to the final result. We note that, therefore, our classification results do not surpass the best ones in the literature. Here, we present three types of experiments, one for RBMs, one for VAEs, and the other for discriminative models. All the samples for the negative pairs were constructed from all the data having different class labels in the current mini batch.

On the Generative Learning of Deep Belief Networks. First, we built a deep belief network (DBNs) [20] on the Cifar-10 dataset [18]. The dataset contains tiny (32×32) color pictures from 10 classes, 50,000 for training and 10,000 for testing. We constructed a DBN of 11 layers, each having 500, 300, 200, 150, 100, 80, 60, 50, 30, 20 hidden and 10 output units, respectively. The weights in every layer were pre-trained by an RBM using learning rate 0.01, one-step CD ($k = 1$). The batch-size was set to 10. For DR, we used the formula Eq. 10, and all data in the mini-batch were used to construct the side information (N), which resulted in approximately 35–45 pairs in each batch. The α trade-off parameter of DR was set to 50. For the fine-tuning backpropagation, we used a stochastic gradient descent without any parameter regularization (such as L_2), with learning rate 0.01 and batch size 10. The fine-tuning phases did not employ DR; therefore, we can see the effect of DR on the training of RBMs. All other parameters were left default. Before training of both models (with and without DR), all weight parameters were initialized with exactly the same values. The DBN and RBM codes were downloaded from deeplearning.net.

The results are shown in Fig. 1. The first 10 plots (first five rows) show the pseudo-log likelihoods obtained during training RBM with DR (solid) and without DR (dashed) at every layer. The last two plots in the last row show the cost and the test error during the fine-tuning, when it was applied after regularized (solid) and unregularized (dashed) pre-training. These plots show that the fine-tuning achieved much faster convergence and much better generalization performance when the weights were pre-trained using DR. We explain this

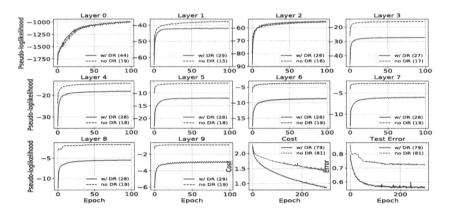

Fig. 1. Learning curves during training of RBM and DBN. The numbers in the parenthesis in the legends indicate the run time in minutes. Cost is defined as cross entropy.

as being due to the fact that DR includes class information about the data, which might not maximize the pure log likelihood but favors solutions where different types of data have more different abstract representations as well. Run times are indicated in the figure legend. Training using DR required more time because side information was generated on the CPU for each mini-batch inline. This possibly could have been accelerated by generating it for each batch before the training procedure, in advance.

On the Generative Learning of VAE. Next, we examined the impact of DR, defined as in Eq. 15, on training of VAE. We employed the classic MNIST dataset [21], which contains (28×28) gray-scaled pictures of handwritten digits: 50,000 of them for training and 10,000 for testing. The VAE consisted one hidden layer with 600 hidden units on the encoding and decoding layer, respectively, while the latent space was 2 dimensional. The training was carried out using the reparameterization trick [17]. The prior and the posterior were defined as $P(H) = N(0, I)$ and $Q_x^\phi = N(z|\mu(x), \sigma^2(x)I)$, respectively, where $\mu = \mu(x)$ and $\sigma = \sigma^2(x)$ are the outputs of the encoding neural network trained on data. Training run 100 epochs with learning rate 0.05 and batch-size 20.

For the visualizations of the learned manifolds, the linearly spaced coordinates of the $[-6, -6] \times [6, 6]$ square were used as the latent variables H. For each of these values we plotted the corresponding generative $P(x|H)$ with the learned parameters. The sampled images are shown in Fig. 2. The panel (A) shows the sampled images from the VAE trained without DR, while panel (B) shows sampled images from the VAE trained using DR. In our opinion, the decoder layer of the VAE trained using DR can produce sharper images.

On the Discriminative Learning of DNN. Finally, we examined the impact of DR, defined in Eq. 16, on training deep, fully connected sigmoid networks. We employed the MNIST dataset. We constructed a six-layer DNN, with hidden layers

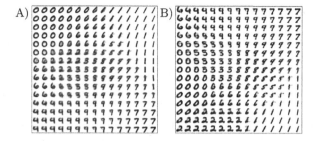

Fig. 2. Learned MNIST manifold (A) without regularization and (B) using DR. DR helps generate sharper images.

containing 30, 30, 30, 20, 20 units and the output layer containing 10 units. Training was carried out via a gradient descent algorithm without any parameter regularization (such as L_2), with learning rate 1.0. The α trade-off parameter of DR was set to 50.0, and it was reduced in each epoch by 10%; therefore, its effect was strong at the beginning of the training but over the time it gradually vanished. For side information, 202,770 data pairs were randomly chosen at the beginning of the training. Before training of both models (with and without DR), all weight parameters were initialized with exactly the same values.

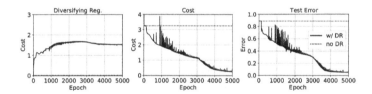

Fig. 3. Curves during training fully connected sigmoid DNN. Cost is defined as cross entropy. Middle plot does not include DR.

The results and the learning curves are shown in Fig. 3. The plots show the value of the regularization, the value of the cost function (without regularization) during training, and the error obtained on the test data, respectively. These plots show that the standard backpropagation algorithm (dashed line) was not able to train the model on this structure, and it gave an 88.65% classification error. We note that the tricky part here is the first hidden layer because the information compression is very high (784 → 30), so the backpropagation must train the corresponding weights very well. However, in practice the backpropagation failed to propagate the error from the output layer to the first layer. However, DR (solid line) is applied directly on every layer providing useful class-label-related information locally. The backpropagation method using DR (solid line) during the first 127 epochs did not change the cost and the test error but it rapidly increased the diversification until it reached a critical point from where the gradients of the cost function led to a good (local) minimum. This resulted in a 5.44% classification error.

5 Conclusions

In this article we have introduced a novel regularization method, termed diversifying regularization (DR), to help train deep generative and discriminative models for classification tasks. The main idea behind DR is that different types of data should have different abstract representation in the model's hidden layers. Therefore, DR applies a penalty on hidden units if they obtain similar features on data belonging to different classes. In the experimental results on deep belief networks, we have shown that DR is capable of including essential information about class labels implicitly and propagating them through layers in the greedy layer-wise pretraining. Therefore, the regularized weight initialization already includes some information about the class labels, and the subsequent fine-tuning phase can obtain smaller generalization errors and faster convergence. On the other hand, discriminative training of deep models can also benefit from DR, because it can provide good gradients at low layers, directly helping coping with vanishing gradient problem.

Our method, DR, is not limited to DNNs, VAEs, RBMs, and DBNs. It can be included in many generative models and methods, such as Gaussian mixture models and wake-sleep algorithms as well.

Acknowledgments. We gratefully acknowledge the support of NVIDIA Corporation with the donation of the GTX Titan X GPU used for this research.

References

1. Andrieu, C., De Freitas, N., Doucet, A., Jordan, M.I.: An introduction to MCMC for machine learning. Mach. Learn. **50**(1–2), 5–43 (2003)
2. Arnold, L., Ollivier, Y.: Layer-wise learning of deep generative models. CoRR abs/1212.1524 (2012). http://arxiv.org/abs/1212.1524
3. Bengio, Y., Lamblin, P., Popovici, D., Larochelle, H.: Greedy layer-wise training of deep networks. Adv. Neural Inf. Process. Syst. **19**, 153 (2007)
4. Bengio, Y., et al.: Learning deep architectures for AI. Found. Trends® Mach. Learn. **2**(1), 1–127 (2009)
5. Choi, E., Lee, C.: Feature extraction based on the Bhattacharyya distance. Pattern Recognit. **36**(8), 1703–1709 (2003)
6. Csiszár, I.: Information-type measures of difference of probability distributions and indirect observations. Stud. Sci. Math. Hung. **2**, 299–318 (1967)
7. Doersch, C.: Tutorial on variational autoencoders. arXiv preprint arXiv:1606.05908 (2016)
8. Erhan, D., Manzagol, P.A., Bengio, Y., Bengio, S., Vincent, P.: The difficulty of training deep architectures and the effect of unsupervised pre-training. In: AISTATS, vol. 5, pp. 153–160 (2009)
9. Fukushima, K., Miyake, S.: Neocognitron: a self-organizing neural network model for a mechanism of visual pattern recognition. In: Amari, S., Arbib, M.A. (eds.) Competition and Cooperation in Neural Nets. LNBM, vol. 45, pp. 267–285. Springer, Heidelberg (1982). https://doi.org/10.1007/978-3-642-46466-9_18
10. Goodfellow, I., Bengio, Y., Courville, A.: Deep Learning. MIT Press (2016). http://www.deeplearningbook.org

11. Goudail, F., Réfrégier, P., Delyon, G.: Bhattacharyya distance as a contrast parameter for statistical processing of noisy optical images. JOSA A **21**(7), 1231–1240 (2004)
12. Hartman, E.J., Keeler, J.D., Kowalski, J.M.: Layered neural networks with Gaussian hidden units as universal approximations. Neural Comput. **2**(2), 210–215 (1990)
13. Hornik, K.: Approximation capabilities of multilayer feedforward networks. Neural Netw. **4**(2), 251–257 (1991)
14. Hubel, D.H., Wiesel, T.N.: Receptive fields, binocular interaction and functional architecture in the cat's visual cortex. J. Physiol. **160**(1), 106–154 (1962)
15. Jonschkowski, R., Höfer, S., Brock, O.: Contextual learning. CoRR abs/1511.06429 (2015). http://arxiv.org/abs/1511.06429
16. Jordan, M.I., Ghahramani, Z., Jaakkola, T.S., Saul, L.K.: An introduction to variational methods for graphical models. Mach. Learn. **37**(2), 183–233 (1999)
17. Kingma, D.P., Welling, M.: Auto-encoding variational bayes. arXiv preprint arXiv:1312.6114 (2013)
18. Krizhevsky, A., Hinton, G.: Learning multiple layers of features from tiny images (2009)
19. Larochelle, H., Bengio, Y.: Classification using discriminative restricted Boltzmann machines. In: Proceedings of the 25th International Conference on Machine learning, pp. 536–543. ACM (2008)
20. Larochelle, H., Bengio, Y., Louradour, J., Lamblin, P.: Exploring strategies for training deep neural networks. J. Mach. Learn. Res. **10**, 1–40 (2009)
21. LeCun, Y., Bottou, L., Bengio, Y., Haffner, P.: Gradient-based learning applied to document recognition. Proc. IEEE **86**(11), 2278–2324 (1998)
22. LeCun, Y., et al.: Generalization and network design strategies. Connect. Perspect. 143–155 (1989)
23. Neal, R.M.: Connectionist learning of belief networks. Artif. Intell. **56**(1), 71–113 (1992). https://doi.org/10.1016/0004-3702(92)90065-6
24. Salakhutdinov, R., Hinton, G.E.: Deep boltzmann machines. In: AISTATS, vol. 1, p. 3 (2009)
25. Salakhutdinov, R., Larochelle, H.: Efficient learning of deep Boltzmann machines. In: AISTATs, vol. 9, pp. 693–700 (2010)
26. Sutskever, I., Hinton, G.E.: Deep, narrow sigmoid belief networks are universal approximators. Neural Comput. **20**, 2629–2636 (2008)
27. Tieleman, T.: Training restricted Boltzmann machines using approximations to the likelihood gradient. In: Proceedings of the 25th International Conference on Machine Learning, pp. 1064–1071. ACM (2008)
28. Xing, E.P., Ng, A.Y., Jordan, M.I., Russell, S.: Distance metric learning with application to clustering with side-information. In: NIPS, vol. 15, p. 12 (2002)
29. You, C.H., Lee, K.A., Li, H.: An SVM kernel with GMM-supervector based on the Bhattacharyya distance for speaker recognition. IEEE Signal Process. Lett. **16**(1), 49–52 (2009)
30. Yuille, A.L.: The convergence of contrastive divergences. In: Saul, L.K., Weiss, Y., Bottou, L. (eds.) Advances in Neural Information Processing Systems 17, pp. 1593–1600. MIT Press (2005). http://papers.nips.cc/paper/2617-the-convergence-of-contrastive-divergences.pdf

Multi-instance Learning for Structure-Activity Modeling for Molecular Properties

Dmitry V. Zankov[1]([✉]) [iD], Maxim D. Shevelev[1], Alexandra V. Nikonenko[1,2],
Pavel G. Polishchuk[3], Asima I. Rakhimbekova[1], and Timur I. Madzhidov[1]

[1] Kazan Federal University, Kazan, Russia
dvzankov@gmail.com
[2] Volgograd State University, Volgograd, Russia
[3] Palacký University Olomouc, Olomouc, Czech Republic

Abstract. In this paper, the approach of multi-instance learning is used for modeling the biological properties of molecules. We have proposed two approaches for the implementation of multi-instance learning. Both approaches are based on the idea of representing the features describing the molecule as a one vector, which is produced from different representations (instances) of the molecule. Models based on the approach of multi-instance learning were compared with classical modeling methods. Also, it is shown that in some cases, the approach of multi-instance learning allows to achieve greater accuracy in predicting the properties of molecules.

Keywords: Multi-instance learning · Neural networks · QSAR

1 Introduction

Chemoinformatics is a modern interdisciplinary science at the intersection of computer science and chemistry [1]. It involves the application of different computational and information processing techniques to solving a variety of problems in the field of chemistry. One of these important problems is establishing the relationship between the structure of a chemical compound and its biological activity, or its any other desired property. Machine learning approaches are widely utilized to build highly predictive quantitative structure-activity (property) models (QSAR/QSPR) [2].

One of the key limitations of conventional structure-property modeling is the requirement that each molecule has to be represented by a single instance [3]. In other words, a molecule has to be associated with a single set of features. Despite a molecule is often considered as a set of atoms and bonds and represented frequently as a planar (2D) graph, it is indeed a dynamic 3D object and simultaneously exists in many 3D forms (or instances), that are different by spatial arrangement of atoms, called conformations (Fig. 1). Every conformation has associated energy. Low-energy conformations are usually formed due to rotation of molecular fragments around single bonds (Fig. 1). For some applications 2D molecular graph features (often called 2D descriptors) are not enough and lack important information about molecular structure. In this case, one has

© Springer Nature Switzerland AG 2020
W. M. P. van der Aalst et al. (Eds.): AIST 2019, CCIS 1086, pp. 62–71, 2020.
https://doi.org/10.1007/978-3-030-39575-9_7

to resort usage of features associated with molecular 3D geometry. However, selection of the geometry used for feature calculation is artificial and influence model quality a lot. "Bioactive" conformation that binds to protein (Fig. 1) could neither be lowest energy nor even belong to potential energy minima.

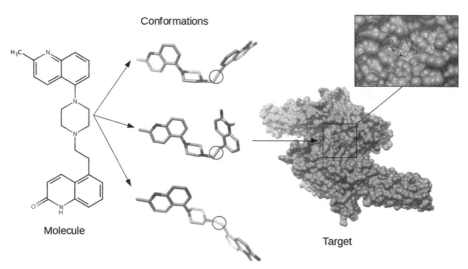

Fig. 1. 2D and 3D representations of the molecule. The rotatable bond is circled by the red line. (Color figure online)

The concept of multi-instance learning (MIL) naturally treats peculiarities of chemical objects [3], which are often difficult to represent by a single molecular form. In MIL, supervised learner receives several feature vectors of instances of the same object and the objects' labels. Despite the fact that multi-instance learning approach was originally proposed specifically to solve chemical problems [4] it did not receive much development in chemistry: there are only two publications in which the properties of small molecules were modeled [5]. The multi-instance learning approach was never compared to conventional QSAR approaches.

There are many implementations of multi-instance modeling approaches based on modifications of Support Vector Machine (SVM) algorithm: MI-SVM and mi-SVM [6], NSK and STK [7], MissSVM [8], MICA [9], sMIL, stMIL, and sbMIL [10]. They use different kernels: instance-based kernels (MI-SVM, mi-SVM, MICA, MissSVM), set-based kernels (NSK) or hybrid approaches (sMIL, stMIL, and sbMIL). Approaches using instance-based kernels produce models with relatively high accuracy, however their high computationally complexity and memory consumption makes them impractical to train on big datasets. Approaches using set-based kernels are less susceptible to these issues. However their accuracy may depend on a distribution of instances within objects [11]. Other MIL approaches use proportionalization or averaging of features of instances to get a feature vector of an object [12, 13]. The class conditional log likelihood ratio (MICCLLR) algorithm represents an object by a feature vector containing statistics computed based on the class conditional log-likelihood ratios of the attribute values of the instances of the object [14]. Therefore this algorithm relies on the conditional independence assumptions which are not always true. The abovementioned xMaP approach is somewhat similar to this approach, it also convert input instances (conformations) in a distribution of variables representing a whole object (molecule). The decision tree algorithm was adapted to multi-instance learning. however if instances are represented by numerical variables it is equivalent to the simple proportionalization minimax scheme [15, 16]. The developed SVM algorithm based on minimax kernel is also equivalent to the corresponding proportionalization scheme. Other approaches are Citation-kNN and Bayesian-kNN based on modified Hausdorff distance measure [17]. Recently a new promising algorithm based on deep convolution neural networks was proposed [18]. It was successfully applied for image annotation and its architecture is not suitable for solving chemical problems.

In this work we propose multi-instance learning approaches adapted for chemical objects modeling that are based on conventional machine learning approaches and the neural nets, and benchmark them for chemical properties modeling. Molecular conformations are considered as instances of molecules, encoded by their own feature vectors. The property of the chemical object is associated with one or more instances from the entire set, but it is not known with which one. This will help to overcome the compound representation shortcomings caused by the use of the conventional structure-property modeling practices, making it possible to utilize multiple molecular forms/instances, which may in turn lead to an improvement of the model's predictive abilities due to more complete description of molecular features. Classification task is considered in this work. The MIL approaches proposed will be compared with conventional QSAR techniques based on application of features of 2D molecular graph and features of lowest energy 3D structure.

2 Modeling

2.1 Data

To build the models, 3 data sets were taken from the paper [19]. Data sets contain molecules that can exhibit biological activity against various diseases.

A data set of 114 angiotensin converting enzyme (ACE) inhibitors consisted of 65 active and 49 inactive molecules. In the original paper, the data set was divided into the train (76 molecules) and test (38 molecules) sets. ACE inhibitors are used as pharmaceutical drugs for treatment of cardiovascular diseases [20].

A data set of 361 dihydrofolate reductase inhibitors (DHFR) consisted of 203 active and 158 inactive molecules. The data set was divided into the train (237 molecules) and test (124 molecules) sets as described in the original paper. DHFR inhibitors can be useful in the treatment of cancer and as a potential target against bacterial infections [21].

A data set of 111 acetylcholinesterase inhibitors (ACHE) consisted of 86 active and 25 inactive molecules. The data set was divided into the train (74 molecules) and test (37 molecules) sets as described in the original paper. Acetylcholinesterase inhibitors are used for treatment of postural tachycardia syndrome, Alzheimer's disease and a lot of other diseases [22].

The test sets in the paper [19] were selected with the maximum dissimilarity to the training sets. The sets were structured this way to examine the predictive accuracy of models when extrapolating outside the training set space.

2.2 Feature Calculation

In this work the RDKit (Open source toolkit for cheminformatics in Python) [23] was used to generate features of 2D molecular graphs and 3D conformations.

Molecular conformations (the number depends on the selected parameter) for each molecule were generated using in house scripts[1]. All possible spatial conformations in which RMSD (structure diversity) threshold parameter was 0.5 and energy cutoff was 100 were generated.

There were 200 features for 2D-structure description of molecules, such as molecular weight, the number of different types of atoms and other. Also there were 914 features (16 types) for 3D conformation representation, such as principal moments of inertia, radius of gyration etc.

2.3 Methods

The general idea behind the proposed approach is that one or more molecular forms (instances) of many active compounds responsible for compound activity should have similar feature vector. Thus, instances can be clustered by similarity. It is assumed that some of the identified groups (clusters) will be enriched with molecular forms of active compounds responsible for compound activity. This stage will largely determine the model quality. Molecule is encoded by a new feature vector whose length is equal to the number of identified clusters. The values of this vector are the number of molecular forms which fell into a particular cluster. Alternatively, this vector can be represented as a bit string containing the information about to which clusters instances of the compound were assigned. This procedure can be considered as a conversion of the original feature space into a latent space and the latent vector encodes a compound represented by many instances with a single feature vector. The resulting feature vectors are used to develop models using conventional machine learning approaches.

Unsupervised Clustering-Based Approach. This approach is based on application of clustering at the first step. The molecular forms are grouped using such conventional clustering method as k-means algorithm. Each molecule is represented by a feature vector. Instances of the same molecule can be grouped several times in the same clusters. Latent vector is generated by counting occupancy of every cluster by instances of some molecule. Latent vectors are used for building model by Random Forest method. This approach is easy-to-implement and adjust and it uses only conventional machine learning techniques. The drawback of the approach is the unsupervised nature of clusterization algorithm which can hardly be optimized for greater model performance.

Supervised Clustering-Based Approach. The second approach is based on neural networks, first similar MIL approach was proposed by Ramon and De Raedt [24]. In this

---

[1] https://github.com/DrrDom/rdkit-scripts.

case, two neural networks are designed and stacked together – the output of the first network is the input to the second one. The task of the first network is to create the latent space representation of an input object represented by multiple instances. It takes an instance feature vector and, passing it through the net, attributes it to output nodes (analogs of clusters) with some probabilities. The probabilities of individual instances of a compound are summed for each node to get a feature vector representing this compound. This is equivalent to mean pooling in convolution neural networks. The obtained feature vector is used as input for the second neural network which establishes correlation with endpoint values of compounds. Since all operations are differentiable both models can be trained simultaneously. Therefore, the latent space representation will be adapted to the modeled property during model training. The supervised approach of latent variable generation may produce models with higher predictive performance. Recently, different tricks (dropout, residual connections, etc.) to enhance neural net-based MIL model performance were explored [25], effect was visible but low. Thus, we used aforementioned two nets-based model.

The basic architecture of the neural network used for multi-instance learning is shown in Fig. 2. Each molecule was represented by an array, where each row was a feature vector calculated for each conformation of the molecule. Thus, the data set of molecules and their conformations was represented as a three-dimensional array of the form the number of molecules, the number of conformations, the number of features.

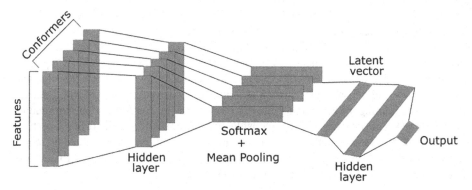

Fig. 2. Architecture of neural network used in supervised clustering-based approach

2.4 Validation

The optimization of the hyperparameters of the models was carried out using the cross-validation procedure with dividing data into 5 folds. The selection of the best models was performed using the ROC-AUC value.

For the Random Forest algorithm the following parameters were used: 250 trees were used, maximum features was selected from 0.1 to 0.3 with step 0.1, log2 or square root of number of features. The number of clusters was varied from 2 to 1000.

During building models based on neural network (Fig. 2), the number of neurons in hidden layers was optimized. The number of neurons in hidden layers was selected from

2 to 1024, as powers of 2. Also during the tests it had been found that the number of learning epochs has a significant influence on the quality of the models. In this regard, the number of learning epochs was also the optimized hyperparameter and was selected from the range from 1 to 1000.

3 Results and Discussion

3.1 Influence of Number of Conformations

For molecules from the mentioned data sets, different sets of conformations (instances) were generated having 1 or up to 5, 10, 15 or 25 conformations. In the case of a single conformation representation conformations with the lowest energy calculated within MMFF were considered. It should be noted that in this case no clustering was performed before building of Random Forest models. The mean pooling operation implemented within the neural network approach had no effect on the learning process.

Table 1 compiles model performance on selected datasets. The best models out of many ones characterized by different hyperparameters were selected based on cross-validation results. From the obtained results, it follows that the approach based on neural networks is characterized by a higher accuracy of predicting the activity of molecules on all three data sets. One can notice that cross-validation quality is much better than those on test set, which is due to the fact that test set comprised of molecules most dissimilar to training set.

Table 1. The results of modeling the activity of molecules for different data sets. CV is the metric value obtained on cross-validation. Best test set prediction in row is bolded.

Model	Conformations (instances)									
	1		5		10		15		25	
	ROC AUC value									
	CV	Test	CV	Test	CV	Test	CV	Test	CV	Test
	ACE									
Random Forest	0.85	**0.69**	0.88	0.69	0.88	0.69	0.87	0.69	0.87	0.69
Neural Network	0.96	0.75	0.97	**0.89**	0.97	0.84	0.97	0.83	0.97	0.80
	DHFR									
Random Forest	0.81	**0.79**	0.69	0.60	0.66	0.67	0.69	0.60	0.68	0.66
Neural Network	0.88	**0.88**	0.89	0.88	0.90	0.84	0.90	0.86	0.91	0.87
	ACHE									
Random Forest	0.63	**0.61**	0.64	0.57	0.58	0.52	0.63	0.52	0.56	0.61
Neural Network	0.76	0.72	0.74	0.88	0.69	0.87	0.75	**0.92**	0.74	0.83

In the case of modeling the activity of molecules on the ACE and ACHE data set, the unsupervised clustering-based model built on one conformation (instance) is characterized by the same prediction accuracy on the external test, as models built on 5,

10, 15 and 25 examples. Thus, multi-instance learning does not give any advantages over single-instance case. However, when using the supervised neural network based approach, models built on several conformations (instances) are characterized by higher prediction accuracy than models built on one instance. At the same time, ROC AUC values reach a plateau and do not rise with number of conformations. It seems that model training becomes more complicated with rise of number of instances; and simultaneously the amount of new information introduced by inclusion of additional instance becomes lower and lower. These two effects can explain such saturation and even lowering of model performance with number of instances used.

In the case of the DHFR data set, the models built on one conformation were already highly predictive and the multi-instance learning approach could not improve predictive ability of models even when neural networks are used. This result can be explained by the fact that there is always a chance to randomly select an instance, which for a given data set and the type of modeled property is well described by the particular type of features, i.e. inclusion of 3D information is not necessary for a given data set.

It was noted that neural net-based MIL models with the greatest ROC AUC on cross-validation are not the best on external test set: there are some hyperparameter combinations that result in the models that much better predict external test set but their performance estimated on cross-validation is slightly lower. It can be caused by mentioned inconsistency in training and test set selection: prediction of test set requires more generalized model (that require more training epochs) than cross-validation which approaches good performance using less epochs.

3.2 Comparison with Traditional QSAR

In this section, we summarize the obtained results and compare the classical approach to modeling the properties of molecules and the approach of multi-instance learning. To build models on classical 2D descriptors, the Random Forest method and the multilayer perceptron with one hidden layer (MLP) were used. Table 2 summarizes results obtained. In Table 2, the third column shows the best ROC AUC achieved using multi-instance learning setup (bolded in the Table 1). One can notice that RF model is worse than MLP. It is interesting that RF models on 2D representation-based features are better than on 3D ones.

Comparing to models built on 2D molecule representation - based features unsupervised clustering-based (RF-based) approaches have lower performance. It seems that some adjustment of model, more careful selection of clustering scheme could improve results. Supervised clustering-based neural nets models guarantees better performance than 2D representation based single-instance approach used in traditional QSAR. Only in DHFR case performance of both models is the same. It supports our conclusion that for this data set one instance is sufficient for achieving the best model performance and information about spatial structure is not important for discrimination of active and inactive compounds. Thus, taking into account several representations of the molecule allows us to achieve a higher accuracy of the classification of molecules into active and inactive than traditional approach, based on molecular graph features.

Table 2. ROC AUC metrics for best models built on feature vectors representing 2D molecular graph and feature vector of random 3D conformation.

Model	2D (Traditional QSAR)		3D (Multi-instance learning)	
	CV	Test	CV	Test
ACE				
			5 instances	
Random Forest	0.81	0.76	0.88	0.69
Neural Network	0.91	0.86	0.97	0.89
DHFR				
			5 instances	
Random Forest	0.85	0.81	0.69	0.60
Neural Network	0.92	0.88	0.89	0.88
ACHE				
			15 instances	
Random Forest	0.62	0.60	0.64	0.57
Neural Network	0.84	0.87	0.75	0.92

4   Conclusions

Molecules are complex dynamic objects that can be usually represented by a set of instances - conformations that are different by spatial arrangements of constituting atoms. Multiple-instance learning is a promising approach for modeling chemical object properties since it can tackle strictly intrinsic complexity of molecules. In the work, two simple approaches for MIL have been proposed for chemical structure-biological activity modeling. One approach is based on unsupervised clustering of multiple instances with subsequent generation of latent vectors showing number of instances of the molecule falling into every cluster. The second approach is based on application of specially adapted architecture of neural network (similar to convolutional NN) that implicitly learns optimal clustering.

Proposed classification approaches were first benchmarked versus traditional single instance based QSAR approaches on three biological activity data sets taken from literature. It was shown that neural network-based approach allows generating more performant models than the unsupervised clustering-based one. The model, built on one conformation within neural network MIL architecture shows lower ROC AUC values than models based on several instances (conformations). However, there is some saturation effect when increase of number of instances lowers model performance. Generally supervised clustering-based MIL approach is more performant (or at least has similar performance) than traditional QSAR approach based on 2D molecular graph features. We noticed at certain hyperparameters neural net-based MIL approach show much better performance on test set (such models become substantially better than single-instance ones), but since cross-validation ROC AUC slightly worse, they are not selected. Such inconsistency in cross-validation and test set quality metrics can be explained by fact that test sets were selected by maximal dissimilarity to training set and thus optimality criterions do not coincide in cross-validation and test set.

The study shows that MIL approach, especially based on neural networks are very promising for modeling chemical object properties. Further wide-scope comparison of MIL approaches with traditional QSAR should be performed.

Acknowledgements. This work was supported by Foundation the Ministry of education, youth and sport of Czech Republic, agreement number MSMT-5727/2018-2 and by the Ministry of Science and Higher Education of the Russian Federation, agreement No 14.587.21.0049 (unique identifier RFMEFI58718X0049).

References

1. Varnek, A., Baskin, I.I.: Chemoinformatics as a theoretical chemistry discipline. Mol. Inform. **30**, 20–32 (2011)
2. Dudek, A.Z., Arodz, T., Galvez, J.: Computational methods in developing quantitative structure-activity relationships (QSAR): a review. Comb. Chem. High Throughput Screen. **9**, 213–228 (2006)
3. Varnek, A., Baskin, I.: Machine learning methods for property prediction in chemoinformatics: quo vadis? J. Chem. Inf. Model. **52**, 1413–1437 (2012)
4. Dietterich, T.G., Lathrop, R.H., Lozano-Pérez, T.: Solving the multiple instance problem with axis-parallel rectangles. Artif. Intell. **89**, 31–71 (1997)
5. Bergeron, C., Moore, G., Zaretzki, J., Breneman, C., Bennett, K.: Fast bundle algorithm for multiple-instance learning. IEEE Trans. Pattern Anal. Mach. Intell. **34**, 1068–1079 (2011)
6. Andrews, S., Tsochantaridis, I., Hofmann, T.: Support vector machines for multiple-instance learning. In: Proceedings of the 15th International Conference on Neural Information Processing Systems, pp. 577–584. MIT Press, Cambridge (2002)
7. Kwok, J.T., Cheung, P.-M.: Marginalized multi-instance kernels. In: Proceedings of the 20th International Joint Conference on Artificial Intelligence, pp. 901–906. Morgan Kaufmann Publishers Inc., San Francisco (2007)
8. Zhou, Z.-H., Xu, J.-M.: On the relation between multi-instance learning and semi-supervised learning. In: Proceedings of the 24th International Conference on Machine Learning, pp. 1167–1174. ACM, New York (2007)
9. Mangasarian, O.L., Wild, E.W.: Multiple instance classification via successive linear programming. J. Optim. Theory Appl. **137**, 555–568 (2008)
10. Bunescu, R.C., Mooney, R.J.: Multiple instance learning for sparse positive bags. In: Proceedings of the 24th International Conference on Machine Learning, pp. 105–112. ACM, New York (2007)
11. Doran, G., Ray, S.: A theoretical and empirical analysis of support vector machine methods for multiple-instance classification. Mach. Learn. **97**, 79–102 (2014)
12. Krogel, M.-A., Wrobel, S.: Feature selection for propositionalization. In: Lange, S., Satoh, K., Smith, C.H. (eds.) DS 2002. LNCS, vol. 2534, pp. 430–434. Springer, Heidelberg (2002). https://doi.org/10.1007/3-540-36182-0_45
13. Dong, L.: A comparison of multi-instance learning algorithms (2006). https://hdl.handle.net/10289/2453

14. EL-Manzalawy, Y., Honavar, V.: MICCLLR: multiple-instance learning using class conditional log likelihood ratio. In: Gama, J., Costa, V.S., Jorge, A.M., Brazdil, P.B. (eds.) DS 2009. LNCS (LNAI), vol. 5808, pp. 80–91. Springer, Heidelberg (2009). https://doi.org/10. 1007/978-3-642-04747-3_9

15. Ruffo, G.: Learning single and multiple instance decision trees for computer security applications. University of Turin, Torino (2000)

16. Xu, X.: Statistical learning in multiple instance problems (2003). https://hdl.handle.net/10289/ 2328

17. Wang, J., Zucker, J.-D.: Solving the multiple-instance problem: a lazy learning approach. In: Proceedings of the Seventeenth International Conference on Machine Learning, pp. 1119–1126. Morgan Kaufmann Publishers Inc., San Francisco (2000)

18. Zeng, T., Ji, S.: Deep convolutional neural networks for multi-instance multi-task learning. In: Proceedings of the 2015 IEEE International Conference on Data Mining (ICDM), pp. 579–588. IEEE Computer Society, Washington, DC (2015)

19. Sutherland, J.J., O'Brien, L.A., Weaver, D.F.: A comparison of methods for modeling quantitative structure – activity relationships. J. Med. Chem. **47**, 5541–5554 (2004)

20. Lopez-Sendon, J., et al.: Expert consensus document on angiotensin converting enzyme inhibitors in cardiovascular disease. The Task Force on ACE-inhibitors of the European Society of Cardiology. Eur. Heart J. **25**, 1454–1470 (2004)

21. Li, R., et al.: Three-dimensional structure of M. tuberculosis dihydrofolate reductase reveals opportunities for the design of novel tuberculosis drugs. J. Mol. Biol. **295**, 307–323 (2000)

22. Colovic, M.B., Krstic, D.Z., Lazarevic-Pasti, T.D., Bondzic, A.M., Vasic, V.M.: Acetylcholinesterase inhibitors: pharmacology and toxicology. Curr. Neuropharmacol. **11**, 315–335 (2013)

23. www.rdkit.org, http://www.rdkit.org/

24. Ramon, J., De Raedt, L.: Multi instance neural networks (2000)

25. Wang, X., Yan, Y., Tang, P., Bai, X., Liu, W.: Revisiting multiple instance neural networks. Pattern Recogn. **74**, 15–24 (2018)

Cross-Efficiency of International Sanctions: Application of Data Envelopment Analysis and Network Methodology

Dmitry Zaytsev[(✉)], Valentina Kuskova, Polina Lushnikova,
and Gregory Khvatsky

International Laboratory for Applied Network Research,
National Research University Higher School of Economics,
Moscow, Russia
zaytsevdi2@gmail.com

Abstract. In this paper we provide the methodology for evaluating effectiveness of international sanctions using Data Envelopment Analysis (DEA), which we use for generating the network matrix for further analysis. DEA is a non-parametric technique used to compare performance of similar units, such as departments or organizations. DEA has wide applications in all industries, and has been successfully used to compare performance of hospitals, banks, universities, etc. The most important advantage of this technique is that it can handle multiple input and output variables, even those not generally comparable to each other. We use the "Threat and Imposition of Sanctions (TIES)" Data 4.0 for analysis. This database contains the largest number of cases of international sanctions (1412 from the years 1945–2005) imposed by some countries on others, takes into account simultaneous sanction imposition, and also estimates the cost of all sanctions - both for those who receive and those who impose them. As input variables for DEA model we use the impact of sender commitment, anticipated target and sender economic costs, and actual target and sender economic costs. As the output variable, we use the outcome of sanctions for senders. We describe how to use DEA cross-efficiency outputs to build the network of sanction episodes. Our proposed combination of DEA and network methodology allows us to cluster sanction episodes depending on their outcomes, and provides explanations of higher efficiency of one group of sanction episodes over the others.

1 Introduction

Recently, leading countries have increased the use of sanctions as a strategy of pressure on other countries. Sanctions are a softer tool of pressure compared to military interventions, and usually, they materialize in the form of prohibition of

© Springer Nature Switzerland AG 2020
W. M. P. van der Aalst et al. (Eds.): AIST 2019, CCIS 1086, pp. 72–82, 2020.
https://doi.org/10.1007/978-3-030-39575-9_8

a certain economic activity. Sanctions can be considered one of the elements of hard power, which allows countries to send signals to others and exert pressure on each other without combat. Despite the wider use of sanctions (relative to military action), they still have a number of significant drawbacks. For example, they are difficult to manage and implement, which casts doubt on their effectiveness [1]. This is a critical issue that has received much attention.

Today, economic sanctions are one of the most common instruments of foreign policy [3]. They became a tool in a foreign policy toolkit as an alternative or complement to the use of force or strengthening the bargaining power. They are actively used as a means of forcing individual states to fulfill the political demands of the countries initiating sanctions [4,5]. It is believed that sanctions are an inexpensive alternative to war when diplomacy fails. The number of cases of application of sanctions by the UN Security Council is growing [6]. Since 1966, more than 30 sanctions regimes have been introduced that still exist today. At the same time, receiver states are actively developing and improving the mechanisms for circumventing sanctions and mitigating their impact on the national economy and the stability [7]. Often, the use of sanctions becomes the subject of regulatory and ethical differences between the objects of sanctions (as a rule, these are developing countries) and the initiators of sanctions (usually in this role are the developed countries). The initiators consider sanctions to be one of the legal means of coercion to fulfill international obligations or to comply with certain norms [8–10]. Interest in economic sanctions as an instrument of pressure is currently increasing, since other key instruments of foreign economic impact have their own objective limitations. These include trade, financial and macroeconomic policies, the provision of assistance and the economic sanctions themselves. However, it is always necessary to remember that the potential of economic sanctions is always limited by the question of their effectiveness.

Broad use in political practice is combined with a broad interpretation of the very concept of sanctions. Sometimes, sanctions include measures of trade or economic war. In the expert community, the functions of sanctions are interpreted differently, offering different versions of the typology of sanctions [11–14]. The literature on sanctions is broad and diverse, and the only summarizing comment is that sanctions are different in purpose, objectives, and scales. The key issue in the scientific discourse is the question of the reasons and conditions for the effectiveness of sanctions. Why in some cases sanctions are successful, while others are not, and even lead to results contrary to expectations?

The definition of sanction effectiveness is somewhat vague, but a consensus was reached nonetheless by the scientific community. Researchers believe that the effectiveness of sanctions consists of two components: the degree to which the outcome of the policy to which the sending country aspired was achieved and the contribution to the success achieved by the sanctions [11].

In most cases, when examining the effectiveness of sanctions from a political angle, researchers consider such factors as accompanying policies, international cooperation, international assistance to the target country, previous relations between the sender and the target states, and democracy against autocracy.

From the economic point of view, researchers study the impact of the average cost of sanctions for the target state, the size of the country, trade relations, economic health, political stability in the target country, and the type of sanctions [12]. Most researchers agree that sanctions are a more humane analogue of war, yet one area where there is still little consensus is the assessment of the effectiveness of sanctions [13,14].

When studying the effectiveness of sanctions one of the most common methods of analysis is a case study. Many researchers turn to this method because they already have a clear idea of effective vs. ineffective cases and are considering them directly [15–17]. Another group of researchers addresses linear regression (and in particular, logistic regression). This method allows predicting the effectiveness of sanctions (dependent variable) for a number of independent variables [18,19].

$$y = f(x, \beta) + \epsilon$$

Researchers also use cluster analysis, where cases are grouped by efficiency and scientists get out similarity patterns between them [20]. In the literature [2] researchers have identified a number of political and economic variables that affect the effectiveness of economic sanctions, including international cooperation, the former relationship between the sender and target states, the average cost of sanctions, trade relations, etc. However, to the best of our knowledge, none of the existing studies were actually able to quantify the sanctions' cross-efficiency - in other words, mutual impact on the sender and the receiver - or provide methodology for its evaluation.

In this study, we attempt to close this literature gap by introducing the methodology for quantifying sanctions' cross-efficiency. We identify the most effective imposition of sanctions in the period from 1945 to 2005 using the method of DEA (Data Envelopment Analysis) and describe how to use DEA results for establishing a network of efficient and inefficient sanctions, further analyzed with network methodology.

2   Data and Method

As already described above, some researchers addressed the issue of sanctions' effectiveness, and even identified key variables that influence the "success" of sanctions. However, there is still no agreement which examples to consider effective, and there are some factors that have not yet been studied in the analysis of effectiveness. Thus, the focus of this study is to identify the most effective examples of imposing sanctions and the variables that affect success.

Economic sanctions in the "TIES 4.0" database are defined as actions taken by one or more countries to limit or terminate their economic relations with the target country in an attempt to convince the target to change its policy. By definition, a sanction must: (1) include one or more sender states and a target state, and (2) be implemented by the sender to change the behavior of the receiver. Actions taken by states that restrict economic relations with other countries solely for reasons of domestic economic policy are, therefore, not considered to be political sanctions. Sanctions take various forms, including such actions as tariffs, export controls, embargoes, import bans, travel bans, freezing of assets, reduction assistance and blockade, and the type of sanction is an important variable in our analysis.

Then, we turned to the standard set of variables that were identified at the literature review stage, and also added variables that were not previously considered in the analysis. Thus, in our model, we use the following set as input variables:

- Sender Commitment. This variable takes the values 1 (weak commitment), 2 (moderate), to 3 (strong commitment).
- Anticipated target economic costs. This variable measure minor (1), major (2), and severe (3) potential costs for target state economy.
- Anticipated sender economic costs. This variable measure minor (1), major (2), and severe (3) potential costs for sender state economy.
- Target economic costs. This variable measure minor (1), major (2), and severe (3) ex post economy costs for target state economy as a result of sanctions imposition.
- Sender economic costs. This variable measure minor (1), major (2), and severe (3) economy costs for sender state economy as a result of sanctions imposition.

Output variable:

- Settlement nature for sender state. This variable is designed to capture how the sanctions episode was settled. On a scale of 1–10, it measures how well did the sender fare as a result of the threat/imposition of sanctions.

We analyze the effectiveness of sanctions using Data Envelopment Analysis (DEA). DEA is a non-parametric method based on solving an optimization linear programming problem of maximizing the outputs for a given amount of resources (inputs) or minimizing the resources used (inputs) at a given level of output products (outputs). The unit of analysis is the decision-making unit (DMU); in our case - countries. To solve the linear programming problem, a DEA model formulates a mathematical description of the system of relations of weighted output variables corresponding to the results of the DMU activity to the weighted input variables corresponding to the used DMU resources. From the technical point of view, DEA compares each of the DMUs in pairs from the point of view of their "output-input" relationship in order to obtain relative efficiency estimates. Those DMUs that receive the highest scores are at the frontier of production capability and become benchmarks for other DMUs in the sample [21–23].

Formally, DEA approach can be described as follows. There are K objects, or DMUs, and each has multiple indices:

- A vector m of expended resources, or inputs, I_{km}
- A vector of n results or outputs, O_{kn}
- Each resource has a weight of x_m
- Each result has a weight of y_n

It is important to note that the weights of resources and outcomes are not known *a priori*. DEA solves a system of equations to optimize a combination of weights, deeming the object either efficient or inefficient relative to other objects in the evaluation. The efficiency of each object is the ratio of the sum of all outputs obtained to the sum of all expended resources:

$$\frac{\sum_{n=1}^{N} O_{kn} y_n}{\sum_{m=1}^{M} I_{km} x_m}$$

The optimization problem is then formulated using the objective function, maximizing the efficiency of the evaluated unit, subject to imposed constraints:

$$max z = \frac{\sum_{n=1}^{N} O_{kn} y_n}{\sum_{m=1}^{M} I_{km} x_m},$$

Subject to:

$$\sum_{n=1}^{N} O_{kn} y_n - \sum_{m=1}^{M} I_{km} x_m \leq 0,$$

k = 1, 2, ..., K

$$x_m, y_n \geq 0$$

for all m, n.

In recent years, the use of the DEA approach has become very popular for evaluating the effectiveness of organizations in various industries and countries. Due to the fact that DEA has a small number of restrictions, this assessment method has become more preferable in cases where other assessment methods are dependent on the nature of the relationship between "inputs" and "outputs" [24, 25].

Formally, the DEA models are aimed at exploring the boundaries, rather than central trends. Thus, regression analysis "smoothes" observations to identify central linear dependence, which greatly simplifies research findings, while DEA models are more convenient for identifying peripheral relationships excluded by regressions [26–28].

It remains important to define the concept of "effectiveness" in accordance with the DEA methodology. According to Rhodes and Cooper, the DEA approach assumes two views on efficiency: isolated efficiency in the context of the "inputs" and "outputs" of the DMU (let's call it "full effectiveness") and comparative efficiency. Full efficiency is achieved by a DMU *iff* none of its "inputs" and "outputs" can be improved without degrading the other "inputs" and "outputs" (Pareto-Kupmans efficiency). As for comparative efficiency, DMUs will be evaluated as a fully effective unit based on the available data *iff* the performance of other DMUs does not show that some of the "inputs" and "outputs" of a fully effective DMU can be improved without degrading others [29].

At the heart of these definitions of "efficiency" is the concept of Wilfredo Pareto about the modern "economy of welfare," in which it is impossible to improve the position of some people without deteriorating the situation of others. In 1951, Koopmans used the Pareto criterion in his work "Analysis of Production and Distribution Activities." According to Koopmans, there are "finished products" ("outlets") that can be recognized as the best if their improvement can only be achieved by worsening other "finished products." These "finished products" are compared in pairs with each other many times to get their own efficiency weights - that is, their estimates. However, each "finished product" can be produced using a specific combination of "resources" ("inputs") [30, 31].

3   Model

Based on the data set that we have, we decided to use variables such as sender commitment, anticipated target and sender economic costs, and target and sender economic costs. Since DEA does not work with missing data, we first removed those cases for which some data were missing. Also, we removed cases where outcomes where coded as -9 (unknown) and 0 ("sender completely fails" as DEA does not process zeros). Therefore, from the 1412 cases we went down to 229 cases suitable for the analysis. Given that we started with the census of sanctions, the remaining sample size is appropriately representative.

In this study, we applied DEA using the software package "PIM-DEAsoft V3" [57]. We used the Constant Returns to Scale (CRS) model, because we assume that outcome from sanctions (their effectiveness) proportionally depend on the sender state commitments to force target state, and potential and ex post costs of sanctions for both sender and receiver of sanctions.

In our model, there are 17 cases or sanctions episodes that received 100% efficiency scores based on our analysis (Table 1). The code of the case is in the format of the "year, month, day and sanction number."

The first interesting result is that majority of the efficient sanctions are imposed by the USA. This is not surprising, given that the almost half (48.22%) of all worldwide sanctions are imposed by the USA. Nenetheless, the percentage of USA-imposed efficient sanctions (71%) is higher than could be expected using weighted percentage alone. The second interesting result is that there are no "famous" cases among the efficient, e.g., sanctions against Cuba or North Korea. This could be due to the higher level of counteraction by the target state in these cases. Finally, also efficient is the most recent case (2005) of ECOWAS vs. Togo. This case was described as a rare case of unity among African nations [58], when the late dictator Gnassingbe's son resigned after the African Union had joined the Economic Community of West African States (ECOWAS) to impose sanctions on his regime and, together with the USA and the EU, threatened to cut all diplomatic ties with Togo. More qualitative analysis is needed to understand the nature of the efficient sanctions obtained by the results, but it is beyond the scope of this article.

Table 1. Description of DEA-efficient sanctions

Sanction code	Description
1952010801	USA vs. Egypt
1952010803	USA vs. Iraq
1953021201	USA vs. Italy
1953021202	USA vs. Belgium
1953021205	USA vs. the Netherlands
1972090801	Several countries vs. Zimbabwe
1991100101	USA vs. Haiti
2000051901	Several countries vs. Fiji
2001012202	USA vs. Benin
2001042301	China vs. Japan
2001051503	Iraq vs. Turkey
2001100301	USA vs. CAR
2002041104	USA vs. Cambodia
2002080501	USA vs. Colombia
2002091901	USA vs. Malawi
2002111301	USA vs. Zambia
2005020701	ECOWAS vs. Togo

In addition to building frontiers to identify the most efficient cases, we also constructed the cross-efficiency matrix. This tool for interpreting the results consists of creating a table where the number of rows (i) and columns (j) equals the number of units in the analysis. For each cell (ij), the efficiency of unit ij is computed with weights that are optimal to unit j. The higher the values in a given column j, the more likely it is that the unit ij is an example of more efficiently operating practices.

Next, we use the cross-efficiency matrix to create the network of sanctions' episodes. Doing so allows us to cluster sanctions' episodes depending on the similarity of their efficiency function, and analyze why one group of sanction episodes is more efficient than the others. We implement the network on the second model, since it contains the full set of variables of interest. Node ties represent similarity of the sanction efficiency for the two connected countries with the weight of the tie equal to the cross-efficiency between the two countries.

In order to cluster the network, we used the exploratory graph analysis using Pajek with Fruchterman-Reingold algorithm on all sanctions with cross-efficiency higher than 90%. As is clear from Fig. 1, there are four distinct clusters of sanctions, though the analysis of their qualitative composition is beyond the scope of this article.

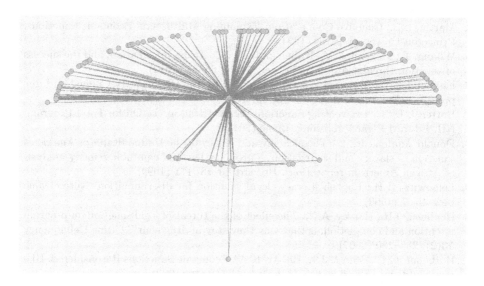

Fig. 1. Cross-efficiency network clusters

4   Conclusion

Despite the multitude of studies on the effectiveness of sanctions on the political arena, the question of the sanctions' effectiveness or the factors that affect it remains on the forefront of international relations research. In this article, we present the methodology that could help shed the light on the sanctions' efficiency by first identifying the most efficient sanctions (relative to each other), and then grouping them in some way to identify the similarities between the efficient cases. To the best of our knowledge, we are the first to propose this methodology in the international relations research, and we hope that further studies can validate our proposed methods with theoretical explanation of the obtained results.

Acknowledgment. The article was prepared within the framework of the Basic Research Program at the National Research University Higher School of Economics (HSE) and supported within the framework of a subsidy by the Russian Academic Excellence Project '5–100.'

References

1. Kerr, W.A., Gaigord, J.D.: A note on increasing the effectiveness of sanctions. J. World Trade **28**(6), 169–176 (1994)
2. Jing, C., Kaempfer, W.H., Lowenberg, A.D.: Instrument choice and the effectiveness of international sanctions: a simultaneous equations approach. J. Peace Res. **40**(5), 519–535 (2003)
3. Selden, Z.A.: Economic Sanctions as Instruments of American Foreign Policy. Greenwood Publishing Group (1999)

4. Martin, L.L.: Coercive Cooperation: Explaining Multilateral Economic Sanctions. Princeton University Press (1994)
5. Whang, T., McLean, E.V., Kuberski, D.W.: Coercion, information, and the success of sanction threats. Am. J. Polit. Sci. **57**(1), 65–81 (2013)
6. Fassbender, B.: Targeted sanctions imposed by the UN Security Council and due process rights. Int. Org. L. Rev. **3**, 437 (2006)
7. Botterill, L.: Circumventing Sanctions Against Iraq in the Oil for Food Program. NH. Edward Elgar Publishing, Lyme (2011)
8. Donald Aquinas Jr., L.: The alchemy and legacy of the United States of America's sanction of slavery and segregation: a property law and equitable remedy analysis of African American reparations. Howard LJ **43**, 171 (1999)
9. Leibowitz, A.H.: English literacy: legal sanction for discrimination. Notre Dame Law **45**, 7 (1969)
10. Hepburn, J.R., Harvey, A.N.: The effect of the threat of legal sanction on program retention and completion: is that why they stay in drug court? Crime Delinquency **53**(2), 255–280 (2007)
11. Hufbauer, G.C., Schott, J.J., Elliott, K.A.: Economic Sanctions Reconsidered: History and Current Policy, vol. 1. Peterson Institute (1990)
12. Figuerola, P.: Economic Sanctions' Effectiveness in a World with Interdependent Networks and Powerful MNCs: The Role of Governance in the Target State (2015)
13. Pape, R.A.: Why economic sanctions do not work. Int. Secur. **22**(2), 90–136 (1997)
14. Pape, R.A.: Why economic sanctions still do not work. Int. Secur. **23**(1), 66–77 (1998)
15. Conlon, P.: United Nations Sanctions Management: A Case Study of the Iraq Sanctions Committee 1990–1994. Brill Nijhoff (2000)
16. Levy, P.I.: Sanctions on South Africa: what did they do? Am. Econ. Rev. **89**(2), 415–420 (1999)
17. Rotunda, R.D.: Judicial comments on pending cases: the ethical restrictions and the sanctions - a case study of the Microsoft litigation. U. Ill. L. Rev. 611 (2001)
18. Ahn, T., Vigdor, J.: The impact of No Child Left Behind's accountability sanctions on school performance: Regression discontinuity evidence from North Carolina. No. w20511. National Bureau of Economic Research (2014)
19. Black, T., Orsagh, T.: New evidence on the efficacy of sanctions as a deterrent to homicide. Soc. Sci. Q. **58**(4), 616–631 (1978)
20. Lektzian, D., Regan, P.M.: Economic sanctions, military interventions, and civil conflict outcomes. J. Peace Res. **53**(4), 554–568 (2016)
21. Cooper, W.W., Seiford, L.M., Tone, K.: Introduction to Data Envelopment Analysis and Its Uses: With DEA-Solver Software and References. Springer, Heidelberg (2006). https://doi.org/10.1007/0-387-29122-9
22. Roll, Y., Hayuth, Y.E.H.U.D.A.: Port performance comparison applying data envelopment analysis (DEA). Maritime Policy Manag. **20**(2), 153–161 (1993)
23. Lertworasirikul, S., et al.: Fuzzy data envelopment analysis (DEA): a possibility approach. Fuzzy Sets Syst. **139**(2), 379–394 (2003)
24. Sherman, H.D., Ladino, G.: Managing bank productivity using data envelopment analysis (DEA). Interfaces **25**(2), 60–73 (1995)
25. Hollingsworth, B., Smith, P.: Use of ratios in data envelopment analysis. Appl. Econ. Lett. **10**(11), 733–735 (2003)
26. Banker, R.D., Charnes, A., Cooper, W.W.: Some models for estimating technical and scale inefficiencies in data envelopment analysis. Manag. Sci. **30**(9), 1078–1092 (1984)

27. Sexton, T.R.: The methodology of data envelopment analysis. New Dir. Program Eval. **32**, 7–29 (1986)

28. Charnes, A., et al. (eds.): Data Envelopment Analysis: Theory, Methodology, and Applications. Springer, Heidelberg (2013). https://doi.org/10.1007/978-94-011-0637-5

29. Zhu, J., Cook, W.D. (eds.): Modeling data Irregularities and Structural Complexities in Data Envelopment Analysis. Springer, Heidelberg (2007). https://doi.org/10.1007/978-0-387-71607-7

30. Ji, Y., Lee, C.: Data envelopment analysis. Stata J. **10**(2), 267–280 (2010)

31. Dajani, M.S., Daoudi, M.: Economic Sanctions, Ideals and Experience. Routledge & Kegan Paul, London, Boston (1983)

32. Tittle, C.R.: Sanctions and social deviance: the question of deterrence, pp. 45–60 (1980)

33. Cooter, R.: Prices and sanctions. Columbia Law Rev. **84**(6), 1523–1560 (1984)

34. Mueller, J., Mueller, K.: Sanctions of mass destruction. Foreign Aff. **78**, 43 (1999)

35. Drury, A.C.: Revisiting economic sanctions reconsidered. J. Peace Res. **35**(4), 497–509 (1998)

36. Lektzian, D., Souva, M.: Institutions and international cooperation: An event history analysis of the effects of economic sanctions. J. Conflict Resolut. **45**(1), 61–79 (2001)

37. Gibbs, J.P.: Sanctions. Soc. Probl. **14**(2), 147–159 (1966)

38. Wollert, R., et al.: Causal attributions, sanctions, and normal mood variations. J. Pers. Soc. Psychol. **45**(5), 1029 (1983)

39. Klein, E.: Sanctions by international organizations and economic communities. Archiv des Völkerrechts **30**(1), 101–113 (1992)

40. Marinov, N.: Do economic sanctions destabilize country leaders? Am. J. Polit. Sci. **49**(3), 564–576 (2005)

41. Cox, D.G., Cooper Drury, A.: Democratic sanctions: connecting the democratic peace and economic sanctions. J. Peace Res. **43**(6), 709–722 (2006)

42. Noland, M.: The (non-) impact of UN sanctions on North Korea. Asia Policy **7**, 61–88 (2009)

43. Yu, J.P., Pysarchik, D.T.: Economic and non-economic factors of Korean manufacturer-retailer relations. Int. Rev. Retail Distrib. Consum. Res. **12**(3), 297–318 (2002)

44. Chambliss, W.J.: Types of deviance and the effectiveness of legal sanctions. Wis. L. Rev. 703 (1967)

45. Giumelli, F.: Coercing, Constraining and Signalling: Explaining UN and EU Sanctions After the Cold War. ECPR Press (2011)

46. Biersteker, T.J., Eckert, S.E., Tourinho, M. (eds.): Targeted Sanctions. Cambridge University Press, Cambridge (2016)

47. Kaempfer, W.H., Lowenberg, A.D.: The theory of international economic sanctions: a public choice approach. Am. Econ. Rev. **78**(4), 786–793 (1988)

48. Torbat, A.E.: Impacts of the US trade and financial sanctions on Iran. World Econ. **28**(3), 407–434 (2005)

49. Gurvich, E., Prilepskiy, I.: The impact of financial sanctions on the Russian economy. Russ. J. Econ. **1**(4), 359–385 (2015)

50. Kittrie, O.F.: New sanctions for a new century: treasury's innovative use of financial sanctions. U. Pa. J. Int. L. **30**, 789 (2008)

51. McGillivray, F., Stam, A.C.: Political institutions, coercive diplomacy, and the duration of economic sanctions. J. Conflict Resolut. **48**(2), 154–172 (2004)

52. Kaempfer, W.H., Lowenberg, A.D.: The political economy of economic sanctions. Handb. Defense Econ. **2**, 867–911 (2007)
53. Cuccia, A.D.: The effects of increased sanctions on paid tax preparers: integrating economic and psychological factors. J. Am. Taxation Assoc. **16**(1), 41 (1994)
54. Drezner, D.W., Drezner, D.W.: The Sanctions Paradox: Economic Statecraft and International Relations, vol. 65. Cambridge University Press (1999)
55. Early, B.R.: Busted Sanctions: Explaining Why Economic Sanctions Fail. Stanford University Press (2015)
56. Cook, W.D., Seiford, L.M.: Data envelopment analysis (DEA)–thirty years on. Eur. J. Oper. Res. **192**(1), 1–17 (2009)
57. Emrouznejad, A., Thanassoulis, E.: Performance Improvement Management Software (PIMsoft): A User Guide. http://www.DEAsoftware.co.uk. Accessed 6 July 2019
58. https://www.theguardian.com/world/2005/feb/27/theobserver1

Natural Language Processing

Using Semantic Information for Coreference Resolution with Neural Networks in Russian

Ilya Azerkovich[(✉)] [iD]

Higher School of Economics, Moscow, Russia
`ilazerkovich@edu.hse.ru`

Abstract. This paper describes an experiment aimed at improving the quality of coreference resolution for Russian by combining one of the most recent developments in the field, employment of neural networks, with benefits of using semantic information. The task of coreference resolution has been the target of intensive research, and the interest at using neural networks, successfully tested in other tasks of natural language processing, has been gradually growing. The role that semantic information plays for the task of coreference resolution has been recognized by researchers, but the impact of semantic features on the performance of neural networks has not been yet described in detail. Here we describe the process of integrating features derived from open-source semantic information into the coreference resolution model based on a neural network, and evaluate its performance in comparison with the base model. The obtained results demonstrate quality on par with state-of-the-art systems, which serves to re-establish the importance of semantic features in coreference resolution, as well as the applicability of neural networks for the task.

Keywords: Natural language processing · Coreference resolution · Neural networks · Semantic relatedness · Russian language

1 Introduction

Coreference resolution, as an important step at machine translation, information extraction, text summarization, etc., is among the most relevant tasks of natural language processing. Two expressions can be considered coreferent if they refer to one and the same real-world entity. Consequently, the goal of automated coreference resolution is to extract chains of mentions, referring to the same entity, from the text.

The algorithms of automated coreference resolution have been created since the middle of XXth century. At first these algorithms have been mostly empiric, based on the rules suggested by its developer, such as the classic algorithm described in [1]. Later the work actively began on the family of algorithms based on machine learning methods and using big data for training [2] or [3], but the rule-based algorithms are also successfully used, for example in the Stanford coreference parser [4]. Recently, followed by the rise of interest towards neural networks, the works aimed at using them for various NLP tasks, including coreference resolution, have started to appear [5, 6].

© Springer Nature Switzerland AG 2020
W. M. P. van der Aalst et al. (Eds.): AIST 2019, CCIS 1086, pp. 85–93, 2020.
https://doi.org/10.1007/978-3-030-39575-9_9

Main types of features used in algorithms of automated coreference resolution include morphological, syntactic, string-based and distance ones. Semantic information, if used, usually is presented as information about named entities or compatibility in terms of top-level ontology nodes (e.g. in [2] or [7]). Features, derived by more detailed analysis, such as semantic relatedness measures, are seldom used despite their potential effectiveness shown in a number of works, such as [8].

Development of algorithms of automated coreference resolution for Russian began later than for English. This may partly be due to algorithms relying upon resources in the corresponding language, which for Russian exist on a much smaller scale. Research, describing systems of coreference resolution in Russian, does exist (see e.g. competition results of [9]), and attempts at using semantic information for analysis are also being made. This paper describes integrating semantic relatedness measures, calculated from open-source data, into a neural network-based algorithm, oriented at Russian language. The achieved results suggest that, while using neural networks for the task of coreference resolution in Russian could be more effective than other methods, by using semantic features further improvement could be achieved.

2 Related Work

Machine learning-based algorithms are the most actively developed class of methods for automated coreference resolution. They can be grouped into several general classes based on structure, the main of them being: the mention-pair models, suggested in the seminal work of Soon et al. [2]; the entity-mention models, described e.g. in [10], which introduces mention clustering and cluster-based features; ranking models, which consider several candidate mentions (e.g. [11]) or mention clusters [12] as possible antecedents. The usage of perceptrons and neural networks has been researched, among others, for a mention ranking model in [13], and for a cluster ranking model in [6].

Russian language-oriented research has begun to develop actively relatively recently, with a breakthrough becoming possible due to the publication of the RuCor corpus, used for the RuEval-2014 competition of coreference resolvers [9]. While most participants of the competition employed different pair-based models for the analysis, clustering approach has been also adopted in [14].

Using semantic information for the task of coreference resolution in English has been studied in several papers. Particularly, the usage of semantic similarity measures as features has been researched in [8] and [15]. For Russian language, the implementation of semantics-based features in the form of hypernym chains and gazetteers has been described in [16], and an attempt at using the data of Wikipedia articles was made in [17].

In this paper we use the semantic information, obtained from the Russian thesaurus RuThes-lite as well as from Wikipedia, to calculate semantic relatedness measures to be used as features in the machine learning algorithm. We attempt to realize a mention-ranking algorithm using neural networks for predicting coreference, based on the one described in [6].

3 System Architecture

In this work we describe a mention-ranking model, derived from the algorithm introduced in [6]. It is based on a feedforward neural network, consisting of two main modules: the mention-pair encoder and the mention-ranking layer. The model was developed, relying on the existing open-source solutions with the use of Keras and Tensorflow libraries for the Python programming language.

3.1 Mention-Pair Encoder

The first module of the network was tasked with transforming the input to its distributed representation. The model receives as input the vector, consisting of embeddings of an antecedent and its potential anaphor and their features as well as several additional pair features (sets of features are described in detail in Sect. 4), and its output is then fed to the mention-ranking model.

Structurally the encoder presents a three-layer fully connected neural network with hidden layers of rectified linear units (ReLU):

$$h_i(a, m) = max(0, W_i h_{i-1}(a, m) + b_i) \tag{1}$$

Here $h_i(a, m)$ is the output of the i-th layer with the input of mention m and its potential antecedent a, W_i is a weight matrix, and b_i is the layer's bias.

3.2 Mention Ranking Model

The second module in the network, the mention-ranking model, estimates the coreference score of the pair of a mention m and its possible antecedent a. As the input it accepts the distributed representation of the pair, the output of the mention-pair encoder. It is represented by a single fully connected layer with the sigmoid activation function:

$$s_m(a, m) = W_m r_m(a, m) + b_m \tag{2}$$

Here s_m is the coreference score of the pair, and r_m is its distributed representation.

3.3 Training the Network

Pretraining the neural network has been determined by [6] among others as an important step in its development. For the pretraining of our network the following function was used:

$$-\sum_{i=1}^{N} \left[\sum_{t \in T(m_i)} \log p(t, m_i) + \sum_{f \in F(m_i)} \log(1 - p(f, m_i)) \right] \tag{3}$$

$T(m_i)$ is the set of all true antecedents of the i-th mention m_i, $F(m_i)$ is the set of all false antecedents of the same mention, and $p(t, m_i) = sigmoid(t, m_i)$.

As the training objective, the slack-rescaled max-margin was used. First, the highest-scoring antecedent of the mention m_i was found:

$$\hat{t}_i = \operatorname{argmax}_{t \in T(m_i)} s_m(t, m_i) \tag{4}$$

Then, the loss function was calculated:

$$\sum_{i=1}^{N} \max_{a \in A(m_i)} \Delta(a, m_i)\left(1 + s_m(a, m_i) - s_m(\hat{t}_i, m_i)\right) \tag{5}$$

$A(m_i)$ here is the set of all possible antecedents of the mention m_i, and $\Delta(a, m_i)$ is the mistake-specific cost function:

$$\Delta(a, m_i) = \begin{cases} \alpha_{FN} \ if \ a = NA \wedge T(m_i) \neq NA \\ \alpha_{FA} \ if \ a \neq NA \wedge T(m_i) = NA \\ \alpha_{WL} \ if \ a \neq NA \wedge \alpha \notin T(m_i) \\ 0 \ if \ a \in T(m_i) \end{cases} \tag{6}$$

Here α_{FN}, α_{FA} and α_{WL} denote costs for different error types: "false new", "false anaphoric" and "wrong link", correspondingly. The values used were {0.5, 0.5, 1}.

For training the model, the Adam optimizer was used. The dropout rate was set to 0.3 for all hidden layers.

4   Feature Sets

For the purposes of this research, the performance of different models with two different feature sets was compared. The default set consisted of string-based, morphological, lexical and distance features, generally used in coreference resolution algorithms. The second feature set also included as features measures of semantic relatedness, calculated from semantic information from two external sources: the Russian Wikipedia and a Russian thesaurus RuThes-Lite.

4.1   Default Model

The default feature set consisted of features, traditionally used for the task of coreference resolution ([2, 6, 7], among others). It combined separate morphologic and lexical features of the mention and the antecedent with features defined for the pair, such as distance between members and matches in strings or POS-tags. As the lexical features of the mentions, the word embeddings were used. The embeddings were obtained from Wikipedia corpus using FastText. If a member of the pair was a noun phrase, the representation of its head was used. An attempt to use the average word embedding of all words in a phrase as well was made, but it yielded worse results.

The complete list of features is given in Table 1 below:

Table 1. The default feature set

Feature class	Features
String-based	Full string match Head string match Partial string match
Distance	Number of NPs between members
Morphological	Number Gender Animacy Number match Gender match Animacy match Both members are proper One of members is a pronoun Both members are pronouns
Lexical	Word embeddings of the NP head

4.2   Semantic Information Extraction

The alternative feature set we used in our research was enriched with measures of semantic relatedness between members of mention-antecedent pairs. To generate these features, the semantic information from two publicly available sources was analyzed. One of them is RuThes-Lite: a thesaurus of Russian, including 55 000 entities that correspond to 158 000 lexical entries [18]. The structure of RuThes-Lite is similar to that of WordNet, with concepts in the thesaurus linked to each other by the set of labeled relations, including IS-A, PART-WHOLE and a number of associative relations. The other source was the Russian segment of Wikipedia. While being smaller than the English one (~1.5 mln articles, compared to ~5 mln articles), it is still one of its largest, making it an important knowledge source. The reason Wikipedia was chosen as a source is its category structure, which can be analyzed in similar terms to a thesaurus: each Wikipedia article is placed within one or several categories that, in their own turn, can be categorized further.

This allowed to analyze the category structure of Wikipedia as a graph, in the same way as Ruthes was analyzed. Categories were considered as graph nodes, and relations of inclusion between them – as edges. Articles belonging to a category were considered terminal nodes of the graph. This representation also made it possible to apply the same semantic relatedness metrics to both of our selected sources.

The following set of measures of semantic relatedness was used for analysis: the path-based measures, suggested by [19, 20] and [21], and information content-based measure, suggested by [22]. For each pair of mention and antecedent the values of the metrics between the head lemmas of both groups were calculated. If any of them was ambiguous, for the combinations of possible meanings the average and maximum values of the metric were calculated and used as features.

5 Experiment Setup and Evaluation

5.1 Corpus Data

The model was trained and tested on the data of RuCor, the Russian coreference corpus, used in the Ru-Eval-2014 competition of Russian coreference resolvers. The corpus consists of 180 texts, containing 3638 coreferential chains with the total of 16557 coreferential mentions. The texts of the corpus are of various lengths and genres: fiction, news texts, scientific articles, blog posts, etc. All texts are tokenized and morphologically and syntactically tagged, which allows to use the data without additional preprocessing.

For the evaluation procedure the texts of the corpus were split into training, validation and test datasets in the 60/20/20 proportion.

5.2 Evaluation Results

In our research we compared the performance of two models based on two feature sets, described above: the default one (model I) and the one enhanced with semantic relatedness measures (model II). Several versions of the second model were considered: supplemented with semantic features calculated on only one of the resources, and on both at once.

All models were at first pretrained to determine the proper feature weights, and then their performance on the test dataset was evaluated. For evaluation the MUC [23] and the B^3 [24] metrics were used. The comparison was conducted using the gold mentions from the RuCor corpus. The results of evaluation, as well as results of similar research described in [14] and [16] with the highest B^3 score, are presented in Table 2. The table also includes the absolute error counts for evaluated models.

Table 2. Evaluation results

	MUC			B^3			Error counts		
	P	R	F1	P	R	F1	FN	FA	WL
Model I	0.683	0.607	0.643	0.568	0.644	0.604	217	63	4.8 K
Model II, RuThes only	0.693	0.729	0.710	0.571	0.624	0.597	230	60	4.5 K
Model II, Wikipedia only	0.641	0.679	0.660	0.566	0.659	0.609	0	79	5 K
Model II, both sources	0.693	0.730	**0.711**	0.568	0.682	**0.620**	100	2	**4.3 K**
[14], random forest	0.740	0.652	0.693	0.739	0.552	*0.631*			
[16], NamedEntities	0.794	0.637	*0.707*	0.794	0.489	0.605			

As can be seen from the table, the general performance of both our models is comparable to that of state-of-the-art systems by [14] and [16]. The MUC score of the variant of Model II that uses the semantic data from both sources is higher than scores of comparison targets, and its B3 score, while 1% lower than that in [14], exceeds the result of [16] by 1.5%. This variant achieved the highest F-measure of all model II variants compared, showing the improvement that can be gained by using semantic features in the analysis. The improvement is also demonstrated by the decrease in error counts of all error types.

Seeing that the difference of our results from the comparison targets is mostly in the lower precision score, increasing it should become the focus of future work.

5.3   Discussion of Results

The results presented above demonstrate that neural networks are viable as a method of coreference resolution in Russian. Trained upon a similar set of features, they perform on par with state-of-the art systems, and only slightly worse than the system using mention clustering. Apart from that, our results show that features derived from semantic information can be successfully used to boost the quality of system's analysis.

Using the features derived from Wikipedia data improves the recall of the system, which can be attributed to large size of the encyclopedia and its coverage of various phenomena. The features derived from thesaurus data, on the other hand, serve to improve the precision, thus the largest increase in quality being gained by combining the features from both information sources. Still, lack of substantial increase in precision after adding semantic features can be observed, which calls for improvements in the feature generation process.

While features based on thesaurus and encyclopedic data help improve coreference resolution for ontologically related mentions, such as hypernyms or synonyms, other complicated cases of coreference still persist. Among them are:

- Direct speech pronouns. First and second person pronouns can be difficult to resolve for a neural network due to their morphological differences from 3rd person ones.
- Split antecedents. Pairs such as *"Иван Тихонович и Татьяна Финогеновна"* ('Ivan Tikhonovich and Tatiana Finogenovna') and *"они"* ('they') will also have differing morphological features, because both heads of the first mention will be analysed separately.
- Relations, depending on context. Cases such as *"Выходец$_i$ из Нигерии$_i$ решил остаться на ПМЖ в Израиле, поскольку на родине$_i$ его якобы преследует опасный призрак"* ('A native of Nigeria$_i$ decided to remain in Israel, because he was haunted by a ghost in his homeland$_i$') are difficult to resolve, because the understanding of the whole sentence is required to link *"homeland"* to the correct country.

To target such cases, additional improvements need to be made in both the structure of the model and the features used. For example, the context-dependent relations can be possibly resolved by implementing similarity measures between word embeddings.

6 Conclusion

In this paper we presented a neural network, designed for the purpose of coreference resolution in the Russian language, and tested two different feature sets to estimate the importance of semantic features for its performance. The results of evaluation using MUC and B^3 metrics demonstrated that its baseline performance is comparable to that achieved in recent researches on the same topic, and that integration of semantic features helps to increase the quality of analysis to a certain degree.

To target the shortcomings of the system, such as low improvements in precision score, as well as complicated coreference cases, future improvements both in the network architecture and semantic feature extraction process are needed. They include: (i) testing alternative network architectures, including recurrent neural networks; (ii) tuning of hyperparameters; (iii) use of alternative word embeddings and features based on them (the BERT language model is of particular interest); (iv) improving the extraction process of semantic features; (v) testing of other relatedness measures, including between word embeddings. Another important development is integration of clustering and cluster-ranking modules to account for entity-level information.

References

1. Hobbs, J.: Resolving pronoun references. Lingua **44**, 311–338 (1978)
2. Soon, W.M., Lim, D.C.Y., Ng, H.T.: A machine learning approach to coreference resolution of noun phrases. Comput. Linguist. **27**, 521–544 (2001). https://doi.org/10.1162/089120101753342653
3. Chen, C., Ng, V.: Combining the best of two worlds: a hybrid approach to multilingual coreference resolution. In: Joint Conference on EMNLP and CoNLL-Shared Task, pp. 56–63 (2012)
4. Raghunathan, K., et al.: A multi-pass sieve for coreference resolution. In: Proceedings of the 2010 Conference on Empirical Methods in Natural Language Processing, pp. 492–501. Association for Computational Linguistics (2010)
5. Fernandes, E.R., dos Santos, C.N., Milidiú, R.L.: Latent structure perceptron with feature induction for unrestricted coreference resolution. In: Proceedings of the Joint Conference on EMNLP and CoNLL: Shared Task, pp. 41–48 (2012)
6. Clark, K., Manning, C.D.: Improving coreference resolution by learning entity-level distributed representations. In: Proceedings of the 54th Annual Meeting of the Association for Computational Linguistics, Volume 1: Long Papers, pp. 643–653 (2016). https://doi.org/10.18653/v1/P16-1061
7. Bengtson, E., Roth, D.: Understanding the value of features for coreference resolution. In: Proceedings of the Conference on Empirical Methods in Natural Language Processing, pp. 294–303. Association for Computational Linguistics, Honolulu, Hawaii (2008)
8. Ponzetto, S.P., Strube, M.: Exploiting semantic role labeling, WordNet and Wikipedia for coreference resolution. In: Proceedings of the main conference on Human Language Technology Conference of the North American Chapter of the Association of Computational Linguistics, pp. 192–199. Association for Computational Linguistics (2006). https://doi.org/10.3115/1220835.1220860
9. Toldova, S., et al.: RU-EVAL-2014: evaluating anaphora and coreference resolution for Russian. In: Computational Linguistics and Intellectual Technologies: Papers from the Annual International Conference "Dialogue", pp. 681–694 (2014)

10. Luo, X., Ittycheriah, A., Jing, H., Kambhatla, N., Roukos, S.: A mention-synchronous coreference resolution algorithm based on the bell tree. In: Proceedings of the 42nd Annual Meeting of the Association for Computational Linguistics (ACL 2004) (2004)

11. Yang, X., Zhou, G., Su, J., Tan, C.L.: Coreference resolution using competition learning approach. In: Proceedings of the 41st Annual Meeting on Association for Computational Linguistics, ACL 2003, pp. 176–183 (2003). https://doi.org/10.3115/1075096.1075119

12. Rahman, A., Ng, V.: Supervised models for coreference resolution. In: Proceedings of the 2009 Conference on Empirical Methods in Natural Language Processing, EMNLP 2009, vol. 2, pp. 968–977 (2009). https://doi.org/10.3115/1699571.1699639

13. Martschat, S., Strube, M.: Latent Structures for coreference resolution. Trans. Assoc. Comput. Linguist. **3**, 405–418 (2015)

14. Sysoev, A.A., Andrianov, I.A., Khadzhiiskaia, A.Y.: Coreference resolution in Russian: state-of-the-art approaches application and evolvement. In: Computational Linguistics and Intellectual Technologies: Papers from the Annual International Conference "Dialogue", pp. 341–352 (2017)

15. Versley, Y.: Antecedent selection techniques for high-recall coreference resolution. In: Proceedings of the 2007 Joint Conference on Empirical Methods in Natural Language Processing and Computational Natural Language Learning (2007)

16. Toldova, S., Ionov, M.: Coreference Resolution for Russian: Computational Linguistics and Intellectual Technologies: Papers from the Annual International Conference "Dialogue", pp. 339–348 (2017)

17. Azerkovich, I.: Employing Wikipedia data for coreference resolution in Russian. In: Filchenkov, A., Pivovarova, L., Žižka, J. (eds.) AINL 2017. CCIS, vol. 789, pp. 107–112. Springer, Cham (2018). https://doi.org/10.1007/978-3-319-71746-3_9

18. Loukachevitch, N.V., Dobrov, B., Chetviorkin, I.: RuThes-Lite, a publicly available version of thesaurus of Russian language RuThes. In: Computational Linguistics and Intellectual Technologies: Papers from the Annual International Conference "Dialogue" (2014)

19. Rada, R., Mili, H., Bicknell, E., Blettner, M.: Development and application of a metric on semantic nets. IEEE Trans. Syst. Man Cybern. **19**, 17–30 (1989). https://doi.org/10.1109/21.24528

20. Wu, Z., Palmer, M.: Verb semantics and lexical selection. In: ACL, pp. 133–138 (1994). https://doi.org/10.3115/981732.981751

21. Leacock, C., Chodorow, M.: Combining local context with wordnet similarity for word sense identification. In: WordNet: An Electronic Lexical Database, pp. 265–283. MIT Press (1998)

22. Resnik, P.: Using Information Content to Evaluate Semantic Similarity in a Taxonomy. IJCAI 448–453 (1995)

23. Vilain, M., Burger, J.D., Aberdeen, J., Connolly, D., Hirschman, L.: A model-theoretic coreference scoring scheme. In: Proceedings of the 6th Message Understanding Conference (MUC-6), pp. 45–52 (1995). https://doi.org/10.3115/1072399.1072405

24. Baldwin, B., Bagga, A.: Algorithms for scoring coreference chains. In: The First International Conference on Language Resources and Evaluation Workshop on Linguistics Coreference, pp. 563–566 (1998)

A Method of Semantic Change Detection Using Diachronic Corpora Data

Vladimir Bochkarev$^{(\boxtimes)}$, Anna Shevlyakova, and Valery Solovyev

Kazan Federal University, Kazan, Russia
vbochkarev@mail.ru, anna_ling@mail.ru, maki.solovyev@mail.ru

Abstract. The article proposes a method for detecting semantic change using diachronic corpora data. The method is based on the distributional hypothesis. The analysis is performed using frequencies of syntactic bigrams from the English and Russian sub-corpora of Google Books Ngram. To obtain the word co-occurrence profile in its new meaning, syntactic bigrams that contributed most to the word distribution change are selected and their time series are clustered. The method is tested on a group of English and Russian words which gained new meanings in the 20th century. The obtained results show that the proposed method allows one to detect semantics changes, as well as to determine the time of these changes.

Keywords: Semantic changes · Diachronic corpora · Google Books Ngram · Cluster analysis · Bigram frequencies

1 Introduction

It has long been recognized that words may change their meaning. Semantic change has traditionally been looked at from a variety of angles. Thus, there are various classifications of meaning change, for example, nine types of semantic changes are determined in [1].

Semantic changes can be driven by language-internal and language-external factors. It is a very complicated task to predict language change, either internally motivated or caused by changes in history. However, one can trace meaning change performing diachronic linguistic studies.

Traditionally, analysis of semantic changes has been carried out using etymology and explanatory dictionaries and thorough examination of relevant published texts. However, creation of large diachronic text corpora provided a valuable complement to more traditional methods of investigation. Large text corpora are an invaluable tool for lexicographers allowing for automatic text processing and making the studies of semantic changes less time-consuming.

The largest text corpus used in semantic studies is Google Books Ngram. It contains scanned books in eight languages. The texts are written in different genres and tagged. Another widely used corpus is COHA, the Corpus of Historical American English, which is also a large collection of texts written in English. Besides text corpora, other language data arrays are used for semantic studies, such as Twitter data [2].

© Springer Nature Switzerland AG 2020
W. M. P. van der Aalst et al. (Eds.): AIST 2019, CCIS 1086, pp. 94–106, 2020.
https://doi.org/10.1007/978-3-030-39575-9_10

Studies of semantic change is a productive subfield of research in the corpus and computer linguistics. However, according to [3, 4], there are no commonly used methods and terminology in this area, which results in lack of consistency. Besides, most semantic change studies are performed for the English language.

This article proposes a new complex method for identifying semantic changes. The Russian and English sub-corpora of Google Books Ngram are used to demonstrate how the method works. The investigation included selection of the groups of syntactic bigrams, calculation of the distribution distance, comparison of the time series and clustering of the revealed syntactic bigrams.

It is shown that the method allows one not only to detect words that change their semantics, but also to determine the time when this change in meaning occurs, as well as to describe new meanings through a set of the most significant uses of the word.

To test the method, two groups of Russian and English words were selected. The first group included the words 'sputnik' and 'sputnitsa'. These words were selected for the analysis because changes in their semantics are obvious. It is reasonable to test the method for the first time using the word examples which interpretation is simple. After that, the method is tested on more complicated cases. To do this, the second group of words was selected which included such words as 'click' and 'apple'. Changes in their semantics can be understood intuitively. However, the methods proposed earlier failed to detect the meaning change of these words and they are a complicated case for the analysis [2]. The task of this work was to test the method using the easiest and the most difficult words for the analysis.

2 Related Works

Changes in word meaning take place continually in all languages. Corpus-based studies provide evidence that these changes can influence frequency of use of words [5, 6]. One of the methods used for tracing semantic change in text corpora is a method based on distribution hypothesis [7]. According to it, a word semantics is determined by its co-occurrence and context and can be approximated by a vector of frequencies of word combinations that contain this word. This method has been improved in [8], which proposes a word embedding method that compresses vectors of word combination frequencies. Currently, it is the most popular method used in various studies in the field of computer linguistics. However, being used in diachronic research, it encounters certain difficulties due to the element of randomness that it contains. To improve this method, Kulkarni [2] introduced a special alignment procedure which is technically very complicated. As a result, semantics of a word is presented as a vector-valued time series. He also compared several methods for detecting changes in semantics and concluded that the word embedding method with alignment is the most effective one. However, in some cases, a simple frequency approach or a method based on part-of-speech recognition still leads to better results.

Various modifications of the frequency method and the word embedding method have been proposed in several papers [9–12]. However, there is no "gold standard" of testing these methods [3], which makes their comparison problematic. Lack of a commonly recognized method explains the relevance of searching for new approaches and methods of detecting semantic changes.

Usually, arbitrary selected words, which change their meaning, are used to demonstrate efficiency of newly proposed methods. In [10], the authors propose a complex methodology of analysis of syntactic dependencies in the Google Books Ngram corpus detecting appearance of new meanings, their extension and narrowing. The algorithm revealed 21 cases of meaning extension in the intervals 1909–1953 and 2002–2005. However, the experts did not confirm the results in 8 cases. Semantics of three words – 'continuum', 'diagonal' and 'intonation' – was studied in different fields of science. For example, the word 'continuum' has its own meaning in physics and mathematics. The authors believe that the proposed algorithm revealed a new meaning. However, this result may be due to a sharp increase in the corpus size in the second interval (approximately 15 times per year), as well as increase in the scientific literature, a parameter which is not controlled by the authors of the article. The word 'continuum' is detected in the corpus approximately 1000 times per year during the 1st interval and 100 thousand times per year during the second interval. One can offer the following explanation of the result presented in this article. There were different meanings of the word in the earlier period. However, the algorithm failed to distinguish them. Probably, the threshold of distinguishing different meanings lies somewhere between 1 thousand and 100 thousand of the word occurrence frequency. Thus, this article does not provide convincing evidence that the proposed method reveals new meanings of words effectively.

The distributional method for detecting semantics changes is proposed in [2]. The method reveals semantic shifts of such words as 'gay', 'tape', 'recording' very well. However, the method does not work well with the words 'click', 'apple' and some others. In general, this method is designed only to detect changes in semantics, but does not allow one to determine the type of these changes - extension or narrowing of the meaning etc.

The proposed method is efficient when new meanings of words appear and cause growth of the word frequency. If the word frequency decreases, the word co-occurrence vectors become "poorer" and do not detect narrowing of the meaning well. The article [13] is one of the few works in which narrowing of the word meaning was detected at a sharp decrease of its frequency.

3   Method

The proposed algorithm includes the following steps:

1. Some vector representation of the word semantics is chosen. Vectors characterising the studied word in different time intervals are constructed. Two intervals, between which the change in semantics is significant, are selected;
2. The distance between the vectors characterising the given word in the target time intervals are calculated using this or that measure.
3. All syntactic bigrams that contain the given word are extracted from the corpus. After that, individual contribution of the extracted syntactic bigrams to the distance value obtained at the previous step is determined.
4. A small number of bigrams making the greatest contribution to the distance value is selected according to the threshold.

5. Groups of bigrams that simultaneously appear or fall out of use and whose frequencies change synchronously are selected using the methods of the cluster analysis. The clusters whose frequency increases indicate appearance of new word meanings. The clusters which decrease in frequency indicate word meanings that fall out of use.

Let us consider each of these steps in more detail.

It is proposed in several works to represent a word semantics quantitatively by vectors of frequencies of syntactic bigrams that include the given word [10]. Syntactic bigrams are primary units of a syntactic structure, including two words [14]. One of them is the head, another one is its dependent. These words are not usually immediate neighbours. By constructing frequency vectors for the target time intervals and using some measure of distributional differences, it is possible to estimate how significant the changes in the word distribution are between these intervals, as well as to estimate its semantic changes.

In this paper, the Jensen-Shannon divergence is used. It is one of the most well-known measures of difference of frequency distributions [15], which will be written as follows:

$$JSD(p\|q) = \frac{1}{2}\sum_i p_i \log_2 \frac{p_i}{(p_i + q_i)/2} + \frac{1}{2}\sum_i q_i \log_2 \frac{q_i}{(p_i + q_i)/2} \qquad (1)$$

Here, p_i and q_i are two frequency sets. This difference measure was used in several works on word semantics and language evolution [2, 16]. However, by calculating divergence of the vectors of the bigram frequencies in successive time intervals, one can detect only the fact of some changes, but cannot directly assess the direction of the changes that occur. The latter requires additional rather time-consuming procedures for comparing the semantic distances between the given word and other words in different time intervals, as was performed, for example, in [3, 10]. Besides, to reliably estimate the divergence of the frequency vectors, a sufficiently large statistics is needed. This requires averaging of the frequencies over rather large time intervals and complicates the dating of the changes.

When a new meaning of a word appears, the word distribution changes and a new group of syntactic bigrams emerge. Thus, the task is to find groups of syntactic bigrams that contain the given word, began to be widely used synchronously and whose frequencies subsequently changed in a similar way.

The classification method can be applied to the series of frequencies of all syntactic bigrams that contain the given word. However, there may be lots of such bigrams. This increases the probability of random similarity in some pairs of the series. Therefore, it is proposed to use a double criterion, preliminary selecting only a small part of syntactic bigrams that contribute most to the increase of divergence between the vectors of frequencies in the successive time intervals.

The Jensen-Shannon divergence provides only an overall estimation of the difference in the distribution of the word in the target time intervals. However, one can also estimate the contribution that each of the syntactic bigrams, which contains a given word, makes to the resulting divergence value. To do this, the terms corresponding to the individual syntactic bigrams are selected in expression (1):

$$\frac{1}{2} p_i \log_2 \frac{p_i}{(p_i + q_i)/2} + \frac{1}{2} q_i \log_2 \frac{q_i}{(p_i + q_i)/2} \tag{2}$$

Having determined the contribution of each bigram according to (2), the bigrams are ranked by decrease of this contribution. Then, after setting a threshold, the bigrams which form the observed divergence value are selected for further analysis.

Thus, having chosen a relatively small number of syntactic bigrams, we consider the time series of their relative frequencies. The task is to distinguish (among them) groups of bigrams whose frequencies change in a similar way over time. In [17, 18], it was shown that the time series of frequencies of semantically related words change in a similar way. Therefore, it is expected that the series of the bigram frequencies related to one and the same meaning of the considered word will also behave in a similar way. Hierarchical clustering methods are often used for such purposes. The used measure of similarity/difference of the time series of the bigram frequencies can be different. However, since the probability distribution of frequency fluctuations of the words and bigrams is not known precisely, it is preferable to use non-parametric criteria. In this paper, the Spearman correlation distance is used.

It is natural to choose the number of bigrams selected for analysis and the threshold value for clustering based on the statistical significance of the detected clusters (that is, minimizing the p-value). In this paper, these parameters were selected manually in accordance with this criterion.

The question arises how to check statistical significance of the detected effects. It should be noted that the number of the detected clusters can be a good statistical data used to test the significance of the changes. Under most reasonable assumptions about the probabilistic properties of the frequency series, the formation of a cluster including a significant proportion of objects is very unlikely. P-value for this or that amount of statistical data can be calculated using statistical modelling. To do this, we generate sets from M of the time series containing N samples (in our case, M is the number of selected bigrams, and N is the length of the time series of their frequencies). Probabilistic properties of these series are set in a model based on the assumption about the features of the studied system. We apply hierarchical clustering with the same parameters (distance measure, algorithm for computing distance between clusters, threshold distance) to these sets and determine the size of the largest cluster. Performing calculations for many times, we find the p-value as proportion of cases, in which the size of the largest cluster was less than or equal to the empirical value. In this paper, the p-value calculations were performed under the assumption that the time series are a normal Wiener process [19]. In [20], there are arguments that prove the fact that the series of logarithms of frequencies often have properties of a normal Wiener process (since we use the rank criterion for the similarity of the series, logarithmization is not essential).

Thus, if significant changes are detected, the algorithm allows one to distinguish a group of syntactic bigrams whose frequencies change in a similar way and which highly likely form a co-occurrence profile of a word used in its new meaning. In cognitive linguistics, the same idea was developed in [21] based on [22, 23]. The analysis of the time series of this group of bigrams allows one to date the emergence of new meanings of words.

To test the suggested method, data on frequencies of syntactic bigrams taken from the Google Books Ngram [24, 25] corpus are used in this article. Frequencies of only those bigrams that consist of vocabulary 1-grams are analysed. (Vocabulary 1-grams consist only of the letters of the corresponding alphabet and, possibly, of one apostrophe). Bigrams which differ only by the keyboard case are regarded to be identical. The conducted experiments showed that lemmatization leads to deterioration of the results. Therefore, it is not used in this work.

4   Case Study

Let's analyse the emergence of new meanings of the Russian word 'sputnik'. Selection of the target time intervals can be performed, for example, using the approaches discussed in [3, 10]. However, in this paper, these intervals are selected manually. In order to avoid the impact of the spelling reform of the Russian language carried out in 1918 (words occurring before 1918 are not always recognized correctly in Google Books Ngram), the data appeared after 1920 are used for the analysis.

Two intervals were chosen – 1920–1953 and 1987–2009. Overall frequencies of the syntactic bigrams that contain the test word ('sputnik') were calculated in these intervals.

In Russian, the original meaning of the word 'sputnik' is a travel companion. The word gained a new meaning after the creation of the first artificial satellite. The Jensen-Shannon divergence between the obtained frequency distributions is 0.2153, which indicates significant changes in the distribution of the tested word from one time interval to another. We rank the syntactic bigrams according to their contribution to the divergence value, select the bigrams which contribution is the greatest and apply hierarchical clustering to the time series of relative frequencies of the selected bigrams (in the interval 1920–2009). The Spearman correlation distance is used as a measure of the time series difference. The unweighted average distance (UPGMA) is used as an algorithm for computing distance between clusters. To check and compare the obtained results, we perform similar calculations for the word 'sputnitsa'. The meaning of the word 'sputnitsa' is very close to the original meaning of the word 'sputnik', it is the feminine form of the noun 'sputnik' and means a woman who accompanies you during your travel. However, a new meaning of the word 'sputnitsa' didn't appear. The used threshold difference value is 0.25. The clustering results are shown in Fig. 1.

It is seen that the large cluster includes 9 bigrams out of 15. All of them are associated with the use of the word 'sputnik' in the 20th century in the sense of an artificial satellite of the Earth. Besides, the word 'gorod-sputnik' (a satellite town) also appears due to metaphorical shift. It is highly unlikely that such a large cluster appeared randomly when the series of the bigram frequencies change independently. The statistical modelling shows that the p-value is $6.79 \cdot 10^{-5}$ for 15 independent series with a length of 90 (if each series has properties of a normal Wiener process).

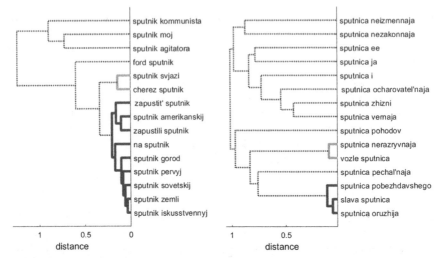

Fig. 1. Clusters of syntactic bigrams that contain the words 'sputnik' and 'sputnitsa'. The dotted line shows the dependencies above the threshold distance, the solid lines show dependencies within the clusters

The second cluster, which includes 2 bigrams out of 15 (they are 'sputnik svyazi' and 'cherez sputnik') is associated with the creation of communication satellites. Four bigrams which were not united in clusters at the given threshold difference value of 3 are the names 'Sputnik Agitatora' and 'Sputnik Komminista' (magazine titles). The third bigram 'ford sputnik' is a fragment of the title of the book [26] published in 1979. Thus, the method revealed three meanings associated with named entities. It should be noted that bigrams associated with named entities can be revealed in advance using other methods (for example, the overview of the modern methods of identifying named entities can be found in [27]) and excluded from the analysis. And only one bigram 'moi sputnik' reflects the main semantics of the test word. The preliminary selection of bigrams according to the values of the contribution to the Jensen-Shannon divergence value (see formula (2)) was intended to exclude bigrams connected with traditional, slowly changing word meanings. Therefore, if there is only one bigram (connected with the main meaning of the word) among the selected bigrams, it means that this technique achieved the goal, which created favorable conditions for finding new meanings of the word 'sputnik'.

Figure 2 shows how the overall frequency of the obtained clusters varies with time. The clusters show a sharp frequency jump with subsequent stabilization or a slight smooth decrease. The most significant frequency peak of the first cluster occurs in 1957–1958 (when the first Earth satellites were launched). The frequency increase of this cluster had begun earlier, in 1954. From the beginning of the 50 s, the term 'sputnik' and the possibilities of space exploration were widely discussed in the USSR scientific popular literature and science fiction. The frequency jump of the second cluster occurs in 1965–1967 (when the first communication satellites were launched, and the first lines of satellite communication were introduced). Thus, for this case, it is possible to accurately date the emergence of the new meanings of the studied words.

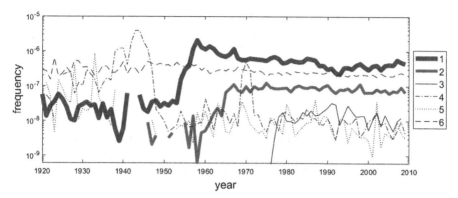

Fig. 2. Change of the overall frequencies of clusters of syntactic bigrams that contain the word 'sputnik'. The thickest line reflects the number of bigrams in the cluster. Line 1 corresponds to the cluster including the syntactic bigrams 'sputnik iskusstvennyj', 'sputnik pervyj', 'sputnik zemli', 'zapustili sputnik', 'na sputnik', 'sputnik sovetskij', 'sputnik gorod', 'zapustit' sputnik', 'sputnik amerikanskij'. Line 2 corresponds to the cluster including 'cherez sputnik' and 'sputnik svjazi'. Lines 3, 4, 5, 5 correspond to the syntactic bigrams 'ford sputnik', 'sputnik agitatora', 'sputnik kommunista' and 'sputnik moj', respectively

The Jensen-Shannon divergence has a moderate value of 0.0877 for the tested word 'sputnitsa' between the frequency distributions of the syntactic bigrams in 1920–1953 and 1987–2009. Analysis of the time series of the bigram frequencies (analogous to the above-described analysis) indicates two small clusters. The first cluster includes the bigrams 'sputnitsa oruzhiya', 'slava sputnitsa' and 'sputnitsa popezhdavshego'. The second one includes the bigrams 'vozle sputnitsa' and 'sputnitsa nerazrivnaya'. Statistical modeling gives a p-value for appearance of the maximum cluster of 3 elements 0.5887 under the assumption that the frequency time series have properties of a normal wiener process. The frequencies of the obtained two clusters significantly decrease from the interval 1920–1953 to the interval 1987–2009.

Figure 3 shows total frequencies of two clusters detected in different years. As can be seen, the frequencies of these clusters decrease in the target interval. The sharpest decrease of the 1st cluster is observed after 1956. To test the hypothesis about the homogeneity of the frequency samples of the first cluster before and after 1956, the Wilcoxon rank-sum test is used. The main hypothesis is rejected at any reasonable level of significance (the p-value is $3.51 \cdot 10^{-6}$). The median frequencies of the first cluster before and after 1956 are $1.91 \cdot 10^{-8}$ and $1.53 \cdot 10^{-9}$, respectively. It means that the frequency has decreased by a factor of 12.5. However, the frequency of the word 'sputnitsa' did not fall after 1956 and even increased by the end of the 20th century.

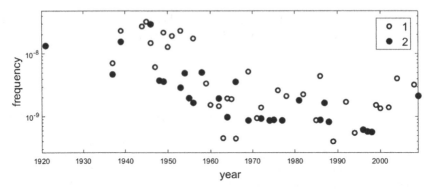

Fig. 3. Change of the overall frequencies of the clusters of syntactic bigrams that contain the word 'sputnitsa'. 1 – the cluster that contains the bigrams 'slava sputnitsa', 'sputnitsa oruzhiia', 'sputnitsa pobezhdavshego'; 2 – the cluster that contains the bigrams 'vozle sputnitsa' and 'sputnitsa nerazrivnaya'

The meaning of the word "sputnitsa" in the bigrams 'slava sputnitsa', 'sputnitsa oruzhiia', 'sputnitsa pobezhdavshego' is metaphorical. The frequency use of this word in these contexts significantly decreased. Thus, the meaning of the word 'sputnitsa' is not extended as the meaning of the word 'sputnik'.

It is noted in [2] that it was a complicated task to trace new meanings of the words 'click' and 'apple' in using the word embedding method. Let us apply the proposed method to analyse semantic changes of these words. To do this, we use the data on frequencies of syntactic bigrams from the English (common) subcorpus of Google Books Ngram and calculate the value of the Jensen-Shannon divergence for the intervals 1920–1970 and 1990–2008. The Spearman correlation distances between the time series of bigram frequencies are be calculated for the interval 1920–2008.

As for the word 'click', the Jensen-Shannon divergence between the vectors in the given time intervals is 0.559, which indicates significant changes in the word distribution. The parameters used for selecting syntactic bigrams and their clustering are the same as in the above-described examples. The selected 15 syntactic bigrams are united in 3 clusters (see Fig. 4 (left)). The first cluster includes 8 bigrams which are associated with the computer field: 'click button', 'click on', 'click ok', 'click then', 'click right', 'click you', 'click tab' and 'select click'. The p-value that corresponds to the appearance of such cluster is $5.34 \cdot 10^{-4}$. The second cluster includes 4 bigrams ('click the', 'click of', 'heard click' and 'with click'). The third cluster consists of 3 bigrams ('click a', 'click double' and 'was click') which are also associated with the computer field. The new meanings are to select something especially in a computer interface by pressing a button on a control device (such as a mouse). The overall frequency of the first cluster bigrams shows rapid growth in the interval 1983–1997 (see Fig. 5A), increasing from $5.6 \cdot 10^{-8}$ to $3.2 \cdot 10^{-5}$ (i.e. approximately 600 times). The overall frequency of the third cluster also increases in the same years. However, its growth is not so significant. Unlike the frequency of the first and third clusters, the overall frequency of the second cluster gradually varies without rapid jumps. The period 1983–1997 is the time of widespread distribution of personal computers and extensive use of the related terminology.

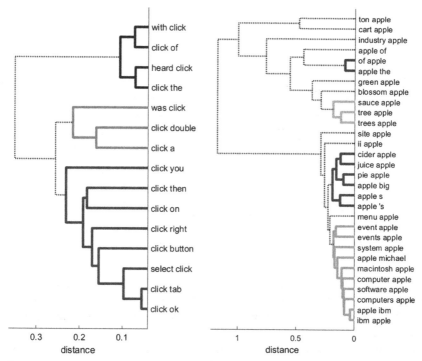

Fig. 4. Clusters of syntactic bigrams that contain the words 'click' and 'apple'. The dotted line shows the dependencies above the threshold distance, the solid lines show dependencies within the clusters

Let us analyse semantics of the word 'apple'. The corpus contains 9173 syntactic bigrams that contain the word 'apple'. This value is the biggest among the considered examples. To reliably distinguish new meanings of this word, we have to increase the number of the clusterized syntactic bigrams from 15 to 20–30 and reduce the threshold value of the correlation distance to 0.2. The Jensen-Shannon divergence between the vectors in the given time intervals is 0.1236. It means that changes in the word distribution is relatively moderate. Ten bigrams (out of 30) unite into a cluster, which corresponds to the p-value $8.6 \cdot 10^{-4}$. This cluster includes the following bigrams: 'computer apple', 'computers apple', 'macintosh apple', 'apple michael', 'ibm apple', 'events apple', 'apple ibm', 'software apple', 'system apple', 'event apple' (see Fig. 4 (right)). All these bigrams, except 'Apple Michael', are associated with the company Apple Inc and its products. Michael Apple is an educational theorist whose first works appeared in the second half of the1980 s of the 20th century. The overall frequency of these bigrams increases rapidly in 1984–1997 (approximately 80 times during these years), the period of the Apple Inc company establishment, and gradually changes later (see Fig. 5B).

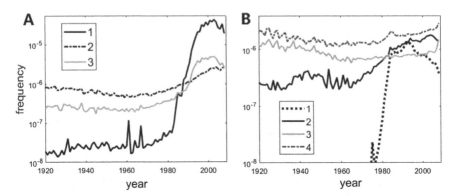

Fig. 5. Change of the overall frequencies of the clusters of syntactic bigrams that contain the words 'click' (A) and 'apple' (B). (A) Line 1 – bigrams 'click you' etc., 2 – bigrams 'was click' etc., 3 – bigrams 'with click' etc.; (B) Line 1 – bigrams 'computer apple' etc., 2 – bigrams 'juice apple' etc., 3 – bigrams 'tree apple' etc., bigrams 'of apple' etc.

There are also three smaller clusters. The first one includes 5 bigrams ('apple's', 'apple big', 'juice apple', 'cider apple', 'pie apple'), the second one includes 3 bi-grams ('trees apple', 'tree apple', 'sauce apple') and the third one includes 2 bigrams ('apple the', 'of apple').

Thus, the proposed method allows one to distinguish emergence of new word meanings and date the time of their widespread use.

5 Conclusion

The method of detecting new word meanings using large diachronic corpora is proposed in this article. The method was tested on several Russian and English words. The data on frequency of the bigrams that contain the tested words were taken from the Google Books Ngram corpus. It is shown that the method allows one to detect emergence of new meanings of words, to identify collocations in which the word is typically used in its new meaning, as well as to date the moment when the words gain new meanings.

To analyze the emergence of new meanings of words fully automatically, it is required to find the way to choose the time interval in which the change in semantics of a word is estimated, as well as to choose the number of analysed syntactic bigrams and the threshold distance value in the process of clusterization explained from the point of view of statistics. It is also needed to analyse the influence of the distance measure used for clusterization of the frequency series on the obtained results. The example of the word 'apple' shows that the series of the bigram frequencies were not always successfully clustered (see Fig. 4, on the right). To detect new values, the distance measure is required that can distinguish the moment of the first appearance of bigrams, even if more significant differences are observed in their frequency changes in the time following. The method has certain limitations on sensitivity since frequency of the word in its new meaning must exceed a certain threshold share among all uses of the word. This question requires further investigation.

Nevertheless, the method application seems to be promising. The examples above show that the method allowed one to reveal new meanings of the English words 'apple' and 'click' though the methods proposed earlier failed to do it (for more details see [2]). Also, unlike other methods, the proposed method allows detecting narrowing of word meaning, which is confirmed by the analysis of the Russian word 'sputnitsa'.

Acknowledgements. This research was financially supported by the Russian Government Program of Competitive Growth of Kazan Federal University, state assignment of Ministry of Education and Science, grant agreement № 2.8303.2017/8.9 and by RFBR, grant № 17-29-09163.

References

1. Bloomfield, L.: Language. Allen & Unwin (1933)
2. Kulkarni, V., Al-Rfou, R., Perozzi, B., Skiena, S.: Statistically significant detection of linguistic change. In: Proceedings of the 24th International Conference on World Wide Web, Florence, Italy, pp. 625–635 (2015)
3. Kutuzov, A., Øvrelid, L., Szymanski, T., Velldal, E.: Diachronic word embeddings and semantic shifts: a survey. In: Proceedings of the 27th International Conference on Computational Linguistics, Santa Fe, New Mexico, USA, pp. 1384–1397 (2018)
4. Tahmasebi, N., Borin, L., Jatowt, A.: Survey of computational approaches to diachronic conceptual change detection. arXiv preprint: arXiv:1811.06278v1 (2018)
5. Juola, P.: The time course of language change. Comput. Humanit. **37**(1), 77–96 (2003)
6. Hilpert, M., Gries, S.: Assessing frequency changes in multistage diachronic corpora: applications for historical corpus linguistics and the study of language acquisition. Lit. Linguist. Comput. **24**(4), 385–401 (2009)
7. Harris, Z.: Distributional structure. Word **10**(23), 146–162 (1954)
8. Mikolov, T., Sutskever, I., Chen, K., Corrado, G., Dean, J.: Distributed representations of words and phrases and their compositionality. In: Advances in Neural Information Processing Systems 26 (NIPS 2013), pp. 3111–3119 (2013)
9. Basile, B., Caputo, A., Semeraro, G.: Analysing word meaning over time by exploiting temporal random indexing. In: Proceedings of the First Italian Conference on Computational Linguistics, Turin, Italy, pp. 38–42 (2014)
10. Mitra, S., Mitra, R., Riedl, R., Biemann, C., Mukherjee, A., Goyal, P.: That's sick dude!: Automatic identification of word sense change across different timescales. In: Proceedings of the 52nd Annual Meeting of the Association for Computational Linguistics, Baltimore, Maryland, pp. 1020–1029 (2014)
11. Kim, Y., Chiu, Yi.-I., Hanaki, K., Hegde, D., Petrov, S.: Temporal analysis of language through neural language models. In: Proceedings of the 52nd Annual Meeting of the Association for Computational Linguistics, Baltimore, USA, pp. 61–65 (2014)
12. Yao, Z., Sun, Y., Ding, W., Rao, H., Xiong, H.: Dynamic word embeddings for evolving semantic discovery. In: Proceedings of the Eleventh ACM International Conference on Web Search and Data Mining, Marina Del Rey, CA, USA, pp. 673–681 (2018)
13. Solovyev, V.: Vozmozhnye mehanizmy izmenenija kognitivnoj struktury sinonimi-cheskih rjadov. V sb. "Jazyk i mysl': Sovremennaja kognitivnaja lingvistika", pp. 478–487. Jazyki slavjanskoj kul'tury, Moskva (2015)
14. Sidorov, G., Velasquez, F., Stamatatos, E., Gelbukh, A., Chanona-Hernández, L.: Syntactic dependency-based N-grams as classification features. In: Batyrshin, I., Mendoza, M.G. (eds.) MICAI 2012. LNCS (LNAI), vol. 7630, pp. 1–11. Springer, Heidelberg (2013). https://doi.org/10.1007/978-3-642-37798-3_1

15. Schütze, H., Manning, C.: Foundations of Statistical Natural Language Processing. MIT Press, Cambridge (1999)
16. Bochkarev, V.V., Solovyev, V.D., Wichmann, S.: Universals versus historical contingencies in lexical evolution. J. R. Soc. Interface **11**, 20140841 (2014)
17. Montemurro, M., Zanette, D.: Coherent oscillations in word-use data from 1700 to 2008. Palgrave Commun. **2**, 16084 (2016)
18. Bochkarev, V., Maslennikova, Yu., Svetovidov, A.: Semantic similarity and analysis of the word frequency dynamics. J. Phys. Conf. Ser. **936**(1), 012067 (2017)
19. Gikhman, I., Skorokhod, A.: Introduction to the Theory of Random Processes. Dover Publications, New York (1996)
20. Cocho, G., Flores, J., Gershenson, C., Pineda, C., Sánchez, S.: Rank diversity of languages: generic behavior in computational linguistics. PLoS ONE **10**(4), e0121898 (2015)
21. Janda, L., Lyashevskaya, O.: Grammatical profiles and the interaction of the lexicon with aspect, tense, and mood in Russian. Cogn. Linguist. **22**(4), 719–763 (2011)
22. Janda, L., Solovyev, V.: What constructional profiles reveal about synonymy: a case study of Russian words for sadness and happiness. Cogn. Linguist. **20**(2), 367–393 (2009)
23. Gries, S., Divjak, D.: Behavioral profiles: a corpus-based approach towards cognitive semantic analysis. In: Evans, V., Pourcel, S. (eds.) New Directions in Cognitive Linguistics, pp. 57–75. John Benjamins, Amsterdam (2009)
24. Michel, J.-B., Shen, Y.K., Aiden, A.P., Veres, A., Gray, M.K., et al.: Quantitative analysis of culture using millions of digitized books. Science **331**(6014), 176–182 (2011)
25. Lin, Y., Michel, J.-B., Aiden, E.L., Orwant, J., Brockman, W., Petrov, S.: Syntactic annotations for the Google Books Ngram Corpus. In: Li, H., Lin, C.-Y., Osborne, M., Lee, G.G., Park, J.C. (eds.) Proceedings of the Conference on 50th Annual Meeting of the Association for Computational Linguistics 2012, Jeju Island, Korea, vol. 2, pp. 238–242. Association for Computational Linguistics (2012)
26. Gordon, A., Ford, R.: Sputnik khimika. Mir, Moskow (1979)
27. Yada, V., Bethard, S.: A survey on recent advances in named entity recognition from deep learning models. In: Proceedings of the 27th International Conference on Computational Linguistics, Santa Fe, New Mexico, USA, pp. 2145–2158. Association for Computational Linguistics (2018)

Automated Approach to Rhythm Figures Search in English Text

Elena Boychuk[1]([✉]) [ID], Inna Vorontsova[1] [ID], Elena Shliakhtina[1] [ID],
Ksenia Lagutina[2] [ID], and Olga Belyaeva[1] [ID]

[1] Yaroslavl State Pedagogical University named after K.D. Ushinsky,
Respublikanskaya Street 108/1, 150000 Yaroslavl, Russia
elena-boychouk@rambler.ru, arinna1@yandex.ru, ElenaV_Yar@mail.ru,
olbelyaeva@yandex.ru
[2] P.G. Demidov Yaroslavl State University,
Sovetskaya Street 14, 150003 Yaroslavl, Russia
lagutinakv@mail.ru

Abstract. Text rhythm is recognized as being one of the most important subject areas of modern linguistic studies. There is a considerable amount of literature on the analysis of rhythm in poetry and literary prose. However, few researchers have addressed the problem of using automated tools for rhythm analysis, whereas automated methods can be of great benefit to this cause, especially when the research is conducted on large text corpora. This paper presents a new automated approach to integrated search of rhythm figures in fiction including anaphora, epiphora, anadiplosis, symploce and simple repetition provided for by an original lexical tool designed within the framework of the research. The ad hoc experiments have proved this approach to be reliable and informative.

Keywords: Text rhythm · Rhythm analysis · Natural Language Processing · Rhythm figures · Automated approach to rhythm analysis

1 Introduction

In terms of linguistics rhythm is a literary device that demonstrates the long and short patterns through stressed and unstressed syllables, particularly in verse form [1]. However, rhythm has a more complex structure that can manifest itself at various linguistic levels in various text types and is characterized by a special "movement" (mouvant) [13]. Fictional text rhythm has so far lacked a clear definition that can be followed without reservation. Nonetheless in this paper we define rhythm as regular repetition of similar and commensurable units of speech that performs structuring, text-forming, and expressive functions [6].

The main purpose of rhythm analysis is deep penetration into the creative method of an author, into their intent, originality of individual creativity and skill. Identifying the specificity of the rhythm of writer's works makes it possible

© Springer Nature Switzerland AG 2020
W. M. P. van der Aalst et al. (Eds.): AIST 2019, CCIS 1086, pp. 107–119, 2020.
https://doi.org/10.1007/978-3-030-39575-9_11

to more successfully solve the problem of determining the authorship of texts. This method is widely used in poetic text analysis, while its application to the research of fiction can be questionable [9]. The problem confronted is large text processing. Therefore, development of automated tools for rhythm analysis in a non-poetic text is among the primary tasks of computational linguistics.

2   Problem Statement

The phonetic aspect of rhythm analysis is usually defined as the sonic periodicity, i.e. change of consonants and vowels and reiteration of the same sounds. It also includes the analysis of accent distribution, pauses and tempos. The lexical aspect of rhythm involves repetition of words, for example, at the beginning or the end of sequences. Within the framework of a grammar aspect of literary stylistics, morphological and syntactic means of rhythmization are distinguished for the analysis of rhythm. For this analysis, such characteristics as the repeated use of words with a common root, the repetition of words in different tenses, rhetorical questions, exclamations, homogeneous sentence members are important.

The paper considers an original software application designed for the automated search for lexical figures of repetition. The application is set to be simple and convenient focusing on finding the figures of repetition and displaying them on the screen.

The following figures of repetition are selected for the research:

– anaphora (repetition of words at the beginning of a sentence or clause);
– epiphora (repetition of words at the end of a sentence or clause);
– symploce (joint use of anaphora and epiphora);
– anadiplosis (repetition of the final word of one clause or sentence at the beginning of the following clause or sentence);
– simple repetition.

Summarily, the application should perform the tasks set above and be easy to use.

3   Existing Tools: State-of-the-Art

Most of works in the field of rhythm analysis look at the definition of phonetic aspects. American researchers Greene et al. [11] analyze the rhythm of a poetic text. They define the analysis as extraction of patterns from existing online poetry corpora. They use these patterns to generate new verses and translate the existing poems.

The "Ritminme" program (http://www.ritminme.ru) allows to evaluate the poetic rhythm selecting a rhyme to the given word and defining a rhythm scheme basing on revealing the stressed and unstressed syllables.

Kishalova [12] proposes a similar approach. She analyzes the rhythm in Russian texts with the "PULSE" program. The tool considers the average factor of unstressed syllables when analyzing the rhythmic structure of the text. Experiments with fiction, news, and scientific texts show the dependency of the rhythm on the text style.

Methods adopted in syntactical analysis are also applicable to rhythm analysis of poetry and prose.

Belousov and Dusakova [4] propose the tool that helps to analyse text rhythm basing on the automatic computation of sentence lengths.

Couranjou and Lachambre [7] analyze the grammar structure of sentences and their rhythm. The syllables are counted and a rhythm curve is built with regards to the syllable number in a text portion, the research is based on a case study of fiction.

Toldova et al. [17] compare systems of anaphora detection. The best results are shown by linguistic algorithms that apply approaches based on rules and ontologies.

Dubremetz and Nivre [8] use a binary logistic regression classifier to extract chiasmuses, anaphoras, and epiphoras from political texts. This system proves to be quite efficient: for epanaphora the accuracy and F-measure are around 55–63%, for epiphora—around 50%, for chiasmus—78.3%.

The best approaches to text rhythm analysis enable experts to make their decisions on the basis of several types of text or word features.

In the area of rhythm analysis in Russian, English and French, the "Rhymes" program (http://rifmovnik.ru) by N. Ketsaris assists in text processing in order to find rhymes for the specified word by phonetical and lexical characteristics.

Russian researchers Boychuk et al. [6] analyze the rhythm of French fiction. Their program allows to do it by applying a number of methods grouped according to the aspect of analysis: phonetic, lexical and grammar.

Balint et al. [2,3] propose a method of English text rhythm evaluation. It analyzes different rhythmic features, including organizational, lexical, grammar, phonetic, and metrical. The program that implements this method uses statistical algorithms that allow to achieve around 80% accuracy in prediction of the text genre by its rhythm features.

Niculescu and Trausan-Matu [14,15] describe a Natural Language Processing application that analyzes the rhythm of English, Romanian, and French texts of different styles: poems, fiction, and political speech. It automatically performs word hyphenation, search of stressed and unstressed syllables using a dictionary, and detection of phonetic, metrical, and grammar rhythm features. The authors show the benefits of their application for the automatization of linguistic research.

The existing instruments prove to be targeted at the analysis of text rhythm at the phonetic and lexical levels and/or at the assessment of sentence length. The novelty of the tool described in this paper is seen in its ability to search and process stylistic devices based on repetition.

4   The Approach

The basis of the rhythmic means of this tool is a repetition that has the following structure: source element + repeating elements. Depending on the means, the elements of repetition can have the initial, final, middle, contact, non-contact, cross positions within a rhythmic unit.

In order to achieve a higher degree of rhythmization with the help of these figures, we should set out the following conditions: (1) presence of several repeating elements, (2) close proximity of the original and repeating elements, (3) high frequency of elements in the text.

To analyze the use of figures of rhythmization, it is necessary to clarify some stylistic terms since they allow for different interpretations in dictionaries and academic literature.

Anaphora is one of the most complex figures. Firstly, it does not have a clear distinction from the concept of deixis, often represented as a category combining anaphora and deixis itself. In this combination, anaphoric relationships indicate the context elements when referring from one word or phrase to another. This relationship is called an associative anaphora but as for rhythm analysis a full (tautological) anaphora is most important.

Secondly, in stylistics there are several types of anaphora as a rhetoric figure depending on the ways of its manifestation: phonetic, lexical and syntactic. At the same time, some dictionaries consider a syntactic anaphora as a lexical subtype, and others as an independent type. In this paper, a lexical anaphora is regarded as most indicative from the rhythm viewpoint. Two elements of anaphora repeated in adjoining sentences are sufficient for text rhythm perception.

Anadiplosis and simple repetition appear to be most confusing in terms of distinction.

Most linguists share an opinion that anadiplosis (Greek "ana"—again and "diploos"—double) is a repetition of the final word of one clause at the beginning of the following clause or the repetition of the final word in a sentence and the beginning of the next sentence [10]. However, a different opinion exists. For example, French linguists Bergez, Robrieux say: "Cette figure de répétition consiste à reprendre dans une phrase (souvent au début) un mot ou un groupe de mots de la phrase précédente, de maniére à établir une liaison" (This stylistic figure consists in repeating a word or a group of words of one sentence at the beginning of the next sentence) [5,16]. Thus, in this definition, the number of elements within anadiplosis is not determined.

The tool discussed in this paper also offers the implementation of a simple retry search. At this level of development, such types of repetitions as reduplication and epanalepsis are not distinguished.

The latter is one of the most controversial in stylistics, since, depending on the conditions of use, it can be identified with anaphora, epiphora, anadiplosis and other stylistic figures. The term "epanalepsis" is used as synonymous with simple repetition, while there is a definition of etymological paraphrase for both figures [10]. The main stylistic function of these figures is the expression of an emotional state, certain feelings: anger, pain, despair, joy, etc.

In this paper we consider epanalepsis as a repetition of a word or a part of a statement interspersed with intermediate words.

Besides, it is important to determine the number of the elements. The smaller the distance between repetitions, the more rhythmic the text is. Epanalepsis is defined as a figure that excludes contact repetitions at the junction of clauses or at the junction of sentences (anadiplosis functions), as well as cases of non-contact repetitions at the beginning (anaphora) or the end (epiphora) of clauses and sentences, and simultaneously at the beginning and at the end of clauses (symploce).

It is also necessary to clearly distinguish between epanalepsis and reduplication (palilogy). Both are based on repetition: the repetition of words in the contact position to accentuate their semantics is called reduplication, while the repetition of words and phrases after intermediate words is considered as epanalepsis.

The algorithm for analyzing the entered text when identifying a lexical anaphora, epiphora, or symploce includes the search for repetitions at the beginning, the end of sentences, as well as at the junction of sentences and clauses. To locate the position of the desired word, a punctuation criterion is applied (comma, period, semicolon, exclamation and question marks, three dots).

To identify anadiplosis, the tool searches for repeating words separated by commas without changing the forms, regardless of the position in a sentence. To reveal the reduplication, the tool searches for a repetition of words separated by commas without changing word forms at the beginning of a sentence. When working with epanalepsis, it is necessary to take into account the repetition of words that are not in a contact position in relation to each other, but are located within a small rhythmic unit.

5   Software Implementation of the Tool

A randomly selected text is uploaded and automated search for the five figures described above (anaphora, epiphora, anadiplosis, symploce and simple repetition) starts. The elements are highlighted with the corresponding colors, and a list of all words in the text and the number of their repetitions are displayed.

As a result, the researcher can receive information about the used lexical figures in the text, propose a theory about the text authorship, evaluate the quality of translation and analyze the rhythm of the text.

The functional tool has a simple interface for displaying the text and the list of all its words with the number of repetitions. Also it finds specific lexical figures in the text, highlights the found elements and display their list separately Fig. 1.

The program was implemented on the JavaScript language using HTML and CSS. It is available at https://github.com/text-processing/html-tool.

The word list with the number of repetitions is formed according to the following algorithm. In the beginning, all non-letter characters characters are replaced with spaces. Extra spaces are removed, i.e. where their number is more than one. Then the whole text is divided into an array, where the elements

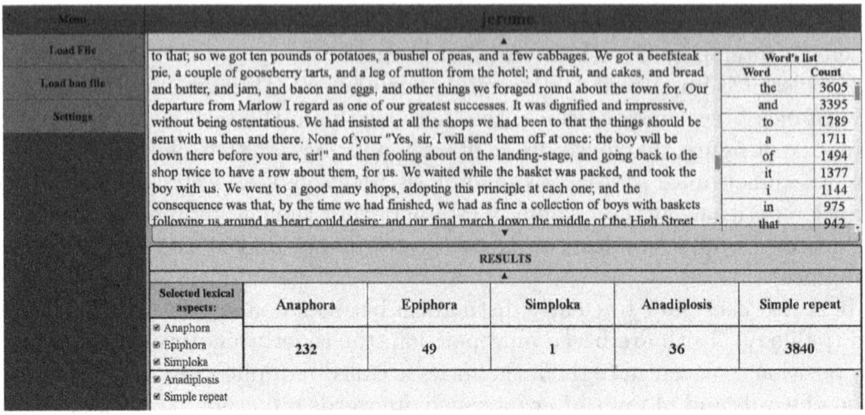

Fig. 1. Main page of the application

are individual words. In the resulting array the algorithm counts the number of repetitions of each element. As a result, we get an associative array of words associated with the number of their repetitions. The elements of the resulting array are sorted in descending order by the number of repetitions and are displayed in a table.

The user can edit the list of displayed words, i.e. delete specific words that are not needed for investigation. The list is formed according to the algorithm described above, but during the element-by-element output of the result, the displayed item is searched for in the stop list (the list of stop words, i.e. ignored words). If the element is found, it is not displayed.

The tool also allows to search such lexical units as anaphora, epiphora, symploce, anadiplosis, and simple repetitions.

The search for anaphora in the text occurs according to the following algorithm. In the beginning, it forms an array of the first words of each sentence in the text. This array counts word repetition number. If it exceeds one, the word is entered into the resulting array. Then a new array is created from phrases the beginning of which is formed of words from the resulting array. In this array we search for repeating elements again. If repetition number is less than two, the element is deleted. The remaining elements form the resulting array by updating the existing elements or adding new ones. In the end, we get a list of anaphoras.

The algorithm for searching epiphoras is similar. The significant difference is that the array is formed from the last words of the sequences. At the same time, the formed phrases contain words from this array not from the beginning, but from the end.

After extraction of these two figures, the application searches symploce comparing lists of anaphoras and epiphoras and choosing cases when they appear together.

The following algorithm is applied to the search of the anadiplosis. The program searches elements by the pattern: "word/phrase + punctuation mark +

word/phrase" and adds them into the resulting array. Then the algorithm compares the left and right parts separated by a punctuation mark. If they are identical, the element is added to the resulting array.

The search for simple repetition is carried out according to the following algorithm. In the beginning, all non-literal characters are replaced with spaces. Then the whole text is reduced to a single register and divided into an array, where the elements are individual words. In the resulting array the algorithm counts the number of repetitions of each element. If this number is less than two, then the element is deleted. In the end, we get an array whose elements are simple repetitions.

6   Experiments

The experiment was conducted by a research team based at the Department of Foreign Languages, Yaroslavl State Pedagogical University named after K.D. Ushinsky. The text corpus was derived from the works of 23 English authors selected with regards to the manifestation of such rhythmic figures as anaphora, epiphora, anadiplosis, symploce and simple repetition. The following writers were chosen for that purpose: K. Atkinson, J. Austen, Ch. Bronte, Ch. Dickens, E.M. Foster, J. Fowels, N. Gaiman, E. Gaskell, Th. Hardy, J. Joyce, D.H. Lawrence, D. Lessing, D. du Maurier, I. McEwan, I. Murdoch, S. Muriel, T. Pratchett, J.K. Rowling, R.L. Stevenson, S. Thomas, J.R.R. Tolkien, O. Wilde, V. Woolf.

The researchers were divided into 2 groups, the first of which processed the text manually, the second - used the Rhythmanalysis tool. The main objectives of the experiment were (1) to determine the efficiency of the tool in terms of time and accuracy of results, (2) to detect errors when working with the tool, (3) to analyze problems and devise ways to solve them.

The first group of 8 researchers worked for 32 days 2 h a day. The second group (also 8 people) coped with the task within 4 days, working 2 h a day. Thus, the efficiency of the application is obvious: the text processing time is reduced by 87.5%, which allows the researcher to be spared from a lot of tedious and monotonous work.

The second stage of the experiment implied the identification of errors, which the tool allows when searching for rhythmic figures (anaphora, epiphora, symploce, anadiplosis, simple repetition) in the texts. At this point, the re-searchers had to spend more time checking on the data provided by the tool, since they had to work both with the tool and with the texts.

The results of the experiments are presented in Table 1. In the analyzed texts the tool found 4407 anaphoras. 89% out of them were detected correctly. For example, *"I **wanted** a miracle job advertisement. I **wanted** someone to come along and say, "Just do what you're good at and we'll give you enough money for your rent, bills, cigarettes and some nice food and clothes"* (Scarlett Thomas).

In the course of the analysis we also found a few disputable cases of anaphora. First of all, not all anaphora elements were detected by the tool. In the following

Table 1. Results of lexical aspects search

Means of rhythmization	Number of occurrences in texts	Accuracy
Anaphora	4407	89
Epiphora	2564	93
Symploce	18	65
Anadiplosis	458	62
Simple repetitions	28542	93

case it is caused by the word number limit in anaphora: *"It asks whether it is possible or even desirable to disrupt this. It asks whether it is possible to find meaning in a world overflowing with it"* (Scarlett Thomas). Another reason for the partial detection of anaphora is the presence of an extra element within the analyzed unit:

"Not in Sylvie's room (they had long ago ceased to think of it as a room that belonged to two parents).

Not in Maurice's room, so generously sized for someone who spent more than half his life living at school." (Kate Atkinson)

However, the displayed text fragment allows to manually detect all elements of the anaphora.

Not all displayed cases of the pronoun **he** can be regarded as anaphora. For example, *"He was a member of a cycling club and every Sunday tried to wheel as far away from Birmingham's smogs as he could, and he took his annual holiday by the sea so that he could breathe hospitable air and think himself an artist for a week.*

He thought he might try to put some figures in his painting, it would give it a bit of life and 'movement', something his night-school teacher (he took an art class) had encouraged him to introduce into his work" (Kate Atkinson).

There is a text fragment of considerable length between the pronouns. Moreover, the first pronoun is in the middle of the paragraph while the second one is at the beginning of a new paragraph.

In the following example the displayed preposition cannot be reckoned as anaphora because the word is used in different meanings: *"At first, being little accustomed to learn by heart, the lessons appeared to me both long and difficult; the frequent change from task to task, too, bewildered me... At that hour most of the others were sewing likewise; but one class still stood round Miss Scatcherd's chair reading..."* (Charlotte Bronte).

In 29 cases simple repetition was displayed as anaphora. For example:

– *Eight years.*
– *Eight years!* (Charlotte Bronte)

In 21% of cases the detected units were not anaphoras. Here, it is important to emphasize that all cases of the definite article usage cannot be referred to as anaphora. For example, *"**The** Librarian jumped it. **The** Luggage, of course, followed them with a noise like someone tapdancing over a bag of crisps"* (Terry Pratchett). Besides, according to the rules of the English language the stress is usually put on meaningful parts of speech. Therefore, the article cannot influence the rhythm of the text in general. The only possible exception is the situation when the article is a part of another rhythm figure, for example, gradation.

The contextual analysis allowed to detect the following number of end-of-a-sequence repetitions provoking a rhetorical effect, for example: 35/46 (76%) in I. Murdoch's "The Bell", 174/187 (93%) in D. Du Maurier's "Rebecca" and 70/74 (94.5%) in V. Wolfe's "To the Lighthouse". We view this statistics to speak for a high sensitivity of the method applied to the search of **epiphora**:

*"Frank **knew**. And Maxim did not know that he **knew**"* (Daphna Du Maurier "Rebecca").

*"The young man was abusing the government. William Bankes, thinking what a relief it was to catch on to something of this sort when private life was disagreeable, heard him say something about "one of the most scandalous acts of the present government." Lily was **listening**; Mrs Ramsay was **listening**; they were all **listening**"* (Virginia Wolfe "To the Lighthouse").

While analyzing the use of **anadiplosis**, the following was revealed:

1. The average number of uses per book is 18–37 units. A total of 458 cases of anadiplosis were investigated. Of these, 62% are the correct search.
2. Most cases of use are classified by the tool as reduplication (repetition of the same words in the contact position in one clause, whereas anadiplosis is used at the junction of the clauses), for example: *"A **very, very** brief time, and you will dismiss the recollection of it, gladly, as an unprofitable dream, from which it happened well that you awoke. My **little, little** child.' cried Bob. The father of a **long, long** line of brilliant laughs."* (Ch. Dickens)
3. The tool considers repetitions that are characterized by a polysyndeton (the use of coordinating conjunctions close together, and more than needed, for stylistic effect) as an anadloplosis, for example: *"Scrooge went to bed again, **and thought, and thought, and thought** it over and over, and could make nothing of it."* (Ch. Dickens). This case must be regarded as a reduplication with a polysyndeton.
4. The most indicative cases of anadiplosis revealed by the instrument are the cases of using it at the junction of clauses: *"It was right to do **it, it** was kind to do **it, it** was benevolent to do it, and he would do it again. It was right to do **it, it** was kind to do **it, it** was benevolent to do it, and he would do it again."* (I. Murdoch). In this case, the example is also combined with the epiphora (to do it at the end of sentences).

While carrying out the experiment the lexical tool identified the following phrases as **symploce**: *"O! O!", "Norbert! Norbert!", "Why? Why?", "And so? And I'm a hundred times more vulnerable than I was. And so?", "Nothing! Nothing!", "No! No!", "The Lighthouse! The Lighthouse!"*. After a thorough analysis, these examples are considered to be just a simple repetition of all the elements of the phrase which can be seen as reduplication, but not symploce.

As for **simple repetition**, this tool has the highest percentage of manifestations in the texts (more than 1000 cases per work). This is primarily due to the fact that simple repetition as a complex means of rhythm involving the division into different types of repetitions, combines the cases of use of reduplication, epanalepsis, and in some cases anadloplosis. It is possible to layer of one means to another.

7   Discussion

To improve **anaphora** detection it is necessary to

- exclude articles from the list of the checked words;
- reduce the number of the words in a search unit to 20;
- define the anaphora parameters more clearly.

For that purpose it is important to limit the number of words to 2–3 during the search and to introduce the punctuation parameter that will help to distinguish anaphora from simple repetition.

The main reason for misidentifying the **epiphora** is a considerable length of sequences (mainly sentences) where new clauses hamper the perception of the stylistically marked epiphora:

"...and he could not help hoping that Toby would sooner or later force such a tête-à-tête upon **him**. *He wished that somehow he could pull out of this mess the atom of good which was in it, crystallizing out his harmless goodwill for Toby, Toby's for* **him**". (Iris Murdoch "The Bell").

"He wanted something else that I could not give him, something he had had **before**. *I thought of the youthful almost hysterical excitement and conceit with which I had gone into this marriage, imagining I would bring happiness to Maxim, who had known much greater happiness* **before**". (Daphna Du Maurier "Rebecca").

In a few cases (not exceeding 3 in the sample) the repeated words, usually personal pronouns, have different referents, which enables us to consider such repetition as accidental and therefore irrelevant:

"Her face was glowing and she put up one hand to hide **it**. *Her cigarette fell on the floor and she abandoned* **it**". (Iris Murdoch "The Bell")

Other examples (10 in the sample) reveal the inability of the method to detect repetition at the end of clauses along with its successful identification of the epiphora in sequential sentences:

"Truthfulness **is enjoined**, *the relief of suffering* **is enjoined**, *adultery* **is forbidden**, *sodomy* **is forbidden**.

And I feel that we ought to think quite simply of these matters, thus: truth is not glorious, it is just **enjoined**; *sodomy is not disgusting, it is just* **forbidden**". (Iris Murdoch "The Bell").

The solution to the problem is expected to be found in designing a feature that would make the tool capable of simultaneously analyzing a sentence-and a clause-long context with the "comma" serving the marker.

As a way of resolving the problem of differentiating between **anadiplosis** and **reduplication** by the tool, three possible ways are proposed:

- formulation of rules for the search for reduplication as a repetition at the beginning of a sequence in steps of 3–4 words;
- exclusion of all adjectives (including a comparative degree) and adverbs from the list of anadiplosis;
- restriction of cases of anadiplosis only to those that are observed at the junction of clauses, separated by semicolon, period, question or exclamation marks, as well as three dots.

The main feature distinguishing between the anadiplosis and reduplication with a polysyndeton should be considered as a union which is not an element of anadiplosis. In order to clearly identify structures with a polysyndeton, it seems appropriate to compile a list of all possible unions used at the junction of clauses.

Due to the fact that cases with the pronoun "it" have 67% of the total number of anadiplosis use, it is proposed to add cases with it to the list of search conditions of the tool when it is considered as a repetition at the junction of clauses separated by commas (it, it).

Taking into account the fact that the lexical tool identifies simple phrases and utterances as **symploce**, but not the repetition of the beginning and the end of the sentence which consists of the subject and/or the predicate, it is considered necessary to apply a restriction (of at least 5 words which should be in a clause/sentence) to the tool while detecting symploce in such texts as fiction.

As a solution to the problem of inaccurate detection of the types of **simple repetitions**, it is necessary to divide these rhythmic figures into reduplication and epanalepsis. As for reduplication, the parameter should be the contact position of words separated by commas (but not at the junction of clauses or sentences), and, as for epanalepsis, the parameter should be the use of identical words at certain intervals.

8 Conclusion

The described tool for the analysis of fiction text rhythm allows to conduct integrated research into lexical rhythm figures including anaphora, epiphora, anadiplosis, symploce and simple repetition. The tool has shown 80.4% reliability which we view as relatively high. Pitfalls in the process of lexical figures search and evaluation are in a few cases explained by the impossibility of excluding certain conditions of the repetition use. If this occurs, the decision whether the case should be attributed to a lexical figure is taken by the researcher. This automated approach holds the lowest number of pitfalls when applied to the search of anaphora, epiphora and simple repetition. To make the detection of the anadiplosis more precise, simple repetition must be divided into reduplication and epanalepsis, which will spare cross search between the anadiplosis and the simple repetition. In terms of symploce, this repetitive figure is rare in

English texts, the search results can be improved by expanding the word range between the repeated clauses. Future work will concentrate on the definition of the author's individual style by means of rhythm figures as well as on the application of the above principles of rhythm analysis to texts written in languages other than English.

Acknowledgements. The reported study was funded by RFBR according to the research project №19-07-00243.

References

1. Literary Devices. Definition and Examples of Literary Terms. Metaphor. http://www.literarydevices.net/metaphor/
2. Balint, M., Dascalu, M., Trausan-Matu, S.: Classifying written texts through rhythmic features. In: Dichev, C., Agre, G. (eds.) AIMSA 2016. LNCS (LNAI), vol. 9883, pp. 121–129. Springer, Cham (2016). https://doi.org/10.1007/978-3-319-44748-3_12
3. Balint, M., Trausan-Matu, S.: A critical comparison of rhythm in music and natural language. Ann. Acad. Rom. Sci. Ser. Sci. Technol. Inf. **9**(1), 43–60 (2016)
4. Belousov, K., Dusakova, G.: Analyzer of the rhythmic structure of the text: attribution of texts based on rhythmical patterns. Cifrovaya gumanitaristika: resursy, metody, issledovaniya **1**, 49–51 (2017). (in Russian)
5. Bergez, D., Géraud, V., Robrieux, J.J.: Vocabulaire de l'analyse littéraire. Armand Colin (2010)
6. Boychuk, E., Paramonov, I., Kozhemyakin, N., Kasatkina, N.: Automated approach for rhythm analysis of French literary texts. In: Proceedings of 15th Conference of Open Innovations Association FRUCT, pp. 15–23. IEEE (2014)
7. Couranjou, P., Lachambre, B.: Pae stylistique informatique. Computer stylistic, in Le Bulletin de l'EPI 56, pp. 24–35 (1989). (in French)
8. Dubremetz, M., Nivre, J.: Rhetorical figure detection: chiasmus, epanaphora, epiphora. Front. Digit. Hum. **5**, 10 (2018)
9. Freyermuth, S.: Poétique de la prose ou prose poétique? le rythme contre le prosaïsme. Questions de style, Vous avez dit prose? pp. 67–80 (2009). (in French)
10. Fromilhague, C., Sancier-Chateau, A.: Introduction à l'analyse stylistique. Bordas (1996)
11. Greene, E., Bodrumlu, T., Knight, K.: Automatic analysis of rhythmic poetry with applications to generation and translation. In: Proceedings of the 2010 Conference on Empirical Methods in Natural Language Processing, pp. 524–533. Association for Computational Linguistics (2010)
12. Kishalova, L.: Analysis of features of rhythmic structure of texts of different styles of the speech, pp. 257–261. Vestnik Bryanskogo gosudarstvennogo universiteta (2016). (in Russian)
13. Meschonnic, H.: Critique du rythme. Anthropologie historique du langage. Verdier: coll. "Verdier Poche" (2009). (in French)
14. Niculescu, I.D., Trausan-Matu, S.: Rhythm analysis of texts using natural language processing. In: RoCHI, pp. 107–112 (2016)
15. Niculescu, I.D., Trausan-Matu, S.: Rhythm analysis in chats using natural language processing. In: RoCHI, pp. 69–74 (2017)

16. Robrieux, J.J.: Les figures de style et de rhétorique. Dunod (1998)
17. Toldova, S., Azerkovich, I., Ladygina, A., Roitberg, A., Vasilyeva, M.: Error analysis for anaphora resolution in Russian: new challenging issues for anaphora resolution task in a morphologically rich language. In: Proceedings of the Workshop on Coreference Resolution Beyond OntoNotes, CORBON 2016, pp. 74–83 (2016)

Adapting the Graph2Vec Approach to Dependency Trees for NLP Tasks

Oleg Durandin ⓘ and Alexey Malafeev$^{(\boxtimes)}$ ⓘ

National Research University Higher School of Economics, Nizhny Novgorod, Russia
oleg.durandin@gmail.com, amalafeev@yandex.ru

Abstract. In recent works on learning representations for graph structures, methods have been proposed both for the representation of nodes and edges for large graphs, and for representation of graphs as a whole. This paper considers the popular graph2vec approach, which shows quite good results for ordinary graphs. In the field of natural language processing, however, a graph structure called a dependency tree is often used to express the connections between words in a sentence. We show that the graph2vec approach applied to dependency trees is unsatisfactory, which is due to the WL Kernel. In this paper, an adaptation of this kernel for dependency trees has been proposed, as well as 3 other types of kernels that take into account the specific features of dependency trees. This new vector representation can be used in NLP tasks where it is important to model syntax (e.g. authorship attribution, intention labeling, targeted sentiment analysis etc.). Universal Dependencies treebanks were clustered to show the consistency and validity of the proposed tree representation methods.

Keywords: Syntax · Graph embeddings · Graph2vec · Representation · Dependency tree · Embeddings · Universal dependencies

1 Introduction

Techniques for vector representation (embeddings) and representation learning have been increasingly attracting attention lately. This is due to the fact that effective representation of input data is key to solving most machine learning and deep learning problems. Vector representations, initially used in natural language processing [1, 2], later saw widespread application.

Of particular interest are vector representations of graphs (graph embeddings). Graphs can be of arbitrary size and structure, but using recently developed methods [1], they can be reduced to a fixed-size vector, which is very convenient for machine learning and deep learning. A dependency tree is a graph with a few distinct properties: it is connected, acyclic and oriented. A special approach to vector representation is needed to accommodate dependency trees, as we will show in more detail in the next section.

© Springer Nature Switzerland AG 2020
W. M. P. van der Aalst et al. (Eds.): AIST 2019, CCIS 1086, pp. 120–131, 2020.
https://doi.org/10.1007/978-3-030-39575-9_12

2 GRAPH2VEC Approach

In [1], a neural approach was proposed for the training of whole-graph representations. Importantly, this approach is completely unsupervised. The learning method is data-driven; it requires a fairly large set of graphs to learn graph representations. An advantage of this approach is the ability to capture structural similarity: vectors for structurally similar graphs are close in the vector space.

Consider the following formal problem statement, as given in [1]. Given a set of graphs $G = \{G_1, G_2, \ldots\}$ and a positive integer δ (the expected size of the vector representation), we want to train a δ-dimensional distributed representation for any graph $G_i \in G$. The matrix of representations of all graphs will be denoted as $\Phi \in \mathbb{R}^{|G| \times \delta}$. Let $G = (N, E, \lambda)$ be a graph, where N is the set of nodes, $E \subset (N \times N)$ is the set of edges. The graph G is labeled if there is a function $\lambda : N \to l$ that assigns a unique label from the alphabet l for each node $n \in N$. In addition, the edges can also be labeled, and in this case there is an edge labeling function: $\eta : E \to l$.

The main idea of the graph2vec approach is to view the entire graph as a document and the rooted subgraphs (that encompass a neighborhood of certain degree) around every node as words, that compose the document. In other words, different subgraphs compose graphs in a similar way that different words compose sentences/documents when used together [1].

We omit the basic graph2vec algorithm, because it is described in the source paper. Note that we modify only the method of representing the graph. In [1], a specific method for extracting rooted subgraphs was described, which is based on the WL [3] relabeling process.

3 Adaptation Graph2vec Approach to Dependency Trees

3.1 Dependency Tree Concept

Let $S = (w_0, w_1, \ldots, w_N)$ denote a sentence of length N. Also, as in most computer linguistics literature [4], we add the "artificial" token ROOT, which is represented as w_0. Similarly, let $A = (a_0, a_1, \ldots, a_N)$ be the representation of the sentence S, where a_i corresponds to the word w_i $(0 \leq i \leq N)$. In addition to encoding information about each word w_i in isolation (for example, its POS-tag), a_i most often also encodes positional information about w_i.

For example, Fig. 1 shows the dependency tree for the sentence *A dog is chasing a cat*.

Fig. 1. Dependency tree example.

The head of the first determinant a is dog, whereas the head of the second a is cat. Additional information a_i about w_i is represented by the POS tag. Also, for each dependency relationship between the head and the dependent word, the type of dependency is specified. Obviously, POS tags and dependency labels are not unique and can be seen in the same sentence more than once.

Each dependency tree is a connected, acyclic and oriented graph with certain functions: $\mu : N \to l_{pos}$, where l_{pos} is the set of POS tags possible in a given language, and the function $\eta : E \to l_{dep}$, where l_{dep} is a set of dependency types that exist in the language in question.

Thus, the dependency tree is represented by the tuple $T_i = (N, E, \mu, \eta)$. Otherwise, the task remains the same – to construct a δ-dimensional vector representation of the set of trees $T = \{T_1, T_2, \ldots\}$, with the property that structurally similar trees will be located near each other in the space of the distributed representation.

3.2 Relaxing of the WL Relabeling Process Constraints

What limitations of the GetWLSubGraph algorithm make it ineffective when applied to dependency trees? First of all, the assumption about the uniqueness of the labels of the nodes and edges of the graph, since the same POS tags can correspond to different words in dependency trees. In addition, the sentence may contain the same dependency types between different words. This is often the case with fairly complex sentences that are quite common in natural language.

Secondly, the standard GetWLSubGraph does not take into account the use of edge labels, which represent, in our case, the type of relationship between a pair of words in a sentence. And finally, the original GetWLSubGraph procedure considers general purpose graphs, in which the order of the nodes is not important. A set of root subgraphs is used, which are then sorted topologically.

In our work, an improvement of the GetWLSubGraph algorithm is proposed, with a relaxation of the above limitations:

```
Algorithm GetWLSubGraphForTrees(n,G,d)
Input: n : Node which acts as the root of the subgraph
G = (N,E,λ,η): Graph from which a subgraph is to be ex-
tracted
d: Degree of neighbors to be considered for extracting
the subgraph
Output: sg_n^(d): Rooted subgraph of degree d around node n
Begin
(1) |  sg_n^(d) = {}
(2) |  if d = 0 then
(3) |  |  sg_n^(d):= λ(n)
(4) |  else
(5) |  |  N_n := {n'|(n,n') ∈ E}
(6) |  |  M_n^(d):=[ η(n,n') ⊕ GetWLSubGraph(n',G,d-1) | n' ∈ N_n]
(7) |  |  sg_n^(d):=sg_n^(d) ∪ GetWLSubGraph(n,G,d-1) ⊕ M_n^(d)
(8) |  return sg_n^(d)
```

As can be seen, small changes are made in the algorithm for more efficient use with dependency trees. In particular, we take into account the edge labels and reject the assumption about node and edge uniqueness.

As an illustration, below are some sequences from the tree shown in Fig. 1, extracted according to the GetWLSubgraphForTree algorithm. Note that the results include single POS tags as well as dependency relationships between two and more words:

ROOT VERB DET NOUN VERB DET NOUN ROOT_root_VERB VERB_aux_VERB VERB_nsubj_NOUN VERB_dobj_NOUN VERB_dobj_NOUN_det_DET VERB_nsubj_NOUN_attr_VERB ROOT_root_VERB_root_VERB_nsubj_NOUN_aux_VERB...

3.3 Relabeling Process Based on Contracted Nodes of the Dependency Tree

The GetWLSubGraphForTrees procedure, however, may generate somewhat "artificial" markup showing good results for general purpose graphs, but meaningless from the linguistic point of view.

Let us use the intuition underlying dependency trees. For example, consider the phrase *beautiful cat*, there is an ADJ (beautiful) <- NOUN (cat) dependency, where the noun is the headword and the adjective is the dependent. In this case the adjective can be seen as a "characteristic" of the noun, that is, we can collapse these nodes into one and work with the single entity *beautiful cat*. This corresponds to the linguistic concepts of noun phrase, verb phrase, etc. We can use this idea to relabel the dependency tree so as to get meaningful representations, which can then be used later in the graph2vec procedure:

```
Algorithm ContractedExtraction(G)
Input:
G = (N,E,λ,η): Graph from which a subgraph has to be ex-
tracted
Output: sgₙ: String representation of the graph
Begin
(1) | sgₙ = [ λ(n) | n ∈ N ]
(2) | nodesZeroOutDegree = {n|degree_out(n) = 0, n ∈ N}
(3) | while(nodesZeroOutDegree)
(4) | | for n ∈ nodesZeroOutDegree:
(5) | | | (n,n') = InputEdge(n)
(6) | | | G = ContractedNode(G,n,n')
(7) | | | λ(nn') := λ(n)+'_'+η(n,n')+'_'+λ(n')
(8) | | sgₙ = sgₙ ∪ [ λ(n) | n ∈ N(G) ]
(9) | | if |Nodes(G)| > 1:
(10) | | | nodesZeroOutDegree = [ n ∈ N(G)| degree_out(n) = 0 ]
(11) | | else:
(12) | | | nodesZeroOutDegree = []
(13) | return sgₙ
```

The ContractedExtraction algorithm starts by searching for graph nodes with no outgoing edges (i.e. the output degree of which is zero). Then each such node is iteratively (4) "collapsed" (6) with the head node, and the label of this new node is formed from the POS tags of the dependent and the headword, accompanied by the dependency type (7). Finally, all node tags of the graph in question (8) are added to the representation list of the graph sg_n. If we have not collapsed the graph to one node (9), then we continue examining the nodes of the graph with zero output degree (10), otherwise we assume that the graph is collapsed and return the resulting relabeling representation (13).

3.4 Relabeling Process Based on a Set of Simple Paths Between the Nodes in a Dependency Tree

Consider another algorithm for the relabeling process, which is based on the concept of path. A walk over a graph is an alternating sequence of vertices and edges: $v_0, e_0, v_1, e_1, v_2, \ldots, e_{k-1}, v_k$. A walk is a path if all vertices in it are distinct. Also, we should take into account that the dependency tree is an oriented graph. In the proposed relabeling process, we consider all the simple paths between all possible pairs of vertices in this graph. As a result of the relabeling process, we simply list the sequence of all paths in the given tree.

```
Algorithm SimplePathExtraction(G)
Input:
G = (N,E,λ,η): Graph from which subgraph has to be extract-
ed
Output: sgₙ: String representation of graph
Begin
(1) |  sgₙ:=[]
(2) |  for n₁ ∈ {n | n ∈ N}
(3) |  |  for n₂ ∈ {n |n ∈ N,n ≠ n₁}
(4) |  |  |  simple_paths = FindAllSimplePaths(n₁,n₂)
(5) |  |  |  for one_path in simple_paths
(6) |  |  |  |  sgₙ = sgₙ ∪ [λ(n)+'_'+η(n,n')+'_'+λ(n')|n₁,n₂ ∈ one_path]
(7) |  return sgₙ
```

The proposed representation method has the following linguistic motivation: we consider how one word in a sentence is related to another. Every word is not usually connected with every other word in the sentence, which is especially relevant for complex sentences, because one entity in a sentence is not always directly connected with another entity. In other words, since the dependency tree is an oriented graph, we cannot 'go' from any one word to any other word. Thus we get a finite number of subpaths that are like 'chains' of syntactically connected words in a sentence. The complex of all syntactic relations within a sentence can be described by its dependency tree (which is obvious), but it also can be described by the set of all possible subpaths in this tree.

4 Experiments: Applying the Graph2vec Approach to Dependency Trees

We have conducted a few experiments to apply the proposed approach to dependency tree representation learning, which showed that our methods are effective and that they yield non-trivial results. In the experiments, we used spaCy [5], which is a modern POS-tagging and parsing tool. SpaCy contains a number of pre-trained models, and in addition, one can use different custom models, as we did for the Russian language [6].

4.1 Clustering Dependency Trees from the Universal Dependencies Treebank

Hopkins Statistics Calculation

The first objective that we had in mind was to draw analogies with classic word embeddings in NLP. Will clustering of dependency trees as vectors result in some grouping of trees that have similar structure to each other? What properties will these clusters have? To eliminate the effect of the syntax analyzer errors, we used the treebanks that are part of the Universal Dependencies project [7], in particular, for Russian and English (SyntagRus [8] and the English Web Treebank v. 2.3 [9]).

First, we used the Hopkins statistics [13] as a cluster structure existence test. The SynTagRus [8] corpus has over 66,000 tagged sentences. Relabeling was performed, followed by the standard graph2vec/doc2vec learning process. The resulting vector representations were normalized and the Hopkins statistics was calculated (Fig. 2).

It can be observed that all kernels yield embeddings that have some clustering structure according to the Hopkins test. Also, the Path Extractor kernel results in the highest value for SynTag Rus.

We used a similar method on the English Web Treebank, resulting in higher Hopkins statistics values, compared to SyntagRus. This may be due to the fact that English

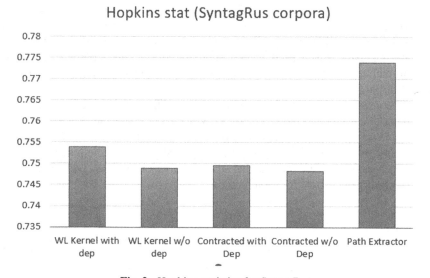

Fig. 2. Hopkins statistics for SyntagRus

has a more fixed word order in comparison with Russian, so the parsing model is less "confused", which results in less variance. It is also interesting that the highest Hopkins statistics value is obtained with Contracted Kernel, with Path Extractor only in the third place (Fig. 3).

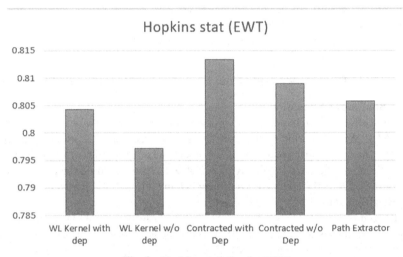

Fig. 3. Hopkins statistics for EWT

Thus, we can conclude that the space of these syntactic embeddings is characterised by some clustering structure, and we will try to explore this structure with classical clustering methods.

Clustering Process

The clustering was performed using the standard k-Means++ algorithm [10] within the Scikit-learn framework [11]. The selection of the optimal number of clusters was performed using the silhouette analysis method [12]. For both English and Russian, five relabeling methods were used: GetWLSubgraphForTrees and ContractedExtraction (for both: with and without taking into account the dependency types), as well as the SimplePathExtraction.

Let us consider the syntactic embeddings obtained from the SynTagRus corpus [8]. The number of clusters was between 2 to 30, with the optimal value selected according to the silhouette. It is worth noting, however, that the silhouette values were higher for smaller numbers of clusters, but linguistic analysis showed that the resulting clusters were too general. Therefore, the number of clusters was increased. Table 1 shows the resulting number of clusters for each method.

Table 1. Methods and number of clusters for SynTagRus trees.

Relabeling method	Number of clusters
GetWLSubgraphForTrees (with dependencies)	14
GetWLSubgraphForTrees (without dependencies)	10
ContractedExtraction (with dependencies)	10
ContractedExtraction (without dependencies)	22
SimplePathExtractor	20

A similar methodology was used on the English Web Treebank [9], see Table 2.

Table 2. Methods and count of clusters for trees from EWT.

Relabeling method	Number of clusters
GetWLSubgraphForTrees (with dependencies)	10
GetWLSubgraphForTrees (without dependencies)	19
ContractedExtraction (with dependencies)	29
ContractedExtraction (without dependencies)	18
SimplePathExtractor	10/14

After performing the clustering, we take the centroids of the clusters and choose a number of nearest embeddings to these centroids. As a metric of similarity, we use cosine similarity, as it is the standard method for embedding evaluation.

During linguistic analysis of the clusters obtained, the following features were noted. While the clusterings differ from one another, they all capture the various structural types in the given language. In other words, the clusterings successfully group structurally similar sentences together. Although it is difficult, if even possible, to obtain a clustering that would be a perfect model of textbook syntax for the given language, all methods result in linguistically sound clusterings with very distinct clusters, that can be considered valid syntax models. All clusterings are freely available online at GitHub[1]. For space constraints, we will not list the interpretations of clusters for all relabeling methods, only for some of them. First, let us consider one relabeling method, SimplePathExtractor, used on SynTagRus. It results in the following clustering:

(1)   Complex sentences with relative clauses;
(2)   Simple sentences with one indirect object;
(3)   Elliptical (incomplete) sentences without verbs;
(4)   Elliptical (incomplete) sentences with verbs;
(5)   Simple sentences with an infinitive and one or more adverb;

---

[1] https://github.com/OlegDurandin/dtree2vec.

(6)   Complex sentences with adverbial clauses and at least one numeral;

(7)   Incomplete sentences with proper nouns;

(8)   Complex sentences that start with a verb and have a noun clause as the object of this verb (importantly, the clause includes a verb and no adjectives, unlike Cluster 11);

(9)   Sentences containing at least one adjective that is a direct dependent of a verb (as in *become silent*);

(10)   Sentences with homogeneous objects (direct or indirect), expressed by nouns;

(11)   Complex sentences that start with a verb and have a noun clause as the object of this verb (the clause includes an adjective and no verbs, cf. Cluster 8);

(12)   Complex sentences with noun clauses that have an infinitive;

(13)   Sentences containing at least one adjective that is a direct dependent of a noun (as in *electronic microscope*);

(14)   Sentences with homogeneous subjects, expressed by nouns;

(15)   Sentences with a noun dependent on another noun (as in *prichiny padeniya*, which is equivalent to *the reasons for the fall*);

(16)   Short simple sentences with an auxiliary verb (e.g. *byl = was*) and an adjective that is its dependent, as well as sentences where the auxiliary verb is omitted (but is assumed);

(17)   Sentences with homogeneous predicates, expressed by verbs;

(18)   Sentences with a participle clause;

(19)   Sentences with adverbial and adverbial participle clauses;

(20)   Sentences with multiple pronouns.

As can be seen, the twenty clusters contain sentences of distinct syntactic structures. Sentences with similar syntactic structures have similar syntactic meanings, or describe different situations in a similar way, so to speak. For instance, consider Cluster 14. In these sentences, the same predicate is shared by multiple subjects. Intuitively, it means that within each sentence, the same statement is made about multiple things (or people). For example, these two sentences from this cluster are completely unrelated semantically, yet they are similar in syntactic structure:

В ход идут коробки из-под фруктов, тележки из супермаркетов, старые ролики, инвалидные коляски, велосипеды. (Fruit boxes, supermarket carts, old rollers, wheelchairs, bicycles are used.).

Посыпались жалобы, протесты (Complaints and protests flowed).

For the sake of comparison, let us also consider the clustering with a smaller number of clusters and a different kernel, GetWLSubgraphForTrees (with dependencies), also on SynTagRus:

(1)   Simple sentences that have adverbs, usually as direct dependents of verbs;

(2)   Simple sentences containing numerals;

3)   Sentences with pronouns and nouns;

(4)   Short incomplete sentences consisting of proper nouns only (as in: *Irina Melnikova.*);

(5)   Sentences with nouns functioning as homogeneous subjects or objects, always expressed by nouns;

(6)   Short incomplete sentences consisting of noun phrases, without proper nouns (as in: *Уроки итальянского.* = *Lessons of Italian.*);
(7)   Sentences with homogeneous attributes expressed by adjectives;
(8)   Complex sentences with noun clauses as objects;
(9)   Sentences with particles, including the negative particle *ne* (=*not*);
(10) Complete sentences that include proper nouns (cf. Cluster 4);
(11) (Mostly) complete sentences consisting only of verbs, adverbs and pronouns (but not nouns like Cluster 3);
(12) Sentences with noun phrases, but no proper nouns;
(13) Sentences with proper nouns and numerals;
(14) Sentences with determiners (e.g. *etot* = *this*).

As can be seen, different clusterings capture different structural types of sentences. A relatively larger number of clusters results in a more detailed syntax model in comparison with a smaller number, as shown by the above clusterings.

Let us also show a clustering for the English language. This time, let us consider an even larger number of clusters and not yet demonstrated clustering method, ContractedExtraction (with dependencies):

(1)   Sentences with a proper noun as the subject and a non-auxiliary verb (e.g. *Kelly hit the nail on the head*);
(2)   Incomplete sentences consisting of proper nouns and having no punctuation (*Karla Ferguson-Granger*);
(3)   Sentences containing pronouns and adverbs (*I will NEVER go here again*);
(4)   Sentences containing particles (*not, to*);
(5)   Sentences with an auxiliary verb and a proper noun as the object or part of the adverbial modifier (*Parking spaces are just big enough for a Mini Cooper*);
(6)   Imperatives usually starting with a verb (*Take her to the vet* or even *Fuck you*);
(7)   Sentences having an adjective with an intensifier dependent (*thoroughly impressed, sooooo beautiful*);
(8)   Sentences having a determiner and adjective combination (*a joint one*, **the** best **little** *motel*, **the** most **common**);
(9)   Sentences with a numeral and an explanation of what it stands for (*22,600 - Number of planes carrying unscreened cargo* or *57 - Percentage of Republican federal judges*) – this is captured by a numeral-punctuation-noun sequence in the dependency tree;
(10) Incomplete sentences without verbs (*Good local bikeshop* or noise sentences like *Groups: alt.animals.cat*);
(11) More complex sentences with proper nouns as subjects and also with noun phrases consisting of an adjective and a noun (*As the leaders like to boast, the Mujahedeen is a family affair*);
(12) Sentences with two or more adjectives describing the same object (*Cheese was falling off, so oily and greasy*);
(13) Shorter sentences containing interjections (*Er, no?* or *Hey guys*);
(14) Sentences with sequences of three connected verbs (a small cluster; e.g. *That* **means making** *yourself* **known**);

(15) Questions with chains of two verbs (usually an auxiliary or modal one and a regular verb, as in *Do you have this information?* or *Should I get a balcony?*);

(16) Sentences with two proper nouns as heterogeneous subjects or objects (*Paul and Judy are on board*);

(17) Incomplete sentences with proper nouns and punctuation (cf. Cluster 2) (*Bill –* or *Dale,*);

(18) Non-interrogative sentences with two verbs linked (cf. Cluster 15) (*We do need a formal plan* or *He will learn to really like this*);

(19) Sentences with chains of two verbs and a lined adverb (*You should reply ASAP*);

(20) Non-standard 'sentences' (http://news.bbc.co.uk/2/hi/programmes/this_world/4446342.stm or *Andrew Edison@ENRON*);

(21) Shorter sentences with a single verb and a dependent adverb (*i completely agree*);

(22) Sentences with a determiner + noun combination (*There is no delivery*);

(23) Incomplete sentences with numerals (*4:00 pm–6:00 pm* or *08/03/2000 11:35* or - *Act 5, Scene 1*);

(24) Complex sentences with adverbial clauses (*Open your image while it's still in color*);

(25) Sentences with prepositional phrases (*He was drinking like a fish* or *put her back in the room*);

(26) Sentences consisting only of nouns (non-proper) and punctuation (*Thanks!!*);

(27) Sentences with pronouns and determiners (*She said a shower would be grand*);

(28) Shorter sentences with pronouns and adjectives (*It is spectacular* or *He's great*);

(29) Sentences that have verbs with pronouns as direct objects (*hurt him* or *hire you*)

As can be seen, the clustering captures many distinct sentence types in English. This results in a fairly detailed syntax model, with sentences having structural similarities grouped together.

5　Conclusion and Future Work

In this paper, an adaptation of the graph2vec model for the processing of dependency trees commonly used in NLP was proposed. In particular, we proposed several modifications of the WL relabeling process taking into account the properties of dependency trees. In the experimental part of our work, a vector space was constructed for the SynTagRus and EWT dependency trees, and clustering of this space was carried out. In the course of linguistic interpretation of the clusterings, some patterns were revealed. Different clusterings capture different structural types of sentences, with relatively larger numbers of clusters resulting in more detailed syntax models. It is also worth mentioning that the proposed methods result in models that completely isolate the syntax layer of the language system, considering it separately from the lexical and semantic layers.

The proposed approach creates new opportunities for further research into syntax. In the future, we plan to identify the common patterns of vector spaces of dependency trees for different languages. Also, a promising research direction is the use of syntactic embeddings in machine translation, paraphrasing, and a number of NLU problems (intention classification, etc.) to prove the efficiency of such embeddings.

References

1. Narayanan, A., et al.: graph2vec: learning distributed representations of graphs. arXiv preprint arXiv:1707.05005 (2017)
2. Collobert, R., et al.: Natural language processing (almost) from scratch. J. Mach. Learn. Res. **12**, 2493–2537 (2011)
3. Shervashidze, N., Schweitzer, P., van Leeuwen, E.J., Mehlhorn, K., Borgwardt, K.M.: Weisfeiler-Lehman graph kernels. J. Mach. Learn. Res. **12**, 2539–2561 (2011)
4. Kubler, S., McDonald, R., Nivre, J., Hirst, G.: Dependency Parsing. Synthesis Lectures on Human Language Technologies, vol. 2. Morgan and Claypool Publishers, San Rafael (2009). https://doi.org/10.2200/S00169ED1V01Y200901HLT002
5. spaCy. https://spacy.io/. Accessed 21 Apr 2019
6. Russian language models for spaCy. https://github.com/buriy/spacy-ru. Accessed 21 Apr 2019
7. De Marneffe, M.-C., et al.: Universal Stanford dependencies: a cross-linguistic typology. In: Proceedings of the 9th International Conference on Language Resources and Evaluation (LREC), pp. 4585–4592 (2014)
8. Dyachenko P.V., et al.: Modern state of the deeply annotated corpus of Russian texts (SynTagRus) [Sovremennoye sostoyaniye gluboko annotirovannogo korpusa tekstov russkogo yazyka (SinTagRus)], National Corpus of the Russian language: 10 years of the project. Proceedings of the V.V. Vinogradov Institute of Russian language. [Sbornik «Natsional'nyy korpus russkogo yazyka: 10 let proyektu». Trudy Instituta russkogo yazyka im. V.V. Vinogradova.], Moscow, vol. 6, pp. 272–299 (2015)
9. Silveira, N., et al.: A gold standard dependency corpus for English. In: Proceedings of the Ninth International Conference on Language Resources and Evaluation, LREC-2014 (2014)
10. Arthur, D., Vassilvitskii, S.: k-means++: the advantages of careful seeding. In: Proceedings of the Eighteenth Annual ACM-SIAM Symposium on Discrete Algorithms. Society for Industrial and Applied Mathematics (2007)
11. Pedregosa, F., et al.: Scikit-learn: machine learning in Python. JMLR **12**, 2825–2830 (2011)
12. Rousseeuw, P.J.: Silhouettes: a graphical aid to the interpretation and validation of cluster analysis. Comput. Appl. Math. **20**, 53–65 (1987). https://doi.org/10.1016/0377-0427(87)90125-7
13. Lawson, R.G., Jurs, P.C.: New index for clustering tendency and its application to chemical problems. J. Chem. Inf. Comput. Sci. (1990). https://doi.org/10.1021/ci00065a010

Recognition of Parts of Speech Using the Vector of Bigram Frequencies

Stanislav Khristoforov, Vladimir Bochkarev[✉], and Anna Shevlyakova

Kazan Federal University,
Kremlyovskaya Street 18, 420008 Kazan, Russian Federation
stnslv91@gmail.com, vbochkarev@mail.ru, anna_ling@mail.ru

Abstract. This paper describes how to automatically recognize parts of speech and other grammatical categories of a word such as gender and number. Unlike some previous works, the vector of syntactic bigram frequencies (including the considered word) is used as the source data for recognition of parts of speech and the grammatical categories. Data on frequencies of syntactic bigrams were obtained from the Russian sub-corpus of Google Books Ngram. We used part–of–speech tags available in Google Books Ngram, as well as data on parts of speech and grammatical categories of words obtained from the electronic dictionary Open Corpora. To train the model, we selected words from the list of 100.000 most frequent words that don't have homonyms and are found in both Google Books Ngram and Open Corpora. A multilayer perceptron with an output layer of the softmax type was used as a recognizer. The vector of frequencies of syntactic bigrams including the test word and one of the 10.000 most frequent words was at the inputs of the network. The neural network was trained by the criterion of minimum cross–entropy. When recognizing parts of speech on the test sample, the average recognition accuracy was 99.1%. Nouns and verbs were recognized best of all (with the accuracy of 99.77% and 99.62%, respectively). The recognition accuracy of the word number was 99.61%. The achieved recognition accuracy of the word gender was substantially lower, it was just 91.9%.

Keywords: Part of speech recognition · Neural networks · Google Books Ngram · Bigram frequency

1 Introduction

Google Books Library is a project developed by Google which offers a digital database of scanned books [1]. It is an extra–large corpus of texts which opened new opportunities for processing data and provided a solid foundation for research studies. It contributed much to linguistics, social and computer science. The updated version of the corpus was created in 2012. The main feature of the new version was part–of–speech tagging [2].

Google Books distinguishes 12 parts of speech: '.' (punctuation marks), ADJ (adjective), ADP (preposition and postposition), ADV (adverb), CONJ (conjunction), DET (determiner and article), NOUN (noun), NUM (numeral), PRON

© Springer Nature Switzerland AG 2020
W. M. P. van der Aalst et al. (Eds.): AIST 2019, CCIS 1086, pp. 132–142, 2020.
https://doi.org/10.1007/978-3-030-39575-9_13

(pronoun), PRT (particle), VERB (verb), X (a catchall for other categories such as abbreviations or foreign words) [2].

The quality of such POS tagging is disputable and was discussed, for example, in [3]. Although POS homonymy is relatively uncommon in English, there are a lot of 1–grams in the corpus tagged as corresponding to different parts of speech. For example, the word "I" is marked up as corresponding to 12 parts of speech. In total, there are 4858 1–grams in the English corpus, tagged as corresponding to 12 parts of speech, 18753 1–grams marked up as corresponding to 11 parts of speech, and 21018, 26508 and 32495 1–grams tagged as corresponding to 10, 9 and 8 parts of speech, respectively.

The Google Books Ngram Russian corpus lacks the tag "pronouns" which is questionable. There are 10 parts of speech in the Russian corpus distinguished by Google Books Ngram. At that, 71 1–grams are tagged as corresponding to 10 parts of speech, 219 1–grams are marked up as referring to 9 parts of speech and 497 1–grams are tagged as corresponding to 8 parts of speech. Thus, Google Books Ngram POS–tagging contains a lot of errors and needs to be improved. It would be useful to check the POS tagging of words important for this or that research. Since the corpus lacks free access to the texts of its books, this can be done only based on information on the frequencies of the studied n–grams.

There were attempts to solve the problems of POS tagging with the help of neural network methods. For example, in [4,5], the authors proposed a model of a morphological analyzer based on a bidirectional recurrent neural network. Words represented at the char–level were used as input data. Convolutional models also proved to be good at solving the POS–tagging problem. In [6], they were applied to text corpora from social networks. Like in previous articles, the authors used symbolic representations of words.

Traditional methods are also used to solve this problem. For example, [7] proposed linguistic rules to improve recognition quality of several POS–taggers. The work objective is to build neural network models performing recognition of parts of speech and grammatical categories of gender and number of Russian words. The main difference of this work from others of that kind is that information on syntactic bigrams [8] including the studied word (i.e. the word distribution) is used as input data for POS recognition. Thus, there are no source texts available for the analysis but there is statistical information that characterizes word distribution. The Russian language was chosen for research because it is a language with developed morphology and the problem of parts of speech classification and recognition has always been contentious in Russian linguistics.

The data on the frequencies of the syntactic bigrams were obtained from the Russian subcorpus of Google Books Ngram [2]. The data on POS, gender and number of words were obtained from the OpenCorpora electronic thesaurus which includes 115 grammemes, 390273 lemmas and a great amount of word forms [9,10]. It is the largest and most detailed Russian thesaurus; therefore, its data were chosen for our research.

The problem related to homonymy disambiguation is formulated differently in our work than in other works concerning POS–taggers. Parts of speech of homonyms can be defined by their distribution in the sentence. There are no source texts in our case. However, there is statistical information about frequencies of bigrams that contain the given word. Since the corpus is very large, the word can have different meanings being used in it and can refer to different parts of speech.

To solve the problem, one should estimate the percentage of use of the word in this or that part of speech. To perform such estimation, it is desirable to have a trained recognizer which allows distinguishing parts of speech of non–homonymic words. It provides an opportunity to obtain some "ideal" frequency distribution of words corresponding to this or that part of speech and obtain more effective estimation of the percentage of use of the homonym in different variants. However, it is beyond the scope of this work.

The problem of POS recognition remains an acute issue in modern theoretical linguistics. The results of this research based on the distributive approach can contribute much to it. Accurate POS recognition yields more reliable results in corpus–based studies.

2   Model Description

To create the model for estimating the above described properties of the word, we choose the architecture of the classical direct distribution network. Each of the networks performing POS recognition and recognition of the words gender and number was a four–layer perceptron which contained 64, 128 and 128 neurons in the first three layers, respectively. Dimension of the last fourth layer is equal to the number of classes of the training set. There are 3 classes to recognize gender, 2 classes to recognize number, 17 and 12 classes to recognize POS using POS tagging in accordance with the electronic thesaurus OpenCorpora or Google Books Ngram, respectively. Relu was selected as a function to activate all hidden layers. Using this activation function allows you to achieve a sparse activation [11] and thus obtain a model of higher approximation power. The output layer was activated by the softmax function [12]. This ensures that the sum taken over all the outputs of the neural network equals 1, which allows us to consider the output data as a probability distribution on the target classes. In addition, since the dimension of the input vector is high (it is 20.000 in our case, as explained below), the number of weights between the input and the first hidden layer is also high, which can lead to overfitting the model, especially if there are few examples in the training sample. To prevent overfitting, a dropout layer with a parameter 0.3 was created before the first hidden layer [13]. Random 30% of data from the input vector is "cut out" at each training iteration. Thus, an analogue of stochastic regularization is created in the training process. In the testing mode, the information is no longer cut out of the data vector. However, the output data of this layer are corrected in a certain way to avoid distortion in the process of their use by subsequent layers.

The model was trained using the backpropagation method based on gradient descent. Implementation of the gradient method is presented in [14]. The Nadam algorithm is a combination of ideas described in [15,16]. It selects the optimal value of the training speed parameter taking into account the previous values of the gradient norm, and also saves some analogue of the inertia of the solution point movement, which allows you to train deep models for a fairly small number of iterations. In addition, during training, the norm of the resulting gradient can be artificially limited from above so that, when it hits the "steep" parts of the target function (this often occurs when moving in ravine functions), the decision point does not move too far from the study area near the local minimum.

There are approximately 70 thousand examples in the training sample. Therefore, stochastic gradient descent is used to ensure high performance while training the models. The entire training sample is divided into a fixedsize set of batches, and the network weights and error gradient aggregation are updated after all the examples from the batches have been provided. To solve the described problems, the size of the batch was chosen to be 128. Thus, there are about 550 updates of the network scales during one epoch.

The training data are a set of frequency vectors of syntactic 2–grams taken from the Google Books Ngram database [2]. Syntactic bigrams are units of syntactic structures denoting a binary relation between a pair of words in a sentence. In each syntactic bigram, one word is called the head, and the other is its dependent [8]. The number of unique 1–grams in the Russian corpus is 4863328. In this case, the size of the full matrix of syntactic 2–grams frequencies will be 48633282. Therefore, the use of all data from the database when training the neural network is a computationally time–consuming task. We use only the frequencies of syntactic bigrams, which are combinations of the test word with one of the sets of reference words. The word is represented by two frequency vectors of the syntactic bigrams of the Wx and xW type, where 'W' is the tested word form and 'x' is one of the reference forms (the word order in the syntactic bigrams shows which word is the head and which is its dependent). In this work, 10 thousand of the most frequent words occurring in the corpus (namely 1–grams consisting of the letters of the Russian alphabet and, probably, one apostrophe) are used as reference words. Thus, the word is described by a pair of vectors of dimension 10000. Both vectors are concatenated to obtain a single 20000–dimensional vector representing the word.

3   Results

To train and test the POS recognizer, we used a sample of the most frequent words found in the Russian subcorpus of Google Books Ngram. We also selected 70 thousand of most frequent words occurring in the electronic thesaurus Open-Corpora. Then, we selected word forms (from the described samples) which occurred in the thesaurus only in one part of speech and didn't have homonyms belonging to other parts of speech The OpenCorpora thesaurus distinguishes 17 parts of speech: ADJF—adjective (long), ADJS—adjective (short), ADVB—adverb, COMP—comparative, CONJ—conjunction, GRND—gerund, INFN—verb (infinitive), INTJ—interjection, NOUN—noun, NPRO—pronoun noun,

NUMR—numeral, PRCL—particle, PRED—predicative, PREP—preposition, PRTF—participle (long), PRTS—participles (short), VERB—verb (finite verb forms). The training and test samples included 62790 word forms; 80% of them were used for training the neural network model, and 20%—for testing. The trained neural network showed an accuracy of 99.10% on the test sample (the error rate was 0.9%). The recognition accuracy of different parts of speech is shown in Fig. 1 and Table 1. It is clear that good results can only be obtained for classes (for parts of speech) for which there are a sufficient number of examples in the training set. Therefore, the figure and the table show the results only for those parts of speech for which there were at least 400 examples in the sample. The table also shows the number of word forms of each part of speech in the sample.

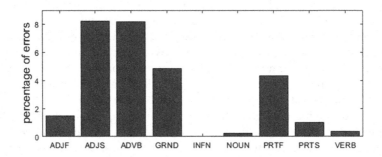

Fig. 1. The percentage of POS recognition errors, %

As it can be seen, the best recognition accuracy is achieved for verbs (infinitives and finite forms) and nouns which represent relatively large groups of words. The worst result was obtained for a small group of words, namely for short adjectives and adverbs. In general, the Spearman correlation coefficient between the number of examples and recognition accuracy was 0.70. On the one hand, this value is relatively large. It indicates a significant influence of the number of examples in the sample on the quality of the model training and the resulting recognition accuracy. On the other hand, it is obvious at this level of correlation that there are other important factors determining the number of POS recognition errors.

Table 2 presents detailed data on error frequencies of various kinds, which allows you to analyze which parts of speech are most often confused.

The errors are unavoidable since the problem of POS recognition remains one of the major contentious issues in linguistic because there are certain limitations and controversial points in the traditional classification of parts of speech. Traditionally, there are three types of grammatically relevant properties of words that differentiate parts of speech: semantic, formal and functional. Sometimes the properties of one part of speech can coincide with the properties of the other part of speech making the classification disputable.

Table 1. The percentage of POS recognition errors

Part of speech	Percentage of Errors, %	Number of examples
ADJF	1.47	14496
ADJS	8.21	686
ADVB	8.16	412
GRND	4.86	745
INFN	0	2039
NOUN	0.23	32348
PRTF	4.34	2312
PRTS	1.02	1032
VERB	0.38	8310

Table 2. Statistics of the POS recognition results on the test sample. The columns show POS from the test sample according to the OpenCorpora Thesaurus, rows show the recognition results in percent

	ADJF	ADJS	ADVB	GRND	INFN	NOUN	PRTF	PRTS	VERB
ADJF	98.53	2.24	3.06	0.69	0	0.14	4.34	0	0.13
ADJS	0.10	91.79	1.02	0	0	0	0	0.51	0
ADVB	0.14	1.49	91.84	0	0	0.05	0	0	0.06
COMP	0	0	0	0	0	0	0	0	0
CONJ	0	0	0	0	0	0	0	0	0
GRND	0.03	0	0	95.14	0	0	0	0	0
INFN	0	0	0	0	100.00	0.02	0	0	0
INTJ	0	0	0	0	0	0.02	0	0	0
NOUN	0.72	4.48	1.02	0.69	0	99.77	0	0.51	0.13
NPRO	0	0	0	0	0	0	0	0	0
NUMR	0	0	0	0	0	0	0	0	0
PRCL	0	0	0	0	0	0	0	0	0
PRED	0	0	0	0	0	0	0	0	0
PREP	0	0	0	0	0	0	0	0	0
PRTF	0.48	0	0	2.08	0	0.02	95.66	0	0
PRTS	0	0	1.02	0	0	0	0	98.98	0.06
VERB	0	0	2.04	1.39	0	0	0	0	99.62

The obtained results showed that the following parts of speech are confused most often in Russian. The gerund in Russian is confused with verbs because it designates action and, thus, can have the same syntactic distribution as the verb. Some linguists even consider the gerund to be a type of predicate [16]. The model also confuses adjectives with participles and vice versa. Participles and adjectives have some common features: they describe object features and are used as a definitive in a sentence. In Russian, participles used attributively agree with the noun gender, number, case, and animateness. Thus, being a morphological form of a verb, participles have such syntactic distribution and such a set of inflexive categories that bring them closer to adjectives. The analysed classification of POS included long and short adjectives. They can be confused because they belong to the category of adjectives and share common properties. However, the percentage of errors made by the model while determining long and short adjectives is relatively low which means that their syntactic distribution has differences and the model can distinguish them.

To train and test the model of word number recognition, 56253 1–grams were selected which occur in the OpenCorpora only in the singular or plural forms. The trained model was tested on the sample of 11250 examples. The number of errors on the test sample was 44 (or 0.39%). According to the results of manual testing, it can be concluded that the tested words occur in the corpus only in the singular or plural since the thesaurus is incomplete. For example, noun forms are confused in the genitive case, homonymous forms of nouns are also confused. In general, it can be said that the model response is correct, but the described examples were recognized incorrectly because the corpus is incomplete.

Gender recognition model was trained and tested on the sample including 32340 1–grams. We selected word forms which occur in the Open Corpora Thesaurus only in one gender. The sample included 47% of masculine, 33.6% of feminine and 19.4% of neuter word forms. The test sample included 6468 examples. In this case, the proportion of erroneously recognized word forms was significantly higher—525 examples (or 8.12%). Figure 2 shows the proportion of recognition errors for word forms of different gender.

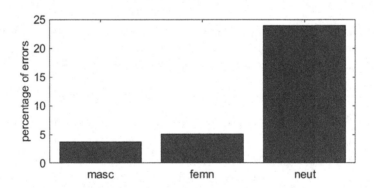

Fig. 2. The percentage of gender recognition errors, %

Masculine word forms are recognized with the best accuracy (8% error). It may seem that gender recognition accuracy depends on the number of words in the training sample (the more words are used, the more accurate the recognition is). However, the situation is more complicated. Table 3 shows details of different types of errors.

Table 3. Statistics of gender recognition results on the test sample. The columns show gender of words from the test sample according to the OpenCorpora Thesaurus, rows show the recognition results in percent

	masc.	fem.	neut.
masc.	96.27	4.47	18.19
fem.	2.64	94.95	5.72
neut.	1.08	0.58	76.09

Taking into account the number of words of each gender, it can be seen that neuter gender is taken for masculine gender and vice versa in most cases. There are 261 of such cases in a test sample, approximately one half of all the errors. Gender variability is typical of a number of lexemes and their classes throughout the history of the Russian literary language which can be expressed both morphologically and syntactically. For example, nouns with certain suffixes, indeclinable nouns, abbreviations, composites can be of both masculine and neuter gender. Gender is considered an inherent quality of nouns. However, it affects the forms of other related words in the process of agreement. In Russian these words are adjectives, pronouns-adjectives, numerals and participles. Past tense verb forms with the suffix -l- are close to this group [17]. Neuter and masculine adjectives have the same endings in some objective cases which can also cause gender recognition errors.

The described method can be used to recognize the grammatical characteristics of words not only by vectors of syntactic bigram frequencies, but also by vectors of frequencies of simple bigrams (pairs of words which are immediate neighbours in a sentence). Comparative testing of such recognizers is of interest since frequencies of syntactic bigrams can be obtained using the corpus markup, but frequencies of bigrams can be obtained without any preliminary analysis of the texts.

The vectors of bigram frequencies for the training and testing of the recognizer were constructed as described above, completely analogous to the vectors of syntactic bigrams. The accuracy of the trained recognizer was 98.74%. Thus, the results are slightly worse (the error probability is 1.4 times higher) than that obtained using the vectors of syntactic bigrams. This is quite expected as the set of syntactic bigrams that include the studied word characterizes it better/more than the set of bigrams that contain this word. Nevertheless, POS recognition can be performed quite effectively using the frequencies of the bigrams. The obtained accuracy of word gender recognition by bigram frequencies was 99.64%, the accuracy of gender recognition was 91.44%.

4   Conclusion

In this paper, we tried to solve the problem of POS tagging and recognition of the grammatical categories of gender and number by word distribution.

Vectors of syntactic bigram frequencies including the given word were used as the source data. The data on the frequencies of the bigrams were obtained from the Russian sub–corpus of Google Books Ngram. The built neural–network models showed an accuracy of 99.1% when recognizing a part of speech, 99.61% when recognizing the word number and 91.88% when recognizing the word gender.

Usually, morphological taggers are tested on specially selected sets of marked texts. For example, the test set for the Russian language was developed at morphoRuEval–2017 [18,19]. The testing results of such morphological taggers as pymorphy2, mystem, UDpipe, rnnmorph are described in [20–25]. According to these results, the accuracy of POS recognition of the modern taggers on the test set is 96–98%.

The recognition accuracy value obtained in our work exceeds these values and equals 99.1%. However, they cannot be directly compared. First, the tagger described in this work and the listed morphological taggers solve fundamentally different problems. In the first case, the decision is made based on the statistical information on the word distribution; in the second case, it is made based on the main function of a word in a sentence. Secondly, high accuracy values of the tagger considered in this work were obtained on a sample words free of homonymy. Nevertheless, the accuracy of the obtained results is good. The proposed POS recognizer can possibly be used in conjunction with traditional morphological taggers to improve recognition accuracy.

These neural network models can be used to improve markup in large diachronic corpora if there is no access to the source texts. Detailed analysis of recognition errors can contribute to linguistic theory.

Acknowledgments. This research was financially supported by the Russian Government Program of Competitive Growth of Kazan Federal University, state assignment of Ministry of Education and Science, grant agreement №34.5517.2017/6.7 and by RFBR, grant №17–29–09163.

References

1. Michel, J.-B., Shen, Y.K., Aiden, A.P., Veres, A., Gray, M.K., et al.: Quantitative analysis of culture using millions of digitized books. Science **331**(6014), 176–182 (2011)
2. Lin, Y., Michel, J.-B., Aiden, E.L., Orwant, J., Brockman, W., Petrov, S.: Syntactic annotations for the Google Books Ngram corpus. In: Li, H., Lin, C.-Y., Osborne, M., Lee, G.G., Park, J.C. (eds.) 50th Annual Meeting of the Association for Computational Linguistics 2012, Proceedings of the Conference, vol. 2, pp. 169–174. Association for Computational Linguistics, Jeju Island (2012)
3. Bochkarev, V.V., Solovyev, V.D., Wichmann, S.: Universals versus historical contingencies in lexical evolution. J. Roy. Soc. Interface **11**, 20140841 (2014)
4. Plank, B., Søgaard, A., Goldberg, Y.: Multilingual part-of-speech tagging with bidirectional long short-term memory models and auxiliary loss. CoRR abs/1604.05529 (2016)
5. Anastasyev, D., Gusev, I., Indenbom, E.: Improving part-of-speech tagging via multi-task learning and character-level word representations. CoRR abs/1807.00818 (2018)

6. Meftah, S., Semmar, N.: A neural network model for part-of-speech tagging of social media texts. In: Calzolari, N., et al. (eds.) LREC, European Language Resources Association (ELRA) (2018)
7. Jatav, V., Teja, R., Bharadwaj, S., Srinivasan, V.: Improving part-of-speech tagging for NLP pipelines. CoRR abs/1708.00241 (2017)
8. Sidorov, G., Velasquez, F., Stamatatos, E., Gelbukh, A., Chanona-Hernández, L.: Syntactic dependency-based N-grams as classification features. In: Batyrshin, I., Mendoza, M.G. (eds.) MICAI 2012. LNCS (LNAI), vol. 7630, pp. 1–11. Springer, Heidelberg (2013). https://doi.org/10.1007/978-3-642-37798-3_1
9. Dictionary OpenCorpora. http://opencorpora.org/dict.php. Accessed 04 May 2019
10. Bocharov, V., Bichineva, S., Granovsky, D., Ostapuk, N., Stepanova, M.: Quality assurance tools in the OpenCorpora project. In: Computational Linguistics and Intellectual Technologies. Papers from the Annual International Conference "Dialogue", vol. 10(17), pp. 107–115. RGGU, Moskow (2011)
11. Glorot, X., Bordes, A., Bengio, Y.: Deep sparse rectifier neural networks. In: Gordon, G., Dunson, D., Dudik, M. (eds.) AISTATS, JMLR Proceedings, vol. 15, pp. 315–323. JMLR.org (2011)
12. Goodfellow, I., Bengio, Y., Courville, A.: Deep Learning. Adaptive Computation and Machine Learning. MIT Press, Cambridge (2016)
13. Srivastava, N., Hinton, G., Krizhevsky, A., Sutskever, I., Salakhutdinov, R.: Dropout: a simple way to prevent neural networks from overfitting. J. Mach. Learn. Res. **15**(1), 1929–1958 (2014)
14. Ruder, S.: An overview of gradient descent optimization algorithms. CoRR abs/1609.04747 (2016)
15. Kingma, D., Ba, J.: Adam: a method for stochastic optimization. arXiv preprint arXiv:14126980 (2014)
16. Botev, A., Lever, G., Barber, D.: Nesterov's accelerated gradient and momentum as approximations to regularised update descent. CoRR abs/1607.01981 (2016)
17. Russkaia korpusnaia grammatika. Rod. http://rusgram.ru/%D0%A0%D0%BE%D0%B4. Accessed 04 May 2019
18. Lyashevskaya, O., Bocharov, V., Sorokin, A., Shavrina, T., Granovsky, D., Alexeeva, S.: Text collections for evaluation of Russian morphological taggers. J. Linguist./Jazykovedný casopis **68**(2), 258–267 (2017). https://doi.org/10.1515/jazcas-2017-0035
19. Sorokin, A., et al.: MorphoRuEval-2017: an evaluation track for the automatic morphological analysis methods for Russian. In: Computational Linguistics and Intellectual Technologies. Proceedings of the International Conference "Dialogue 2017", vol. 16, no. 23, pp. 297–313. RGGU, Moskow (2017)
20. Sboev A., Gudovskikh D., Ivanov, I., Moloshnikov, I., Rybka, R., Voronina, I.: Research of a deep learning neural network effectiveness for a morphological parser of Russian language. In: Computational Linguistics and Intellectual Technologies. Proceedings of the International Conference "Dialogue 2017", vol. 16, no. 23, pp. 234–244. RGGU, Moskow (2017)
21. Anastasyev, D., Andrianov, A., Indenbom, E.: Part-of-speech tagging with rich language description. In: Computational Linguistics and Intellectual Technologies. Proceedings of the International Conference "Dialogue 2017", vol. 16, no. 23, pp. 2–13. RGGU, Moskow (2017)
22. Anastasyev, D., et al.: Improving part-of-speech tagging via multi-task learning and character-level word representations. CoRR. abs/1807.00818 (2018)

23. Morphological analyzer for Russian and English languages based on neural networks and dictionary-lookup systems. https://github.com/IlyaGusev/rnnmorph. Accessed 04 May 2019
24. Straka, M., et al.: UDPipe: trainable pipeline for processing CoNLL-U files performing tokenization, morphological analysis, POS tagging and parsing. In: Calzolari, N., et al. (eds.) LREC. European Language Resources Association (ELRA) (2016)
25. Korobov, M.: Morphological Analyzer and Generator for Russian and Ukrainian Languages. CoRR. abs/1503.07283 (2015)

Formalization of Medical Records Using an Ontology: Patient Complaints

Eduard Klyshinsky[1]([✉])(iD), Valeriya V. Gribova[2](iD), Carina Shakhgeldyan[3,4](iD),
Elena A. Shalfeeva[2](iD), Dmitry B. Okun[2](iD), Boris I. Geltser[4](iD),
Tatiana A. Gorbach[4](iD), and Olesia D. Karpik[5](iD)

[1] National Research University "Higher School of Economics", Moscow, Russia
`klyshinsky@hse.ru`
[2] Institute of Automation and Control Processes, FEB, RAS, Vladivostok, Russia
`{gribova,shalf}@iacp.dvo.ru, okdm@dvo.ru`
[3] Vladivostok State University of Economics and Service, Vladivostok, Russia
`carinashakh@gmail.com`
[4] Far East Federal University, Vladivostok, Russia
`boris.geltser@vvsu.ru, tagorbachdv@gmail.com`
[5] Keldysh Institute of Applied Mathematics, RAS, Moscow, Russia

Abstract. Medical records contain a textual description of such important information as patients' complaints, diseases progress and therapy. An extraction of this information could allow starting with processing information stored in medical databases. In this article we introduce a short description of a medical ontology storing information on patients' complaints. We also describe an algorithm that uses this ontology for extraction of claims from texts of medical records. The algorithm combines both syntactic properties, and peculiarities, of a text and connections between diseases' properties and their values. The algorithm corrects syntactical mistakes according to the hierarchical information from the ontology. The resulting algorithm was proved on 3000 clinical records of Department of Neurosurgery of FEFU.

Keywords: Medical record · Term extraction · Information retrieval

1 Introduction

Modern medical information systems allow storing structured information on patient diseases, disease flow and outcome, therapy etc. The stored information could be divided into non-, weakly and strongly structured. Some of medical information systems divide a medical record into a set of formalized fields of numerical nature or defined by dictionaries; there couldn't be fields with text descriptions in free textual form. Such information could be processed using mathematical methods: statistics, machine and deep learning, time series analysis, etc. Weakly structured information - images, electroencephalograms, cardiograms, ultrasonography, MRI - could be also stored in such medical information

The reported study was funded by RFBR according to the research project # 18-29-03131.

© Springer Nature Switzerland AG 2020
W. M. P. van der Aalst et al. (Eds.): AIST 2019, CCIS 1086, pp. 143–153, 2020.
https://doi.org/10.1007/978-3-030-39575-9_14

systems and mathematically processed. E.g. using Fourier and wavelet analysis, neural networks - for purposes of feature extraction and further analysis. That is why weakly and strongly structured medical information could be used as a source for data mining in such areas as causal relations among functional, metabolic and genetic status of patients, their way of living, therapy, morbidity and survival probability, etc.

However, the most part of medical records are stored in unstructured text form - fortunately, in data media. That is the case of patients' complaints, medical background, patient diary, etc. Russian medical information systems formalize the basic clinical records structure with the content of mentioned fields being tables or text in a free form. An extreme case of electronic medical records is a folder with documents stored by a doctor. Such information cannot be processed directly without preprocessing. The mentioned files or fields of information system should be processed by a fact extractor in order to construct a formalized representation.

This paper describes a new method for formalization of patient complaints written in a textual form. The method constructs a preliminary representation of a complaint text on the base of syntactic analysis; the final representation is refined using an ontology that describes relations among notions of a selected domain - neurological diseases.

2   Overview

In 1986 US National Library of Medicine started a project of Unified Medical Language System (UMLS) - a comprehensive publicly accessible collection of electronic dictionaries, thesauri and ontology. (For more historical information on UMLS consult [1].) It consists of Metathesaurus (hierarchy of terms collected from many vocabularies), Semantic Network (relationships among these terms and their categories), and SPECIALIST Lexicon and Lexical Tools (a large syntactic lexicon of biomedical and general English combined with natural language processing tools). The 2018AB Metathesaurus release (November 2018) contains approximately 3.82 million concepts and 14 million unique concept names from 207 source vocabularies [2]. Metathesaurus vocabulary (Medical Subject Headings - MeSH) was translated into 15 languages including Russian [3].

Currently, this collection is used in several big projects, e.g., MetaMap - a program for information extraction from medical texts [4]. The MetaMap method of a medical text processing consists of two stages: (1) natural language processing of the medical text and fact extraction, and (2) notions refinement. The first stage starts with tokenization and finishes with a syntactic analysis. It includes an acronym/abbreviation identification, multi-word terms extraction, and their identification in dictionaries. One word or phrase could have several entries in thesaurus or dictionaries. That's why MetaMap provides word and multi-word term sense disambiguation. It maps terms combinations and then filters improbable combinations out. However, some terms keep an unresolved ambiguity. The result of the medical text processing is a tagged text with a link to Metathesaurus.

MetaMap system allows extracting and indexing such terms as pharmacy names, their quantity, disease names, body parts, etc. Using these results, one could conduct such processing as a text clustering and classification, text indexing, search results ranking, word sense disambiguation, logical inference, historical information extraction and processing, etc. Authors claim that in 2014 the quality of Medical Text Indexer, based on MetaMap, was as high as 0.6 for precision, 0.56 for recall, and 0.58 for F1 [5].

The MetaMap system was adopted for the Russian language [6]. Authors use Exactus system for the natural language processing, Russian translation of UMLS dictionary, the State Register of Medicinal Remedies and some other local resources. The main purpose of this project is a logical inference for diagnosis of chronic diseases. Using of machine learning algorithms allows the authors to increase the precision of fact extraction up to 82% for a severity of disease and 99% for a flow of disease.

Another big system here is cTAKES [7]. Apart of UMLS, this system uses such extra corpora as SHARP and ShARe. This project also aimed in detection of body part and severity of disease. Authors use SVM method for increasing quality of extraction; however, the final quality is not much better than MetaMap [8].

Currently, the machine learning approach to fact extraction is very common in the medical text processing for different languages. Authors of MedInX system [9] proclaim about 95% for precision and recall in extraction of medical terms from Portuguese text. The same concept is used in TAKELAB system presented at SemEval 2015 devoted to medical texts processing [10]. The authors of [11] use the information extraction approach for a pictorial visualization of an electronic medical record. They extract names of disease and sick body parts, and draw this information on an abstract image of body.

In our project we could not use machine learning techniques [12] since the aim of the project is to find and connect terms stored in the Database of Terms and Observations, described below. That is why we are not using methods of Named Entity Recognition [13] but extracting terms from dictionary and then trying to find correct connections among them. We also could not use common sence ontology [14] or thesauri [15] since they do not containing successfull terminology.

3   Database of Terms and Observations

The main part of our system is the Database of Terms and Observations [16]. It is formed on the basis of the ontology with the same name, designed according to the best modern practice [17,18]. This ontology contains definitions of all concepts classes and consists of two main types of medical terms descriptions – symptoms and factors. Symptoms characterize the current functional state of a patient, and factors are used to describe the risks of various diseases. Symptoms and factors can be combined into logically related groups to make them easier to navigate. Symptoms can be simple or composite. The first ones are described by name and a set of qualitative, numeric, or interval values. Composite symptoms have a name and characteristics. Each characteristic is also described by its name and a set of possible values (qualitative, numerical or interval). Each medical term may have several synonyms. The Database consists of about 1500 Symptoms, about 1300 Features and more than 25500 Values.

The "Symptom" section of this database contains several groups of symptoms: "Complaints", "Objective examination", "Laboratory and instrumental examination". In this article, we use only a group of symptoms of "Complaint", describing the subjective feelings of the patient, characterizing its current functional status and the state of individual systems: digestive, respiratory, circulatory, nervous system, etc. This group contains a subgroup "General complaints", which includes those that occur in many diseases (dizziness, weakness, nausea, sweating, etc.). The subgroup "Pain" is a part of the subgroup "General complaints", it includes the symptoms: headache, back pain, neck pain, sore throat, etc. For most of composite symptoms of the group "Complaints" are used characteristics such as "localization", "severity", "cause", "time of occurrence", "intensity", "frequency", etc. The characteristics of the group "Pain" also include the additional characteristics: "irradiation", "increasing", "increasing", etc.

A fragment of the Database of Terms is presented at Fig. 1. The group of symptoms Pains includes Back Pain that has synonyms Spinal Pain and Lumbodynia. The symptom Back Pain has such characteristics as Localization (possible values are Lumbar Region and Lumbar Spine) and Amplification (possible values are Deep Breath and In a Strong Position).

We have also designed an ontology for description of a medical record. It includes personal information, patient complaints, disease flow, patient history, results of general examination, clinical diaries, and diagnosis. Patient complaints are described as symptoms and their values. When forming a medical record, the symptoms and their values defined in the Database of Terms and Observations are used.

The section of the Database in neurology was created according to 3000 anonymized medical records from the Department of Neurosurgery of Far Eastern Federal University. The information resources described above are stored in a heterogeneous repository developed by the authors [16].

Therefore, the aim of the current project is to create a software tool for information retrieval from patients' medical records. The output is a fragment of the Database of Terms and Observations describing the current state of the patient according to the analysed text of complaints.

4   Algorithm of Term Extraction and Connection

The main idea of the algorithm for patients' complaints extraction is to run a syntactic analysis and correct its results according to the hierarchy of the Database of Terms and Observations. During syntactic analysis, every extracted terms is considered as a single syntactic unit. The algorithm consists of two stages. The Stage 1 conducts the pre-syntactic analysis of a medical record and consists of patient's complaints extraction, tagging, and terms extraction. The Stage 2 conducts syntactic analysis of extracted terms and the whole text of the complaint, and correction of resulting dependency tree.

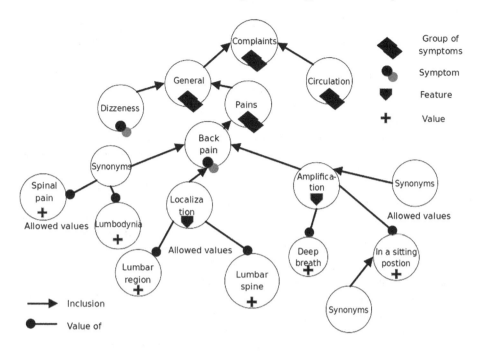

Fig. 1. A fragment of the Database of Terms and Observations

Now we will consider the algorithm in details starting with **Stage 1**.

Step 1 – Patients' Complaints Extraction. There are several options here. The first is to load a text of a complaint from a specific field of a healthcare information system or a file in a specialized format. However, some medical records are stored in a plain text format, therefore, we should extract complaints from such text. Patient's complaints are usually placed close to the beginning of a document and start with such phrases as *Complaints, Chief Complaints* (*Жалобы, Жалобы при поступлении, На момент осмотра жалобы*) etc. In some cases a patient doesn't present any problems. Such situation is described by phrases like *No complaints, Doesn't present any problems, Doesn't present age-related problems* (*жалоб нет, жалоб не предъявляет, жалоб по возрасту не предъявляет, жалоб не предъявляет в связи с тяжестью*) etc. Thus, we should exclude paragraphs with the former formulae.

Step 2 – Complaint Tagging. The extracted text of a complaint is tagged according to a selected dictionary. Here we consider a word as a sequence of Cyrillic characters. We used PyMorphy2 library [19] with OpenCorpora dictionary [20]. Thus, every token of a text is converted into its most frequent initial form.

Step 3 – Terms Extraction. As it was mentioned before, the Database of Terms and Observations stores a hierarchy of such medical terms as Symptoms (P_i),

Group of Symptoms (P_i), names of Features (C_{ik}) and their Values (H_{ijk}). The aim of Step 3 is to find such entities and their synonyms in a complaint.

An entity consists of one or several words. In order to find them in a text, the Database of Terms and Observations was converted into a prefix tree; all words are converted to the most frequent initial form using PyMorphy. Thus, the task of terms extraction could be reformulated as finding of the longest sequence of words from text in the prefix tree. Note that a term in a text can be ambiguous, i.e. it's stored in the Database several times in several branches. E.g., a *significant pain* could belong to quite any body part. Thus, such term should be connected to several branches in the Database. Depending on the extracted branch or branches, a word or a multi-word term will be marked as a Symptom, Feature, Value or their combination.

The result of Stage 1 is a sequence of words and word combinations, some of them are tagged as terms and store a list of connected entities from the Database of Terms and Observations. The aim of the **Stage 2** is to convert such sequence into a hierarchy according to our Database and to disambiguate the list of connections with branches in the Database.

The Russian language has a lot of peculiarities, therefore, there could be mistakes in connecting words using a parser. E.g., a value could be connected to a wrong feature or symptom since such syntactical connection is more probable than a correct one. The opposite problem is the syntactical incorrectness of some connections between features and their values. We cannot follow [4] and construct a Cartesian product of all possible connections filtering them out according to some common sense rules because of complexity of such calculations. Thus, at the Stage 2 we conduct a syntactic analysis and gradually correct its results according to the Database of Terms and Observations.

Step 1 – Replacing Multiword Expressions. Here we are going to reduce the complexity of a sentence by joining multiword terms into one token. Selected multiword terms are parsed by UDPipe parser [21]. The parser returns a dependency tree with a main word as a root. We copy all tags of this word to the new constructed word. It makes a sentence shorter, thus, the parser processes it faster and more correctly on the step 2. All non-term or one-word terms are also tagged by UDPipe by reasons of uniformity. All term nodes store a list of routes from the root node of the Database to an entity node with this term; because of ambiguous nature of terms, there could be several routes in the list.

Step 2 – Parsing. At this step we parse the shortened sentence consisting of one- and multiword terms, non-terms, and service words. The parsing allows us not to consider all possible connections, but syntactically reasonable ones only. However, the parser makes some mistakes in syntactic connections, thus we have to correct them at

Step 3 – Correction of the Tree. The parser fails since a sentence could be syntactically ambiguous. The Russian language of medical records has a very specific structure; some sentences don't contain a verb in their structure but have a long sequence of patient complaints. Therefore, a constructed dependency tree could have mistakes in terms hierarchy; two connected Values, a Feature that is child of its Value etc. Thus, we have to correct the dependency tree constructed at Step 2 according to the Database of Terms and Observations. We state here that two terms could be connected in a dependency tree only if they have a direct path in the Database. Otherwise, a tree should be corrected according to the following rules.

Rule 1. If a parent node has a lower level in the Database hierarchy that its child node, we should swap these two nodes. E.g., if a Feature becomes a child of a Value, we should exchange them in the tree.

Rule 2. If there is no direct path from a child node to its parent node, we should move the child node to the parent's level of a dependency tree. This situation is possible if a Value was incorrectly subordinated to another Feature or Symptom.

Rule 3. Both parent and child nodes are Features but the child node has an upper level of hierarchy in the Database. In this case we should swap these two nodes.

Rule 4. There are two nodes connected to the same parent; one node has a lower level of hierarchy in the Database than the other one and there is a direct path between them. In this case we should subordinate the first node to the second one.

These four rules should be applied to a tree until the process converges. The rules move an incorrectly subordinated node to an upper level or subordinate it to a node with a higher level of hierarchy that could be a parent for this node. A node could be moved up several times until it finds a possible parent. Otherwise, it will be moved to the highest level and stays here. Such situation is possible if our Database is incomplete and this term should have at least one extra reference.

We can use the resulting tree to filter out ambiguous connections of terms with the Database of Terms and Observations. As it was mentioned above, a term could be placed in several nodes of the Database. We should leave a connection if it satisfies one of the following conditions.

Condition 1. The parent node has a route to a node in the Database that has a direct path to the connected node, i.e., the stored route starts from one of the routes of the parent node.

Condition 2. A child node has a route to a node in the Database that has a direct path to the connected node, i.e., the stored root starts on of the child's route.

Condition 3. There is at least one neighbouring node with a route that coincides with the stored route.

These conditions state that we could leave only those routes which construct a correct hierarchy. All other routes should be eliminated.

The resulting dependency tree could be used to construct a subtree from the Database of Terms and Observations. According to Conditions 1–3, the resulting tree contains only such routes which can be connected: a parent contains a root (ideally, one root only) that points to a parent node for a child's route. The resulting subtree could be used as a formal description of a patient complaints.

5   An Example of Processing

Let us consider the following example, that was correctly analyzed: «*При поступлении жалобы на выраженную боль в поясничном отделе позвоночника, ограничение движений в пояснично-крестцовом отделе, нарушение ходьбы, нарушение функции тазовых органов*» (*At admission to hospital complains to an significant pain in lumbar spine, restraint of movement in lumbosacral spine, ambulation disorder, dysfunction of pelvic organs*). After selecting of terms, the sentence has the following structure: *При поступлении жалобы на [выраженную] [боль] в [поясничном отделе позвоночника], [ограничение движений] в [пояснично-крестцовом отделе], [нарушение ходьбы], [нарушение функции тазовых органов]*» (*At admission to hospital complains to an [significant] [pain] in [lumbar spine], [restraint of movement] in [lumbosacral spine], [ambulation disorder], [dysfunction of pelvic organs]*). Once terms are joined and parsed, the dependency tree looks like following (translations are given in parenthesis).

```
2 поступлении (admission to hospital)
  1 при (at)
    3 жалобы (complaints)
      6 боль (pain)
        4 на (to)
        5 выраженную (significant)
      8 поясничном отделе позвоночника (lumbar spine)
        7 в (in)
      10 ограничение движений (restraint of movement)
        12 пояснично-крестцовом отделе (lumbosacral spine)
          11 в (in)
        14 нарушение ходьбы (ambulation disorder)
        16 нарушение функции тазовых органов (dysfunction of pelvic organs)
```
Nodes 3, 6 and 8 have the following connected routes:

3 – Жалобы

6 – Жалобы|Общие|Боли

8 – Жалобы|Общие|Боли|Боль в спине|Составной признак|Локализация|
Тип возможных значений|Качественные значения|поясничный отдел позвоночника

Nodes 3 and 8 are at the same level of the tree, route 8 starts with route 6. Thus, the node 8 should become a child of 3.

In the same way nodes 14 and 16 should be moved to one level up. The resulting tree looks as below.

```
2 поступлении (admission to hospital)
  1 при (at)
    3 жалобы (complaints)
      6 боль (pain)
        4 на (to)
        5 выраженную (significant)
      8 поясничном отделе позвоночника (lumbar spine)
        7 в (in)
     10 ограничение движений (restraint of movement)
       12 пояснично-крестцовом отделе (lumbosacral spine)
         11 в (in)
       14 нарушение ходьбы (ambulation disorder)
         16 нарушение функции тазовых органов (dysfunction of pelvic organs)
```

The resulting subtree of the Database of Terms and Observations is following.

```
Жалобы
  Боли
    Боль в спине
      Выраженность|выраженная
      Локализация|поясничный отдел позвоночника
    Опорно-двигательная система
      Снижение подвижности сустава
      Локализация|пояснично-крестцовый отдел
    Нервная система
      Нарушение походки
      Нарушение функции тазовых органов
```

6 Results of Experiments

For our experiments, we have used 3000 of medical records describing a flow of such neurological diseases as microplasia, brain concussion, stenosis etc. Records were sampled and anonymized by neurologists of Department of Neurosurgery of FEFU. All of these records containing patient complaints part, however, some of them were short, did not contain or deny any complaints: "Вялость, отсутствие аппетита, *тошноту*", "Жалоб не предъявляет". An average size of a clinical record is about 1500 word tokens.

100 records of our collection were randomly sampled, processed by our algorithm, and manually checked. The selected records contain 1610 word tokens of patient complaints, 1093 of these tokens were recognized as 711 terms. The examination of selected medical records demonstrates that precision of our algorithm for terms extraction is 0.96 while recall is 0.82; resulting f1-measure is 0.79. The precision was calculated as a relation of number of extracted terms

to number of correctly extracted terms; recall was calculated as the ratio of the number of extracted terms to the number of manually tagged terms in a text.

The resulting precision for terms connection was as low as about 0.5. By precision of terms connection we understand the ratio of the number of correctly connected terms (syntactically connected in text, hyerarcically connected in the Database, with a term type and meaning being properly define) to the number of connections in the resulting graphs. Such low results could be equally explained by incompleteness of the Database and the set of Conditions changing an hierarchy of nodes in a tree. Some terms could be presented in the Database, however they are not liked to any possible position in the Database. E.g Localization characteristics could be connected to any kind of pain; entering a new kind of pain a doctor could forget to link this new pain to the Localization.

For example, let us consider a sentence "жалобы: на боль в поясничном отделе позвоночника, иррадиирущую в левую ногу, онемение левой стопы, усиление боли при физической нагрузке" which has a misspelling "иррадиирущую", that is why one of the word wasn't properly detected as a term. The resulting graph is presented below.

```
жалобы
  боль
    на
    поясничном отделе позвоночника
      в
    иррадиирущую
      левую ногу
        в
    усиление
      боли
    при физической нагрузке
```

The term "при физической нагрузке" should be connected with the token "боли" but not "боль". Moreover, the term "усиление" was not disambiguated since the parent token have no semantic tags. Thus, there are 4 correctly attributed tokens out of 7 connections in the sentence (we are not considering prepositions here), i.e. the precision of connections is 4/7. In some cases our algorithm connects a node to a wrong parent if it's ambiguous. Thus, our algorithm for building subtree of the Database needs further improvements.

References

1. Selden, C.R., Humphreys, B.L.: Current bibliographies in medicine. https://www.nlm.nih.gov/archive/20040831/pubs/cbm/umlscbm.html. Accessed 25 Apr 2019
2. 2018AB UMLS Release Notes and Bugs Page. https://www.nlm.nih.gov/research/umls/knowledge_sources/metathesaurus/release/notes.html. Accessed 25 Apr 2019
3. MSHRUS (MeSH Russian) - Statistics. https://www.nlm.nih.gov/research/umls/sourcereleasedocs/current/MSHRUS/stats.html. Accessed 25 Apr 2019

4. Aronson, A.R., Lang, F.M.: An overview of MetaMap: historical perspective and recent advances. J. Am. Med. Inform. Assoc. **17**(3), 229–236 (2010)
5. Mork, J., Aronson, A., Demner-Fushman, D.: 12 years on - is the NLM medical text indexer still useful and relevant? J. Biomed. Semant. **8**, 8 (2017)
6. Shelmanov, A.O., Smirnov, I.V., Vishneva, E.A.: Information extraction from clinical texts in Russian. In: Computational Linguistics and Intellectual Technologies. International Conference «Dialogue», vol. 13, pp. 560–572 (2015)
7. Dligach, D., Bethard, S., Becker, L., Miller, T.A., Savova, G.K.: Discovering body site and severity modifiers in clinical texts. J. Am. Med. Inf. Assoc. (JAMIA) **21**, 448–454 (2014)
8. Reatugui, R., Ratte, S.: Comparision of MetaMap and cTAKES for entity extraction in clinical notes. BMC Med. Inform. Decis. Mak. **18**(Suppl. 3), 74 (2018)
9. Ferreira, L.S.: Medical information extraction in European Portuguese - abstract of Ph.D. thesis. Universidade de Aveiro, Departamento de Electrónica, Telecomunicações e Informática, Aveiro (2011). https://www.researchgate.net/publication/220009587_Medical_Information_Extraction_in_European_Portuguese
10. Glavaš, G.: TAKELAB: medical information extraction and linking with MINERAL. In: Proceedings of of the 9th International Workshop on Semantic Evaluation (SemEval 2015) (2015)
11. Ruan, W., Appasani, N., Kim, K., Vincelli, J., Kim, H., Lee, W.-S.: Pictorial visualization of EMR summary interface and medical information extraction of clinical notes. In: Proceedings of 2018 IEEE International Conference on Computational Intelligence and Virtual Environments for Measurement Systems and Applications (CIVEMSA) (2018). https://doi.org/10.1109/CIVEMSA.2018.8439958
12. Tutubalina, E., Nikolenko, S.: Combination of deep recurrent neural networks and conditional random fields for extracting adverse drug reactions from user reviews. J. Healthc. Eng. **2017**, 9 (2017)
13. Giorgi, J., Bader, G.: Towards reliable named entity recognition in the biomedical domain. BioRxiv, p. 526244 (2019)
14. Fellbaum, C.: WordNet: An Electronic Lexical Database. MIT Press, Cambridge
15. Loukachevitch, N., Dobrov, B.: RuThes linguistic ontology vs. Russian wordnets. In: Proceedings of Global WordNet Conference GWC-2014, Tartu (2014)
16. Грибова В. В., Москаленко Ф. М., Шахгельдян К.И. и др.: Концепция гетерогенного хранилища биомедицинской информации. In: Информационные технологии, Том 25, №2, М.: "Новые технологии", pp. 97–106 (2019)
17. Sowa, J.F.: Building, sharing, and merging ontologies. http://www.jfsowa.com/ontology/ontoshar.htm. Accessed 18 Jan 2009
18. Лукашевич Н. В., Б. В. Добров. "Проектирование лингвистических онтологий для информационных систем в широких предметных областях." Онтология проектирования, **5**, 1 (15)
19. Korobov, M.: Morphological analyzer and generator for Russian and Ukrainian languages. In: Khachay, M.Y., Konstantinova, N., Panchenko, A., Ignatov, D.I., Labunets, V.G. (eds.) AIST 2015. CCIS, vol. 542, pp. 320–332. Springer, Cham (2015). https://doi.org/10.1007/978-3-319-26123-2_31
20. Бочаров В. В., Алексеева С. В., Грановский Д. В., Остапук Н. А., Степанова М.Е., Суриков А. В.: Сегментация текста в проекте "Открытый корпус". In:Proceedings of "Компьютерная лингвистика и интеллектуальные технологии" Материалы конференции «Диалог» 2012, Вып, 11, no. 1, pp. 51–60 (2012)
21. Straka, M., Straková, J.: Tokenizing, POS tagging, lemmatizing and parsing UD 2.0 with UDPipe. In: Proceedings of the CoNLL: Shared Task: Multilingual Parsing from Raw Text to Universal Dependencies, Vancouver, Canada, August 2017

A Deep Learning Method Study of User Interest Classification

Alexey Malafeev$^{(\boxtimes)}$ ⓘ and Kirill Nikolaev ⓘ

National Research University Higher School of Economics, Nizhny Novgorod, Russia
aumalafeev@hse.ru, kinikolaev@edu.hse.ru

Abstract. In this paper, a deep learning method study is conducted to solve a new multiclass text classification problem, identifying user interests by text messages. We used an original dataset of almost 90 thousand forum text messages, labelled for ten interests. We experimented with different modern neural network architectures: recurrent and convolutional, as well as simpler feedforward networks. Classification accuracy was evaluated for different architectures, text representations and sets of miscellaneous parameters.

Keywords: Text classification · User interests · Machine learning · Neural networks · Deep learning · Feedforward neural networks · Convolutional Neural Networks · Long short term memory

1 Introduction

As social networks have experienced a recent growth in popularity, analysis of user data is of utmost importance. In particular, extracted information about user interests and preferences can be used in recommendation systems and for target advertising.

Existing systems modeling user interests or preferences typically make use of personal data (e.g. location, marital status, age, online communities, etc.) or metadata (search or watch history, user ratings). However, if such data is inaccessible, one can take advantage of various NLP techniques to find out what users like by extracting this information from user-authored texts, such as blog posts, forum or social network messages. This is the task that will be tackled in this paper. It can be formally stated as follows:

Given a set of documents $D = \{d_1, \ldots, d_n\}$ and a finite set of classes $C = \{c_1, \ldots, c_m\}$ corresponding to interests that are expressed in these documents, with a constraint that each document d has only one corresponding class c, find a classification function f that can determine for any given pair $< d, c >$ whether the document corresponds to the interest: $f : D \times C \rightarrow \{0, 1\}$

As can be observed, we have made several simplifying assumptions. For instance, real-life user messages can contain none or, contrarily, more than one interest. Moreover, we consider a finite set of ten interests (anime, food, art, games, books, music, nature, travel, films, and football), whereas in reality there might be a larger or even infinite set of classes. However, these assumptions are made only at the current stage; later, when the problem is solved with satisfactory results, the task can be made more difficult.

© Springer Nature Switzerland AG 2020
W. M. P. van der Aalst et al. (Eds.): AIST 2019, CCIS 1086, pp. 154–159, 2020.
https://doi.org/10.1007/978-3-030-39575-9_15

2 Related Work

Since the task considered in this paper is standard multiclass classification, let us review the existing methods of text classification used on a range of tasks, such as spam filtering, sentiment analysis, authorship attribution, etc. The current state-of-the-art is deep learning-based text classification, with methods like Convolutional Neural Networks (Zhang et al. 2015), Recurrent Convolutional Neural Networks (Lai et al. 2015), Long-Term Short Memory (LSTM), C-LSTM and Attention-Based LSTM (Zhou et al. 2015; Zhou et al. 2016) architectures. Our work focuses primarily on the deep learning methods.

For text representation, embeddings such as doc2vec (Le, Mikolov 2014) have been extensively used in recent years. In general, distributed representations perform better than traditional features in the sense that embeddings are of significantly smaller dimensionality and are modelling semantic relations between language units, although these representations are not interpretable.

As for methods of inferring user interests or preferences described in literature, there are some that are based on analyzing web bookmarks [Jung and Jo 2003], search query logs (Limam 2010), user profiles (Cantador and Castells 2011), and mobile device logs (Zhu et al. 2015). However, there have been very few works that attempt to apply NLP techniques to extracting user interests. For example, Stone and Choi (2013) used a SVM sentiment classifier on Twitter messages about a particular smartphone design. An older paper (Paik et al. 2001) considers extracting user profiles from emails, which also includes some information on interests and preferences. In (Liu et al. 2013), an attempt is made to improve the accuracy of a recommendation system by extracting opinions from customer reviews written in Chinese. Our work differs from existing research in the following: (1) the task is formulated as a multiclass text classification problem with a set of 10 classes, using a large new dataset of user messages written in Russian; (2) modern deep learning methods are used.

3 Dataset and Text Representation

We automatically extracted a collection of 89,844 text documents not shorter than 150 characters from web forums on different topics (forum.kinopoisk.ru and www.livelib.ru). Each of the documents was assigned a label of one of the ten classes, or "interests": anime, food, art, games, books, music, nature, travel, films, and football. This set of categories had been built in accordance with the empirical analysis of user forum activity. The dataset is very unbalanced: there are more common classes like books (33%) or football (26%) and rare classes like nature and art (about 1% each). Obviously, this poses an additional challenge for our task.

We held out a validation set and a test set of 1000 texts each (uniform distribution, 100 texts of each class). All initial experiments were conducted on the validation set. We also applied several stages of preprocessing: removing stop words and non-Cyrillic tokens, then text lemmatization with MyStem (https://tech.yandex.ru/mystem/).

The following types of text representation were used. Firstly, on the 87,844 texts of the training set, we learned a doc2vec (Le, Mikolov 2014) model. Secondly, we developed two sets of 10 complex features each, based upon most informative words and character trigrams for each class using the Positive Pointwise Mutual Information (PPMI) (Bouma 2009) metric, which allows selecting the most pertinent words and character trigrams that can be used in class prediction. Some examples are listed below:

FOOD: *аппетит* (appetite), *кофе* (coffee), *блюдо* (dish), *гарнир* (garnish); *'ыр '*, *'кеф'*, *'ощ '*, *'сыр'* (parts of food product names)

BOOKS: *слог* (author style), *паланик* (palahniuk), *книжный* (book), *роман* (novel); *'афк'*, *'гюг'*, *'фка'*, *'руэ'*, *'дюм'*, *'элш'*, *'амю'*, *'юма'*. (parts of famous writers' names)

The actual feature values are computed as the proportion of class-specific elements among all words/trigrams in a given text, for every class. Thus, we obtain 20 values: 10 for words predicting each class and 10 for character trigrams. At the next stage, we used the different feature sets and their combinations to determine the contribution of each type of feature and find the best configuration.

4   Experiments

We tested the following methods on our text classification problem: Feedforward NN, CNN, and LSTM. For comparison, we also used some classic machine learning algorithms: Naïve Bayes and Random Forest. The code was implemented in Python using the scikit-learn and Keras frameworks. The best results for a number of methods can be observed in Table 1. We used micro accuracy on the test set as the performance metric.

Table 1. Performance of classic machine learning and deep learning algorithms

Model	Classification accuracy (micro)
Random forest classifier – 100 e	0.595
Gaussian Naïve Bayes	0.627
CNN: Conv1D 26	0.761
Feedforward: Dense 32 – Dense 32	0.771
Bidirectional LSTM 100 – Dense 100 – Dropout 0.2	0.785
LSTM 100 – Dense 200 – Dense 100 – Dropout 0.2	**0.786**

As expected, the deep learning algorithms outperformed the classic machine learning ones by a large margin. The best accuracy of 0.786 was shown by the LSTM algorithm. This result can be considered fairly good, because there are ten classes and the data is quite noisy and unbalanced. It can also be seen that the CNN model performed about 2% worse than LSTM. This can be explained by the fact that our representation ignores structure: all features used are structurally unrelated.

We also attempted to compensate for dataset imbalance by weighting the classes. As expected, this significantly improved accuracy, as shown in Table 2 for two models.

Table 2. The effect of class weighting on classification accuracy

Model	Weighted	Unweighted
LSTM	**0.786**	0.743
Feedforward	**0.771**	0.723

As far as text representation is concerned, we experimented with different feature combinations: the 300-dimensional doc2vec, the same vector plus 10 word-based features or 10 character trigram-based features, as well as the concatenation of all 320 feature values. The results are shown in Table 3. It can be seen that the combination of all features yields the best accuracy. Also, informative words are more effective than character trigrams.

Table 3. The effect of different text representations

Model	Doc2vec	Doc2vec, words	Doc2vec, character trigrams	Doc2vec, words, character trigrams
LSTM	0.657	0.747	0.717	**0.786**
Feedforward	0.640	0.740	0.735	**0.771**

We also investigated the effect of restricting the number of class-specific words to consider when extracting the features. As in the case of class weighting, a general trend was observed for all models: the best result was given by fewer words (see Table 4).

Table 4. Number of class-predicting words by PPMI for each class and classification accuracy

Model	100	200	300
LSTM	**0.786**	0.757	0.749
Feedforward	**0.771**	0.741	0.738

Finally, we also built a confusion matrix to see which interests were most frequently misclassified. It can be seen from Table 5 that the most challenging categories were music and nature. Nature is often confused with art or travel, while music can be sometimes mistaken for almost any other class.

Table 5. Confusion matrix for the best-performing LSTM model

	Anime	Art	Books	Films	Food	Football	Games	Music	Nature	Travel
Anime	80	1	5	3	1	1	3	3	0	2
Art	3	74	2	1	0	0	0	3	5	4
Books	3	4	80	2	1	2	1	4	1	1
Films	3	8	4	78	0	0	1	2	0	0
Food	1	1	0	1	84	0	0	3	9	4
Football	1	0	1	2	0	88	3	1	0	2
Games	4	2	2	5	0	2	80	8	0	2
Music	4	1	2	5	2	4	5	72	3	1
Nature	0	7	1	0	6	0	3	0	72	7
Travel	1	2	3	3	6	3	4	3	10	77

Below is an example of a text that was misclassified:

Интересно всё. Каньоны - впечатляют. А Гранд Кэньон, естественно, больше всего. Необыкновенное ощущение от Леса Гигантов в парке Секвойя. И запах - секвойи.

(Everything is interesting. The canyons are impressive. And the Grand Canyon is, of course, the most. An extraordinary sensation from the Forest of Giants in Sequoia Park. And the smell of redwoods).

The ground truth label for this text was travel, but the system predicted nature. Indeed, this is a fairly ambiguous example, as the author writes about nature at the places he/she has traveled to. There are other similarly ambiguous texts in our data.

5   Conclusion and Future Work

In this paper, we have tackled the text classification problem of identifying user interests by text messages. We have conducted a number of experiments using the modern deep learning architectures. The highest classification accuracy (0.786) was achieved by a LSTM model using the doc2vec representation and twenty complex features based on class-predicting words and character trigrams. Additionally, we experimented with different ways to represent the data, and measured the impact of a few other parameters. Our dataset and code are freely available[1] for the community.

We believe that it should be possible to improve our results by further expanding the training and test data sets, especially for the underrepresented classes; by using more complex text representations and more advanced deep learning algorithms, as well as conducting detailed error analysis. Our solution can be used as a standalone module or as part of a more complex user interest identification algorithm.

---

[1] https://github.com/Pythonimous/forum-classifier.

Acknowledgements. This research was prepared within the framework of the Academic Fund Program at the National Research University Higher School of Economics (HSE University) in 2019 (grant No. 19-04-004) and of the Russian Academic Excellence Project "5–100".

References

Bouma, G.: Normalized (pointwise) mutual information in collocation extraction. In: Proceedings of GSCL, pp. 31–40 (2009)

Cantador, I., Castells, P.: Extracting multilayered communities of interest from semantic user profiles: application to group modeling and hybrid recommendations. Comput. Hum. Behav. **27**(4), 1321–1336 (2011)

Jung, J.J., Jo, G.-S.: Extracting user interests from bookmarks on the web. In: Whang, K.-Y., Jeon, J., Shim, K., Srivastava, J. (eds.) PAKDD 2003. LNCS (LNAI), vol. 2637, pp. 203–208. Springer, Heidelberg (2003). https://doi.org/10.1007/3-540-36175-8_20

Lai, S., Xu, L., Liu, K., Zhao, J.: Recurrent convolutional neural networks for text classification. In: Twenty-Ninth AAAI Conference on Artificial Intelligence (2015)

Le, Q., Mikolov, T.: Distributed representations of sentences and documents. In: International Conference on Machine Learning, pp. 1188–1196 (2014)

Limam, L., Coquil, D., Kosch, H., Brunie, L.: Extracting user interests from search query logs: a clustering approach. In: Workshops on Database and Expert Systems Applications, pp. 5–9. IEEE (2010)

Liu, H., He, J., Wang, T., Song, W., Du, X.: Combining user preferences and user opinions for accurate recommendation. Electron. Commer. Res. Appl. **12**(1), 14–23 (2013)

Paik, W., Yilmazel, S., Brown, E., Poulin, M., Dubon, S., Amice, C.: Applying natural language processing (NLP) based metadata extraction to automatically acquire user preferences. In: Proceedings of the 1st International Conference on Knowledge Capture 2001, October 22, pp. 116–122. ACM (2001)

Stone, T., Choi, S.K.: Extracting consumer preference from user-generated content sources using classification. In: American Society of Mechanical Engineers Conference (ASME 2013, August 4) (2013)

Zhang, X., Zhao, J., LeCun, Y.: Character-level convolutional networks for text classification. In: Advances in Neural Information Processing Systems, pp. 649–657 (2015)

Zhou, C., Sun, C., Liu, Z., Lau, F.: A C-LSTM neural network for text classification. arXiv preprint. arXiv:1511.08630 (2015)

Zhou, P., et al.: Attention-based bidirectional long short-term memory networks for relation classification. In: Proceedings of the 54th Annual Meeting of the Association for Computational Linguistics (Volume 2: Short Papers), vol. 2, pp. 207–212 (2016)

Zhu, H., Chen, E., Xiong, H., Yu, K., Cao, H., Tian, J.: Mining mobile user preferences for personalized context-aware recommendation. ACM Trans. Intell. Syst. Technol. (TIST) **5**(4), 58 (2015)

Morpheme Segmentation for Russian: Evaluation of Convolutional Neural Network Models

Lyudmila Maltina and Alexey Malafeev(✉)

National Research University Higher School of Economics, Nizhny Novgorod, Russia
lpmaltina@gmail.com, aumalafeev@hse.ru

Abstract. This paper is aimed at evaluating the performance of existing models of morphemic analysis for Russian based on convolutional neural networks. The models were trained on a relatively small amount of annotated training data (38,368 words). We tuned the hyperparameters to accommodate the harder task setting, which helped improve the accuracy of the model. In addition to testing 15 different configurations on the available test set, a new sample of 800 words containing roots that are missing in the training sample (e.g. neologisms and recent loan words) was manually created and annotated for morphemic structure (the new dataset is made available to the community). The effectiveness of the models was evaluated on this sample, and it turned out that the performance of the CNN models was much worse on this set (an almost 30% drop in word accuracy). We performed a classification of errors made by the best model both on the standard test set and the new one.

Keywords: Morpheme segmentation for Russian · Convolutional neural network · Model evaluation · Words with out-of-vocabulary roots · Error analysis · Parameter tuning

1 Introduction

A morpheme is the minimum significant unit of a language. Roots and derivational affixes contain information about the lexical meaning of a word, and relational affixes express grammatical meanings. This means that morphemic analysis can be used in tasks related to lexical semantics and morphology. Morphemic analysis allows one to obtain a vector representation (word embedding) of an out-of-vocabulary word using a combination of vector representations of its morphs (morpheme embeddings) [1–3]. Morphemic analysis helps to reduce the vocabulary size and is used in machine translation [4] and speech recognition [5]. Other areas of application of morphemic analysis are morphemic and derivational mark-up of corpora [6], search query expansion by using cognate words in information retrieval [7], and testing of morpheme parses carried out by students [8].

Many different methods have been used to tackle the task of automatic morphemic analysis of Russian words. Very recently, in 2018, convolutional neural networks were applied to the task, with results that can be considered state-of-the-art for Russian [16].

© Springer Nature Switzerland AG 2020
W. M. P. van der Aalst et al. (Eds.): AIST 2019, CCIS 1086, pp. 160–166, 2020.
https://doi.org/10.1007/978-3-030-39575-9_16

The authors tested various configurations and ideas, but the best performance was achieved by an ensemble of three CNN models with different random initializations: F1-score = 97.16 F1-score and 87.53 word accuracy (percentage of words analyzed correctly without taking into account the type of morphs).

In our paper, we use the models by Sorokin and Kravtsova on a much smaller amount of training data in order to see how much it will affect the performance. We use the hyperparameter values proposed by the authors of [16], but also tune the parameters to see if this can partly compensate for an almost twofold decrease in training data. These experiment results can be useful for working with languages with similar morphemic structures, but with less available labeled data than Russian (e.g. other Slavic languages). Additionally, we perform an error analysis of the best-performing model to try and gain insight into ways to improve it. Finally, we also evaluate the model on a new dataset that we created, which consists of words with roots that are not seen in the training set (neologisms, loan words, terms and other words not included in the morphemic dictionary), which is also followed by an error analysis.

2 Evaluation of Convolutional Neural Network Models

We tested the CNN models from [16] using data [15] made available by the authors. The dataset, compiled on the basis of Tikhonov's dictionary [9], is a large collection of 95,922 Russian words segmented for morphemes with indication of the type of each morph. We made sure the words in the training and test samples were not repeated, which was sometimes the case in the original dataset. The dataset was randomly divided into training, validation and test samples in the ratio of 40/30/30 (38, 368/28, 777/28, 777 words). As can be seen, we make the proportion of the training data much smaller than 75% used in [16], thus making the task significantly more difficult. All data and materials are available at [17]. Some characteristics of data are given in Table 1. We can observe that train, validation and test samples have similar characteristics, while the sample of words containing previously unseen roots shows some differences.

Table 1. Dataset characteristics

Sample	Prefixes	Roots	Suffixes	Endings	Linking morphs	Postfixes	Average number of morphs per word
Train	0.114	0.319	0.367	0.137	0.036	0.028	3.824
Validation	0.116	0.318	0.367	0.135	0.036	0.029	3.836
Test	0.116	0.318	0.366	0.136	0.036	0.028	3.829
Previously unseen roots	0.022	0.436	0.377	0.145	0.012	0.006	2.726

Firstly, we performed parameter tuning. The *nepochs* hyperparameter value was reduced to 10 epochs (instead of 75 by default), and *early_stopping* – to 5 (instead of 10) in order to speed up training. The following hyperparameters were chosen as the ones to be customized: the number of convolutional layers, the width of the filter, the number of filters, the number of dense output units, dropout rate and the number of ensembled models with different random initializations. We considered 15 combinations of values, including two combinations proposed by the authors of [16].

For size constraints, we will not list the evaluation results for all 15 configurations, but only for two proposed by the above authors (#1 and #4), and for two that scored best during our evaluation (#13 and #15). These results are given in Table 2, while the complete tables for all 15 hyperparameter combinations and their evaluation results are available at [17]. Note that for each performance metric (precision, recall, F1, word accuracy), we give three values: the first is the result on the validation sample, the second – on the test sample, and the third – on the sample with previously unseen roots (see Sect. 4 for details).

Thus, the best performance was shown by our two models, resulting in a 2% increase in word accuracy compared to the baseline. This was achieved by increasing the number of convolutional layers, reducing the dropout rate and using ensembles of 3 or 5 neural networks. Lower scores than those in [16] are explained by the use of a much smaller amount of training data and fewer epochs. However, it is interesting to see that with almost twice less data than in [16], the drop in performance is only about 5% (87.53

Table 2. Evaluation results

Model	Hyperparameters	Precision	Recall	F1-score	Word accuracy
#1	Convolutional layers: 3 width of filters: [5] filters: 192 dense output units: 64 dropout rate: 0.2 ensembled models with different random initializations: 1	0.943/0.943/0.764	0.957/0.956/**0.819**	0.95/0.949/0.79	0.787/0.783/0.516
#4	Convolutional layers: 3 width of filters: [5] filters: 192 dense output units: 64 dropout rate: 0.2 ensembled models with different random initializations: 3	0.956/0.956/0.786	0.954/0.954/0.802	0.955/0.955/0.794	0.804/0.805/0.531
#13	Convolutional layers: 4 width of filters: [5] filters: 192 dense output units: 64 dropout rate: 0.1 ensembled models with different random initializations: 3	0.962/**0.963**/0.784	0.956/0.956/0.809	0.959/**0.959**/0.796	0.823/**0.824/0.544**
#15	Convolutional layers: 4 width of filters: [5] filters: 192 dense output units: 64 dropout rate: 0.1 ensembled models with different random initializations: 5	0.962/0.962/**0.792**	0.956/0.956/0.804	0.959/**0.959**/0.798	0.822/0.823/0.536

word accuracy in [16] and 82.4 with our best model). Discussion of the results on the new dataset of words with 'unfamiliar' roots is given in Sect. 4.

3   Error Analysis

We also performed error analysis to gain insight into the possible ways to improve the performance of the system. From the words in the test sample that our best model made mistakes in, 100 words were randomly sampled. These 100 words were then divided into groups according to the presumed cause of the error. The most common ones are the following: influence of more frequent morphs, presence of low-frequency morphs, de-etymologization, abbreviations and morphological alternation. When analyzing the types of errors associated with historical changes in the morphemic structure of the word, an etymology dictionary [10] was used. The results of the error analysis are presented in Table 3.

As expected, it can be seen that most errors probably result from a lack of data. Indeed, the morphemic system of Russian is highly variable and inconsistent, so having sufficient training data becomes all the more crucial.

Table 3. Types of errors with comments and examples

Cause of the error, number of such errors (in parentheses)	Example (the correct segmentation is shown in parentheses)	Comment
influence of more frequent morphs (34)	*том/ат* (*томат*) *tom/at (tomat)*	The frequency of the morphs-*том*- (-*tom*-) and-*am*- (-*at* -) is greater than that of -*томат*- (-*tomat*-)
unseen or low-frequency morphs (under 15 entries) (28)	*спринтер* (*спринт/ер*) *sprinter (sprint/er)*	The root-*спринт*- (-*sprint* -) is not found in the training set
de-etymologization (16)	*о/град/и/ть/ся* (*оград/и/ть/ся*) o/grad/i/t/sya (ograd/i/t/sya)	Historically, this word used to have the root-*град*- *(-grad-)*, but now it is -*оград*- *(-ograd-)*
roots are abbreviations (5)	*тюз/ов/ец* (*т/ю/з/ов/ец*) *tyuz/ov/ets (t/yu/z/ov/ets)*	The word is derived from *ТЮЗ* (*TYuZ*), which is an abbreviation, so each letter represents a separate root
morphological alternation (3)	*лине/еч/н/ый* (*линееч/н/ый*) *line/ech/n/yy* (*lineyech/n/yy*)	The morph-*лин*- (-*lin*-) (*разлиновать*) (*razlinovat*) has allomorphs-*лине*- (-*line*-) and *лини*- (-*lini*-), which confuses the model
other (14)	*про/гулоч/н/ый* (*про/гул/оч/н/ый*) *pro/guloch/n/yy* (*pro/gul/och/n/yy*)	The morphs-*гул*- (-*gul*-) and-*оч*- (-*och*-) have high frequency, yet the model fails to segment them

4 Evaluation on Words with Previously Unseen Roots

It was also interesting to see how the CNN models would perform on a set of new words (neologisms, loan words etc.) with roots previously unseen in training. To this end, we created a new dataset. From the dictionaries [11, 12, 14] we selected words for which at least one of the roots was not seen in the training set. For these words, cognate words were added that were chosen using the service [13]. Using this method, a sample of 800 words was obtained, which included loan words (e.g. *буккроссинг, bukkrossing, bookcrossing*), terms (*аденозинтрифосфорный, adenozintrifosforniy*, adenosine-triphosphate), neologisms (*загуглиться, zaguglitsya*, be googled), words derived from proper names (*неогумбольдтианство, neogumboldtianstvo*, neo-Humboldtism). Since these words did not have gold standard morpheme analyses, we manually decomposed them. The new dataset is also available at [17].

The test results on this dataset were already shown in Table 1. It is worth noting that the performance of all models is much worse on this dataset. The best model (with the hyperparameters tuned by us) achieved 0.796 F1-score compared with 0.959 on the 'old' test set, and 0.544 (!) word accuracy versus 0.824.

The tested models were able to parse words with unfamiliar roots if affixes have high frequency. For example, words containing the following affixes were often analyzed correctly: postfix *-ся (-sya-)*, infinitive suffix *-ть- (-t-)*, participle suffix *-вш-(-vsh-)*, verb suffix *-и- (-i-)* (*аутсорсить, autsorsit*, to outsource), suffixes *-ств- (-stv-)* (*тьюторство, tyutorstvo*, tutorship), *-изм- (алармизм, alarmism*, alarmism), *-ист- (-ist-)* (*шекспирист, shekspirist*, researcher of Shakespeare), and *-ова-(-ova-)* (*сканировать, skanirovat, to* scan), prefixes*рас- (ras-)* (*распарсить, rasparsit*, to parse up) and *за-(za-)* (*забаговать, zabagovat*, to introduce bugs in code). On the other hand, the English-borrowed prefix *ре- (re-)* and suffix *-инг(-ing-)* were not recognized (as in *ремейк, remeyk*, remake, and *дайвинг, dayving*, diving), as these affixes have a low frequency in the training set. Upon the whole, we can conclude that the model works only satisfactorily (0.544 word accuracy) with words whose roots are not seen in the training set.

5 Conclusion and Further Work

Thus, we tested the existing CNN models with new parameter values to find that they are quite effective for an almost twice smaller amount of labeled training data (about 1% worse by F1-score and about 5% – by word accuracy). Interestingly, the results are much worse on a large sample of 800 words that we created, with 'unfamiliar' roots (loan words, neologisms, etc.), with an almost 30% drop in word accuracy. Also, we performed a linguistic classification of the errors made by the best model, both on the standard test set and on the new one. Prospects for research include using new architectures of neural networks, as well as applying automatic morphemic analysis (as well as morpheme-based embeddings) to more general NLP problems such as various text classification tasks.

Acknowledgements. This research was prepared within the framework of the Academic Fund Program at the National Research University Higher School of Economics (HSE University) in 2019 (grant No. 19-04-004) and of the Russian Academic Excellence Project "5–100".

References

1. Galinsky, R., Kovalenko, T., Yakovleva, J., Filchenkov, A.: Morpheme level word embedding. In: Filchenkov, A., Pivovarova, L., Žižka, J. (eds.) AINL 2017. CCIS, vol. 789, pp. 143–155. Springer, Cham (2018). https://doi.org/10.1007/978-3-319-71746-3_13
2. Sadov, M.A., Kutuzov, A.B.: Use of morphology in distributional word embedding models: Russian language case. In: Computational Linguistics and Intellectual Technologies: Proceedings of the International Conference "Dialogue 2018" (2018). http://www.dialog-21.ru/media/4554/sadovmapluskutuzovab.pdf
3. Arefyev, N.V., Gratsianova, T.Y., Popov, K.P.: Morphological segmentation with sequence to sequence neural network. In: Computational Linguistics and Intellectual Technologies: Proceedings of the International Conference "Dialogue 2018", pp. 85–95 (2018)
4. Fritzinger, F., Fraser, A.: How to avoid burning ducks: combining linguistic analysis and corpus statistics for German compound processing. In: Proceedings of the Joint Fifth Workshop on Statistical Machine Translation and Metrics MATR, pp. 224–234 (2010)
5. Karpov, A.A.: Models and program realization of Russian speech recognition based on morphemic analysis [Modeli I programmnaya realizatsiya raspoznavaniya russkoy rechi na osnove morfemnogo analiza], a Ph.D. thesis, Saint-Petersburg, 129 p. (2007)
6. Tagabileva, M.G., Berezutskaya, Yu.N.: Word-formation annotation of the Russian national corpus: aims and methods [Slovoobrazovatel'naya razmetka Natsional'nogo korpusa russkogo yazyka: zadachi I metody]. In: Computational Linguistics and Intellectual Technologies Papers from the Annual International Conference "Dialogue" [Komp'yuternaya lingvistika i intellektual'nye tekhnologii: Po materialam ezhegodnoy Mezhdunarodnoy konferentsii «Dialog»], vol. 9, no. 16, pp. 499–507. Russian State University for the Humanities [Rossiyskiy gosudarstvennyy gumanitarnyy universitet], Moscow (2010)
7. Bernhard, D.: Unsupervised morphological segmentation based on segment predictability and word segments alignment. In: Kurimo, M., Creutz, M., Lagus, K. (eds.) Proceedings of the Pascal Challenges Workshop on the Unsupervised Segmentation of Words into Morphemes, pp. 19–23 (2006)
8. Decomposition of Words [Razbor slov po sostavu]. http://www.morphemeonline
9. Dictionaries of the Russian language for downloading. The archives of the forum "Speak Russian" [Slovari russkogo yazyka dlya skachivaniya. Arkhivy foruma «Govorim po-russki»]. http://www.speakrus.ru/dict/
10. Fasmer, M.: Etymological dictionary of the Russian language [Etimologicheskiy slovar' russkogo yazyka]. In: 4 Volumes: Translated from German, 4th edn. Editorial Astrel', Moscow (2007)
11. Morpheme Segmentation for the Russian Language. https://github.com/kpopov94/morpheme_seq2seq
12. The Dictionary of Neologisms. Neologisms of the century [Slovar' neologizmov. Neologizmy veka]. https://russkiiyazyk.ru/leksika/slovar-neologizmov.html
13. Cognate Words [Odnokorennye slova] https://wordroot.ru
14. Dictionaries and encyclopedias. Orthographic dictionary by Lopatin, V.V. [Slovari I entsiklopedii. Orfograficheskiy slovar' Lopatina, V.V.]. https://gufo.me/dict/orthography_lopatin

15. Code for AINL2018 Paper Deep Convolutional Networks for Supervised Morpheme Segmentation of Russian Language. https://github.com/AlexeySorokin/NeuralMorphemeSegmentation
16. Sorokin, A., Kravtsova, A.: Deep convolutional networks for supervised morpheme segmentation of Russian language. In: Ustalov, D., Filchenkov, A., Pivovarova, L., Žižka, J. (eds.) AINL 2018. CCIS, vol. 930, pp. 3–10. Springer, Cham (2018). https://doi.org/10.1007/978-3-030-01204-5_1
17. Morpheme Segmentation for Russian: Evaluation of Convolutional Neural Network Model. https://yadi.sk/d/L3YrwGZAmW3Cug

Using Pre-trained Deeply Contextual Model BERT for Russian Named Entity Recognition

Eugeny Mukhin[✉]

Chelyabinsk State University, Chelyabinsk, Russia
eugeny.muhin@gmail.com

Abstract. Named entity recognition is an important part of the Information Extraction process (extracting structured data from unstructured or semi-structured computer-readable documents). To highlight in the text of people, organizations, geographical locations, etc., many approaches are used. Although, well-known bidirectional LSTM neural networks, show good results, there are points for improvement. Usually, the word embedding vector are used as the input layer, but the main disadvantage of the last vector models (word2vec, GLOVe, FastText) is that they do not consider the context of documents.

In this paper we present the effective neural network based on the deeply pre-trained bidirectional BERT model, which was introduced in the fall of 2018, in the task of named entity recognition for the Russian language. The BERT model, trained for a long time on large unannotated corpuses of texts, were used in our work in two modes: feature extraction and fine-tuning for the NER task. Evaluation of the results was carried out on the FactRuEval dataset and the BiLSTM network (FastText + CNN + extra) was taken as the baseline. Our model, built on fine-tuned deep contextual BERT model, shows good results.

Keywords: Named Entity Recognition (NER) · NLP · BERT · Deeply contextual model · Russian language

1 Introduction

1.1 Related Works

Information extraction (IE) is the task of automatic extraction of structured data from unstructured or semi-structured machine-readable documents. One of the important sub-tasks of IE is NER - Named Entity Recognition - the labeling mentions of people, organizations, geographical locations, etc.

Rule-Based and Statistical Models. The first approaches used to extract information from texts are the definition of rule systems (for example, regular expressions, filters). Systems typically use the GLR algorithm and additionally include tools for intermediate text processing (tokenizer, morphological analyzer). Showing rather high accuracy results (~95%), they have rather low recall rate and are laborious for creating and supporting. Also previously, hidden Markov models (HMM, MEMM) were widely used and further development into the CRF classification method [1]. Nowadays CRF is often used as the top layer of deep models.

© Springer Nature Switzerland AG 2020
W. M. P. van der Aalst et al. (Eds.): AIST 2019, CCIS 1086, pp. 167–173, 2020.
https://doi.org/10.1007/978-3-030-39575-9_17

Bi-directional LTSM. Recurrent neural networks for sequence processing were proposed in 1996 [2]. To solve the vanishing gradient problems on long sequences, a model of special memory cells and gates was introduced, which tracks the state of the important features through the entire sequence (LSTM) [3]. For the first time, to solve the NER problem, this method was proposed in 2003 [4], the Effective neural network model was introduced in 2008 [5] and was further developed in the work of 2011 [6]. In practice, these models require almost no hard manual work to feature engineering. Since 2015, a variety of network architectures have been introduced: BRNN + CNN [7], BLSTM-CRF [8], BLSTM-CNN [9]. The main component of these models is Bidirectional recurrent layer. To identify morphological-like features, a character-based convolutional layer is introduced. As a rule, all these models additionally use word embeddings which enhance their quality by using distributional semantics [10]. Mikolov proposed [11], trained on the GoogleNews corpus (about 100 billion words), the word2vec model built on the methods of CBOW (Continuous Bag of Word) and Skip-Gramm. Another recent work, GloVe [12], used SVD decomposition of word co-occurrence matrix, trained on the Wikipedia texts and web pages corpus (6 billion words). Based on these architectures, models achieved F1 scores close to 92% in dataset CONLL2003 [9].

1.2 Deeply Contextual Representation

The key disadvantage of this approach is that word vectors have a single context, and as a result, homonyms obtain the same vectors. To solve this problem researchers proposed various options for extracting features from context [13, 14]. Thus, in 2018, the work of ELMo was published, in which the output vectors are computed over the bidirectional LSTM network with convolutions of characters [15]. The model was trained on a corpus of 1 billion words and showed a significant increase in results in many tasks on the natural language processing.

In 2018, there appeared works based on a fine-tuning approach, as opposed to the extraction of features (feature-based) approach. With this method, first a model has a long preliminary training on a large corpus of unsupervised data, then the layers marked up for different tasks are added to it and there is an additional training of the entire model. Further work is based on the Transformer [16] architecture, which showed a significant improvement in performance in most NLP tasks, especially in machine translation. OpenAI GPT (Generative Pre-Training) [17] is a multi-layer model containing 12 Transformer blocks. The connections between the blocks are lined up from left to right. The model was trained on the BookCorpus for a month using 8GPU.

BERT. As you can (see Fig. 1), the GPT model has only one-way communication between Transformer blocks, while in the most token-level tasks, the context needs to be taken into account both on the left and on the right. Researchers at Google have corrected this flaw by releasing the model BERT (Bidirectional Encoder Representations from Transformers) in the fall of 2018 [18]. Two versions of the bidirectional encoder were released (see Table 1), while the large model is more powerful than GPT, the base one is the same size as the GPT. This allows you to compare two architectures. The model was trained on two tasks. In the first approach - 15% of all tokens was replaced as follows: in 80% of the MASK token, in 10% by a random word, and in the remaining 10% the token was left unchanged. The model should have predicted hidden tokens.

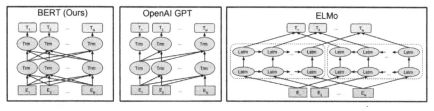

Fig. 1. Modern deeply contextual models.

Table 1. Size of models.

Model	L	H	A	P
GPT	12	768	12	110M
BERT base	12	768	12	110M
BERT large	24	1024	16	340M

The second task was to predict the next sentence, was the continuation of the first or not. In 50% of cases sentence there was the following sentence in the text, in the rest of cases the sentence was the randomly chosen from the corpus. BERT can be used both in the feature extraction mode (FE - feature extraction) and in the fine-tuning mode (FT - fine-tuning) to train the models for specific tasks. For the English language in the NER problem a new SOTA F1 92.6 (CONLL) was obtained.

In this paper, we will consider the use of BERT model for named entity recognition task for the Russian language.

2 Experiments

2.1 Dataset

We test our models on FactRuEval dataset [19]. In 2016, in the framework of the Dialog conference, a competition was held on the allocation of FactRuEval entities. For this, a separate dataset was prepared, which contained marked texts from the websites "Private Correspondent" and "Wikinews" in Russian. Dataset parameters are given in the Table 2.

2.2 Preprocessing

Data for training models was loaded using the built-in splitting of sentences into tokens. Lemmatization and stemming were not used to preserve morphological features. In variants of classic word embeddings (word2vec, GloVe, FastText), the replacement of

Table 2. FactRuEval dataset.

Set	Documents	Sentences	Tokens
Demo set	122	1769	30940
Test set	133	3138	59382

digital sequences with a special token DGT was used. For models that do not use a register, lower-case tokens were used (in order not to lose significant information, it is required to allocate the register into a separate attribute). To build a model, the sentences were padded to a fixed size with a token PAD. For compound tokens, the main label was assigned the first token, followed by the label X.

2.3 Evaluation

For evaluation our models we use the standard approach for information retrieval – Precision, Recall, F-score.

$$Precision = TP/(TP + FP) \tag{1}$$

$$Recall = TP/(TP + FN) \tag{2}$$

$$F = (2 * (Precision * Recall))/(Precision + Recall) \tag{3}$$

The results of the work were evaluated by python-version of the CONLL script [20] (originally Perl).

3 Results

As part of the work, we carried out experiments on various algorithms using neural network models. The model CNN + BiDirectional LSTM + emb was taken as the baseline (Base).

For the BERT model, we used both fine-tunning (BERT FT) and feature extraction (BERT FE) approaches. In our experiments, we use multilingual cased base model (104 languages, 12-layer, 768-hidden, 12-heads, 110M parameters). Architectures of models are given in the Table 3.

The peculiarity of the deep contextual models of ELMo and BERT is that in order to obtain a vector representation of words (feature extraction), it is required to submit to the model input not one word, but an entire sentence. At the output we get the vector of tokens that make up the sentence. These vectors were further used as the input layer of the model.

Base and BERT FE models were trained for 50 epochs, while BERT FT for 20. The results of our experiments are given in the Table 4.

Table 3. Architectures of models.

Layers\Model	Base	BERT FE	BERT FT
Word embeddings	tayga[a]	BERT	BERT
Additional layers	POS + Character CNN		
Hidden layers	BiLSTM		
Classification layer	FC or CRF		FC

[a]https://rusvectores.org/ru/models/

Table 4. FactRuEval results.

Model	Precision	Recall	F1
Bi-LSTM + CRF + Lenta [21]	83.8	80.84	82.10
Bi-LSTM-CRF + ELMo + fine-tuning [22]	83.19	85.41	84.29
CNN + emb (our baseline) + Bi-LSTM + CRF	82.31	77.33	79.74
Bi-LSTM + BERT FE	84.61	79.33	81.86
BERT FT	81.52	85.63	83.52

Analysis of common mistakes shows, that:

- It's difficult for the model to recognize coreferences, for instance «правящей правой **коалицией**», «все то же **правительство**», «целостности **страны**», «Сотрудники **газеты**»;
- The model misses geographical adjectives as LOC entity, for example in sentence «**Санкт-Петербургского** экономического форума»;
- The omission of words like «компания», «оператор», «сервис» in the phrases «компанией PepsiCo», «оператора телерекламы Видео Интернешнл», «сервис YouTube»;
- Confusion in geopolitical objects, the assignment of LOC entities to ORG and vice versa;
- Model also cannot recognize nominal names, for instance «большой семерки», «большой двадцатки».

4   Conclusion

The Named entity recognition task is one of the many practical applications of natural language processing and is used as a preliminary step in the task of facts extracting. Like in many other machine learning domains, high standards are set by neural networks. Until 2018, the main dominant architectures were various combinations of LSTM networks using vector word embeddings. Since the release of the ELMo model, the era of pre-trained deep context models has begun. This paper considered the use of the BERT model for named entity recognition applied to the Russian language.

As in many other downstreamed tasks of the NLP, the BERT model shows good results in Named Entity Recognition.

References

1. Lafferty, J., McCallum, A., Pereira, F.: Conditional random fields: probabilistic models for segmenting and labeling sequence data. In: Proceedings of the 18th International Conference on Machine Learning, pp. 282–289 (2001)
2. Goller, C., Kuchler, A.: Learning task-dependent distributed representations by back-propagation through structure. In: 1996 IEEE International Conference on Neural Networks, vol. 1, pp. 347–352. IEEE (1996)
3. Gers, F.A., Schmidhuber, J., Cummins, F.: Learning to forget: continual prediction with LSTM. Neural Comput. **12**(10), 2451–2471 (2000)
4. Hammerton, J.: Named entity recognition with long short-term memory. In: 2003 Proceedings of the Seventh Conference on Natural Language Learning at HLT-NAACL, pp. 172–175. Association for Computational Linguistics (2003)
5. Collobert, R., Weston, J.; A unified architecture for natural language processing: deep neural networks with multitask learning. In: Proceedings of the 25th International Conference on Machine Learning, pp. 160–167. ACM (2008)
6. Collobert, R., Weston, J., Bottou, L., Karlen, M., Kavukcuoglu, K., Kuksa, P.: Natural language processing (almost) from scratch. J. Mach. Learn. Res. **12**, 2493–2537 (2011)
7. Labeau, M., Loser, K., Allauzen, A.: Non-lexical neural architecture for fine-grained POS tagging. In: Proceedings of the 2015 Conference on Empirical Methods in Natural Language Processing, pp. 232–237. Association for Computational Linguistics (2015)
8. Huang, Z., Xu, W., Yu, K.: Bidirectional LSTM-CRF models for sequence tagging. CoRR, abs/1508.01991 (2015)
9. Chiu, J.P.C., Nichols, E.: Named entity recognition with bidirectional LSTM-CNNs. arXiv preprint arXiv:1511.08308 (2015)
10. Turian, J., Ratinov, L., Bengio, Y.: Word representations: a simple and general method for semi-supervised learning. In: Proceedings of the 48th Annual Meeting of the Association for Computational Linguistics, ACL 2010, pp. 384–394 (2010)
11. Mikolov, T., Sutskever, I., Chen, K., Corrado, G.S., Dean, J.: Distributed representations of words and phrases and their compositionality. In: Proceedings of the Twenty-Seventh Annual Conference on Advances in Neural Information Processing Systems, pp. 3111–3119 (2013)
12. Pennington, J., Socher, R., Manning, C.D.: GloVe: global vectors for word representation. In: Proceedings of the 2014 Conference on Empirical Methods in Natural Language Processing, pp. 1532–1543 (2014)
13. Melamud, O., Goldberger, J., Dagan, I.: context2vec: learning generic context embedding with bidirectional LSTM. In: CoNLL (2016)
14. McCann, B., Bradbury, J., Xiong, C., Socher, R.: Learned in translation: contextualized word vectors. In: NIPS 2017 (2017)
15. Peters, M.E., et al.: Deep contextualized word representations. arXiv:1802.05365
16. Vaswani, A., et al.: Attention is all you need. In: Advances in Neural Information Processing Systems, pp. 6000–6010 (2017)
17. Radford, A., Narasimhan, K., Salimans, T., Sutskever, I.: Improving language understanding with unsupervised learning. Technical report, OpenAI (2018)

18. Devlin, J., Chang, M.-W., Lee, K., Toutanova, K.: BERT: Pre-training of Deep Bidirectional Transformers for Language Understanding. arXiv:1810.04805 (2018)
19. Starostin, A.S., et al.: FactRuEval 2016: evaluation of named entity recognition and fact extraction systems for Russian. In: Proceedings of the Annual International Conference on Computational Linguistics and Intellectual Technologies, Dialogue No. 15, pp. 720–738 (2016)
20. https://github.com/mayhewsw/conlleval.py/blob/master/conlleval.py
21. Anh, L.T., Arkhipov, M.Y., Burtsev, M.S.: Application of a Hybrid Bi-LSTM-CRF model to the task of Russian Named Entity Recognition. arXiv:1709.09686 (2017)
22. Konoplich, G., Putin, E., Filchenkov, A., Rybka, R.: Named entity recognition in Russian with word representation learned by a bidirectional language model. In: Ustalov, D., Filchenkov, A., Pivovarova, L., Žižka, J. (eds.) AINL 2018. CCIS, vol. 930, pp. 48–58. Springer, Cham (2018). https://doi.org/10.1007/978-3-030-01204-5_5

Expert Assessment of Synonymic Rows in RuWordNet

Valery Solovyev⬤, Gulnara Gimaletdinova⬤, Liliia Khalitova(✉)⬤,
and Liliya Usmanova⬤

Kazan Federal University, Tatarstan 2, 420021 Kazan, Russia
gulnara.gimaletdinova@kpfu.ru, lilia_khalitova@mail.ru

Abstract. This article explores the principles of synsets in the RuWord-Net thesaurus and synonyms in the classical dictionaries of Russian synonyms (N = 10) to identify discrepancies and improve the principles of organising synsets in RuWordNet. The relevance of the study is determined by the demand for WordNet resources in natural language processing tasks. The authors selected 102 RuWordNet thesaurus synsets, including nouns (N = 34), adjectives (N = 34) and verbs (N = 34). The meanings of the lexemes were correlated according to the data given in Russian language thesauri (N = 2). The comparative method and an independent expert assessment of RuWordNet revealed a number of discrepancies and inaccuracies in the representation of synsets concerning polysemy, hypo-hyperonymic relationships, lexical meanings of words and parts-of-speech synonymy. On the basis of this study, the authors recommend the elimination of individual shortcomings in the construction of the RuWord-Net synsets, in particular the polysemy and parts-of-speech synonymy.

Keywords: Computer lexicography · Synonymy · Dictionaries of synonyms · RuWordNet thesaurus · Hypo-hyperonymic relationships

1 Introduction

For the modern stage of linguistic research, the creation of large-scale linguistic resources for information retrieval systems is relevant. The development of such resources is carried out according to a modern approach of linguistic research, computational linguistics, whose development is based on the knowledge of general linguistics. In particular, the development of thesauri takes into account modern advances in lexicology, lexicography, semantics, pragmatics and cognitive linguistics.

WordNet is one of the most popular linguistic resources available today. Developed at Princeton University (https://wordnet.princeton.edu/), WordNet is the largest electronic lexical database of English nouns, adjectives, verbs and adverbs. The structure and principles of the organisation of language material, as well as the features of WordNet, are described in detail by a number

© Springer Nature Switzerland AG 2020
W. M. P. van der Aalst et al. (Eds.): AIST 2019, CCIS 1086, pp. 174–183, 2020.
https://doi.org/10.1007/978-3-030-39575-9_18

of scientific studies [1,2]. In the Google Scholar web search engine, WordNet is mentioned (as of April 30, 2019) in 105,000 articles, which indicates the demand for a resource for scientific research. WordNet is used in various studies in the field of Natural language processing (NLP): information retrieval, automatic text classification, automatic text typing, etc. The lexical database is often used to determine the degree of semantic proximity of words [3] and in the word-sense disambiguation task [4]. Currently, numerous WordNet interfaces, APIs and data processing tools have been developed (https://wordnet.princeton.edu/related-projects). WordNet analogues are created for many languages of the world. The Global WordNet Association website http://globalwordnet.org/ provides information about WordNet-like thesauri for more than 70 languages. A number of studies have described the principles for creating multilingual thesauri [5].

This article is dedicated to the WordNet analogue for the Russian language called RuWordNet [6]. The RuWordNet thesaurus was created at Moscow State University under the guidance of Loukachevitch [7,8]. Currently, it is the only Russian-language thesaurus created by experts and built on the principles of WordNet (synonymic rows and the semantic relationships linking them).

This article presents the results of an independent assessment.

The main contributions of the study are:

1. We provide an independent expert assessment of RuWordNet aimed at the improvement of the quality of RuWordNet synsets.
2. Following the results of the work, we give recommendations to eliminate the discrepancies and inaccuracies revealed in the RuWordNet thesaurus.

The paper is organized as follows. Section 2 provides a brief survey on the related work. Section 3 indicates the data and related methodology. Section 4 provides data analysis and key results. Section 5 discusses the results and gives recommendations. Section 6 concludes the work.

2 Related Work

For the Russian language, several attempts have been made to create thesauri similar to WordNet. The first attempt occured 20 years ago when the researchers at the philological faculty of St. Petersburg State University launched the Russ-Net project (http://project.phil.spbu.ru/RussNet/indexru.shtml) [9]. RussNet contained 15,000 words and the suggested synsets were described by experts. The project was completed in 2005. According to [10], the project data is not coded uniformly and cannot be used in NLP applications.

The next attempt to create an electronic lexical database was the Russian WordNet resource [11,12] which contained 100,000 words, obtained in a semi-automatic way from various dictionaries. This thesaurus is currently unavailable.

Another project, Wordnet for the Russian language (http://wordnet.ru/) was a thesaurus obtained by the automatic translation of WordNet into Russian. This resource which contains 30,000 words is unverified [13].

The Yarn project (https://russianword.net/) was a thesaurus created by the crowdsourcing method [10]. Yarn currently contains 145,000 words, but it lacks semantic relationships between synsets, including hypo-hyperonymic, which is typical of thesauri. Both the resource itself and the research based on it are actively developing to date [14–16]. Using Yarn, a selective expert qualitative check of synonymic rows was carried out, where 200 synsets were evaluated by four experts according to the evaluation system, which resulted in 103 synsets of Excellent, 70 of Satisfactory and 27 of Bad quality [10].

The RuWordNet project is developed on the basis of an earlier version called RuThes (http://www.labinform.ru/pub/ruthes/). The thesaurus containing 110,000 words and phrases was created by a semi-automatic method on the basis of an extensive corpus of texts with post-editing. While the independent verification of RuWordNet has not been performed, the authors of this paper see this as the purpose of their study. The data on the thesauri is summarized in Table 1.

Table 1. Comparison of WordNet-like thesauri for the Russian language.

Thesaurus	Number of words	Method of creation	Independent verification	Availability	Development stage
RussNet	15,000	Expert-based	No	Partly available	In progress within Yarn project
Russian-WordNet	100,000	Semi-automatic based on dictionaries	No	Unavailable	Completed
Wordnet for the Russian language	30,000	Automatic translation of WordNet into Russian	No	Available	Completed
Yarn	145,000	Crowdsourcing	Yes	Available	In progress
RuWordNet	110,000	Semi-automatic, based on corpus of texts with post-editing	No	Available	In progress

It seems noteworthy that all the thesauri were created by different methods. Accordingly, it is of scientific interest to conduct a comparative analysis of the quality of the created linguistic databases. The analysis of the quality of RuWordNet synsets presented in the article is a step in this direction. The aim of this research is to make an expert assessment of selected RuWordNet synsets with a focus on qualitative analysis.

3   Data and Related Methodology

The current study which presents the analysis of RuWordNet synsets was conducted by independent experts ($N = 4$) in Russian semantics and lexicography [17]. First, the experts selected three semantic groups that are supposed to be the most difficult for semantic analysis: a) feelings and emotions, b) mental and verbal activity, and c) human relationships and social life. The raw data consist of 102 synsets from RuWordNet including noun synsets ($N = 34$), adjective synsets (N = 34) and verb synsets (N = 34). The total number of analysed and compared lexemes is 976 for nouns, 520 for adjectives and 499 for verbs.

Second, the authors chose classical academic dictionaries of Russian synonyms (N = 10) [18–27] with different principles for representing synonymic rows, and used a comparative method to analyse the selected synsets in RuWord-Net and dictionaries of Russian synonyms particularly considering discrepancies. Thus, a relatively small number of analysed synsets were justified by a qualitative rather than a quantitative approach.

Third, the meanings of the lexemes, mainly those that were polysemantic, were refined in Russian language thesauri (N = 2) [28, 29], since numerous cases of discrepancies due to polysemy occurred.

The discrepancies found were summarised and systematised in the form of tables. The full list of words selected for analysis in this study and the tables representing comparative analysis of the synsets are available on the project website (https://kpfu.ru/kompleksnyj-analiz-struktury-i-soderzhaniya-366287.html).

Statistical analysis of raw research data allowed the determination of the features of RuWordNet, improved the quality of synsets, as well as the identification and correction of errors in the thesaurus.

4   Results

The RuWordNet thesaurus presents a hierarchical lexeme treatment principle based on hypo-hyperonymic relationships that are established between the generic and species synsets, which are the main structural elements of this thesaurus. One of the basic principles of this resource is the ability to interchange lexical units in most contexts. Moreover, the basic relationships are supplemented by the following: causation and consequence, domain, word formation (single-root words) and parts-of-speech synonymy.

Similarities between the lexemes in RuWordNet synsets and synonymic rows in dictionaries of Russian synonyms were justified if 50% or more dictionaries supported the same meanings, otherwise the lexemes were fixed as discrepancies. This allowed us to make a number of generalisations regarding:

(1) the description of the polysemy of lexemes;
(2) a presentation of hypo-hyperonymic relationships;
(3) the narrowing and extension of the meanings of words;
(4) a description of parts-of-speech synonymy.

The following are the results of studying the questions above.

4.1   The Polysemy of Lexemes

There are differences in the description of the synsets in RuWordNet and the dictionaries of Russian synonyms. In nine synsets of nouns out of the 34 examined (26%), the lexemes viewed as polysemantic in dictionaries of Russian synonyms were presented in the RuWordNet thesaurus as monosemantic (*strakh* (fear), *radost'* (joy), *skuka* (boredom), etc.).

A comparative analysis of the synsets in the RuWordNet thesaurus and dictionaries of Russian synonyms revealed significant differences in the description of polysemantic adjectives. Thus, 11 of 34 (32%) RuWordNet synsets are presented as monosemantic, while thesauri [28,29] mark them as polysemantic (adjectives *bezlyudnyy* (deserted), *gostepriimnyy* (hospitable), *neozhidannyy* (unexpected), *truslivyy* (cowardly)).

Similar to nouns, nine synsets out of 34 (26%) demonstrated cases of polysemantic verbs marked as monosemantic in the RuWordNet, which contradicts the descriptions provided by dictionaries of Russian synonym thesauri (the verbs *znat'* (know), *grubit'* (be rude), *mechtat'* (dream), etc.).

4.2   Hypo-hyperonymic Relationships

A comparative analysis of synsets in RuWordNet and dictionaries of Russian synonyms revealed discrepancies in the principles of describing synonymic, hyponymic and hyperonymic relationships. Regarding noun synsets ($N = 34$), discrepancies in the interpretation of the synonym status between RuWordNet and most dictionaries were noted in 24 synsets (71%): lexemes, defined as synonymic in dictionaries, referred to hyponyms or hyperonyms in the RuWordNet thesaurus. Regarding the total number of analysed nouns $N = 976$), 63 cases of discrepancies were identified, which was 6%.

Regarding adjective synsets ($N = 34$), 23 synsets (68%) lexemes marked as hyponyms or hyperonyms in the RuWordNet thesaurus were viewed as synonyms in most of the analysed dictionaries. Of the total number of analysed adjectives $N = 520$), there were 43 such cases (8%).

For verbs ($N = 34$), similar discrepancies were found in 20 analysed synsets (59%). Regarding the total number of verb lexemes analysed ($N = 499$), 64 cases of discrepancies per lexeme were identified, which was 11%.

A qualitative analysis of these discrepancies and the interpretation of possible causes are presented in the next section of the article.

4.3   Narrowing and Extension of Meanings

The analysis of the lexemes included in RuWordNet synsets and dictionaries of Russian synonyms revealed some discrepancies in the quantitative and qualitative filling of synonymic rows, which might be explained by different approaches to the interpretation of synonymy and semantic proximity of words in general. We analysed cases of the most significant discrepancies and found the following. First, in the synsets of nouns, adjectives and verbs, there are lexemes which are not represented in the RuWordNet synsets, but are included in the synonymic rows in dictionaries of Russian synonyms. Thus, we can distinguish the narrowing of the lexical meaning when describing lexemes in RuWordNet compared to dictionaries of Russian synonyms. Second, there are also words included in

RuWordNet synsets, that are not presented in the synonymic dictionaries, which we assume to be an extension of the lexical meaning.

The results of a comparative analysis of narrowing and extension of lexical meanings of nouns, adjectives and verbs and statistical data are presented in Fig. 1.

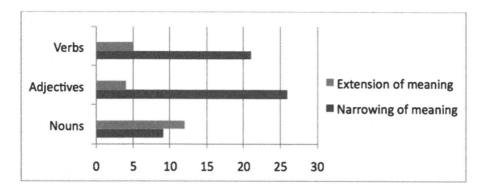

Fig. 1. Narrowing and extension of lexical meanings of nouns, adjectives and verbs.

4.4   Parts-of-Speech Synonymy

As noted earlier, while representing the relationships between words in synsets, RuWordNet suggests the list of part-of-speech synonyms. We were interested in whether parts-of-speech synonymy was presented in all analysed synsets, and whether there were errors or inaccuracies in the description of parts-of-speech synonymy. The analysis revealed the following. Parts-of-speech synonymy in 34 analysed noun synsets was given in 20 cases, which was 59%; however, in the remaining 14 synsets (41%), parts-of-speech synonymy was not included. In the case of adjectives, part-of-speech synonymy was described in 32 analysed synsets (94%) and was not represented in two synsets – only (6%). Of the 34 verb synsets analysed, the parts-of-speech synonyms were given in RuWordNet in 19 cases (56%), while in the remaining 15 verb synsets (44%), parts-of-speech synonyms were not indicated. Inaccuracies in the description of part-of-speech synonymy were found in one noun synset (3%), in four adjective synsets (12%) and in two verbal synsets (6%).

5   Discussion and Recommendations

The question of revising the scope of certain words in RuWordNet remains open, since the problem of hypo-hyperonymic and synonymic relationships between similar lexemes is still debatable and there is no clear answer regarding the semantic status of these units. The analysis revealed significant discrepancies in the description of polysemy: words marked as polysemantic in Russian thesauri are presented as monosemantic in RuWordNet, which requires clarification and adjustment. For example, nouns *strakh* (fear), *radost'* (joy), *skuka* (boredom), *len'* (laziness), *zhalost'* (pity), *toska* (grief), *mechta* (dream), *obman* (deceit) and *mest'* (revenge).

In RuWordNet the verb *obmanut'* (deceive) is presented with only one meaning: "to deceive, mislead", while the Russian thesaurus by Ozhegov identifies five meanings of the verb *obmanut'* (deceive): 1. Mislead. 2. Break the promise. 3. Fail to meet expectations/assumptions. 4. Underpay (when calculating wages). 5. Betray, violate marital fidelity [29].

The following meanings are given in the Russian thesaurus by Efremova: 1. Consciously mislead smb. 2. To commit trickery, fraud towards smb. 3. Fail to fulfil your promises, not keep your word. 4. To show deception in love; betray (wife, husband). 5. Seduce (girl, woman) [28].

Recommendations for expanding the meaning of the listed verbs should be considered casual, especially in those controversial cases concerning hypo-hyperonymic and synonymic relationships between similar lexemes.

Discrepancies in the principles of describing synonymic, hyponymic and hyperonymic relationships could be illustrated by the following examples (for statistics see Sect. 4).

In the description of the synset *vostorg* (delight) the following words are listed as hyponyms in RuWordNet: *upoyeniye* (flush), *ekstaz* (ecstasy), *ekzal'tatsiya* (exaltation); whereas all the analysed dictionaries of Russian synonyms define them as synonyms [18, 19, 22, 26]. Analysing the synonymic row of the adjective *boyazlivyy* (fearful), we find that seven out of ten dictionaries of Russian synonyms define the relationship between the lexemes *boyazlivyy* (fearful) and *robkiy* (timid) as synonymic; whereas in RuWordNet, *robkiy* is marked as a hyperonym.

We revealed some discrepancies between the words represented in RuWordNet and synonymic rows in the dictionaries of Russian synonyms. For example, the synset for the word *obizhat'* (offend) in RuWordNet contains the following set of synonyms: *zatseplyat'* (hook) and *ushchipyvat'* (pinch), while none of the dictionaries identify them as synonyms. Similarly, the dictionaries do not establish synonymic relationships between *obshchat'sya* (communicate) and *povestis'* (be tricked by); *pridirat'sya* (carp) and *shpynyat'* (poke), *pridirat'sya* (carp) and *podkapyvat'sya* (intrigue); *mechtat'* (dream) and *leleyat'* (cherish); *rasskazyvat'* (tell) and *opisat'* (describe).

Some synsets in RuWordNet in the section "part-of-speech synonymy" have included the words of the same parts of speech as synonyms. For example, the adjectives *belen'kiy* (white) and *veselen'kiy* (cheerful) are given as synonyms of different parts of speech (as nouns) to the adjectives *belyy* (white) and *veselyy* (funny), respectively. We assume that this is due to the phenomenon of substantivisation; however, we recommend clarifying the part-of-speech synonyms of these words. Regarding verbs, the words *znat'* (know) and *nadsmekhat'sya* (make fun of) require clarification.

The cases of narrowing of the meanings of synonyms in RuWordNet compared to the dictionaries of Russian synonyms are explained since RuWordNet was based on a news corpus, while the classical dictionaries of synonyms were focused on fiction. However, we believe that, with the expansion of the corpus and the inclusion of fiction, the list of synsets in RuWordNet will increase, while the discrepancies will decrease.

The cases of expansion of the meanings of synonyms in RuWordNet are likely related to a larger number of words (110,000) compared to the dictionaries of Russian synonyms, covering a significantly smaller layer of vocabulary.

Evaluating the quality of hypo-hyperonymic relationships in RuWordNet is an extremely complicated task, primarily due to the lack of a formal (operational) definition of hypo- and hyperonymy in linguistics. Modern computational linguistics also does not provide methods for the automatic detection of hypo-hyperonymic relationships corresponding to "golden standards". Moreover, there is no elaborate Russian language thesaurus compiled by professional lexicographers. This could be beneficial in comparing and analysing ambiguous data. To establish valid hypo-hyperonymic relationships, we recommend carrying out extensive theoretical studies that go far beyond the scope of this article.

6   Conclusions

In this research, we analysed the synsets presented in RuWordNet thesaurus and compared the data with dictionaries of synonyms of the Russian language. The comparative method and an independent expert assessment of RuWord-Net revealed a number of discrepancies and inaccuracies in the representation of synsets concerning polysemy, hypo-hyperonymic relationships, lexical meanings of words and parts-of-speech synonymy. The recommendations would be beneficial in the creation and improvement of similar linguistic databases, and expert assessment seems to be the most appropriate approach in cases when qualitative analysis is needed.

Acknowledgments. This research was financially supported by the Russian Foundation for Basic Research (Grant No. 18-00-01238), and by the Government Program of Competitive Development of Kazan Federal University.

References

1. Miller, G.A.: WordNet: a lexical database for English. Commun. ACM **38**(11), 39–41 (1995)
2. Fellbaum, C.: WordNet and wordnets. In: Brown, K., et al. (eds.) Encyclopedia of Language and Linguistics, 2nd edn, pp. 665–670. Elsevier, Oxford (2005)
3. Pedersen, T., Patwardhan, S., Michelizzi, J.: WordNet::Similarity – measuring the relatedness of concepts. In: Demonstration Papers at HLT-NAACL 2004, Association for Computational Linguistics, Stroudsburg, PA, USA, pp. 38–41 (2004)
4. Navigli, R.: Word sense disambiguation: a survey. ACM Comput. Surv. **41**(2), 1–69 (2009)
5. Vossen, P.: EuroWordNet: building a multilingual database with wordnets for European languages. The ELRA Newsl. **3**(1), 7–10 (1998)

6. Thesaurus of Russian Language RuWordNet. https://ruwordnet.ru/ru. Accessed 26 May 2019
7. Loukachevitch, N.: Thesauri in Information Retrieval Tasks. Moscow University, Moscow (2011)
8. Loukachevitch, N., Lashevich, G., Gerasimova, A., Ivanov, V., Dobrov, B.: Creating Russian WordNet by conversion. In: Proceeding of Conference on Computational Linguistics and Intellectual Technologies Dialogue-2016, RSUH, Moscow, pp. 405–415 (2016)
9. Azarova, I.V., Sinopalnikova, A.A., Yavorskaya, M.V.: Guidelines for RussNet structuring. In: Proceeding of Conference on Computational Linguistics and Intellectual Technologies Dialogue-2004, Nauka, Moscow, pp. 542–547 (2004)
10. Braslavski, P., Ustalov, D., Mukhin, M., Kiselev, Y.: YARN: spinning-in-progress. In: Proceedings of the 8th Global WordNet Conference, Bucharest, Romania, pp. 58–65 (2016)
11. Balkova, V., Sukhonogov, A., Yablonsky, S.: Russian WordNet from UML-notation to internet/intranet database implementation. In: Proceedings of the 2nd International WordNet Conference, Masaryk University, Brno, pp. 31–38 (2004)
12. Sukhonogov, A., Yablonsky, S.: Russian WordNet development. In: Proceedings of the 6th Russian Conference on Digital Libraries: Advanced Methods and Technologies, Digital Collections - RCDL 2004, Pushchino, Russia (2004)
13. Gelfenbeyn, I., Goncharuk, A., Lekhelt, V., et al.: Automatic translation of Word-Net semantic network to Russian language. In: Proceeding of Conference on Computational Linguistics and Intellectual Technologies Dialogue-2003 (2003)
14. Ustalov, D., Chernoskutov, M., Biemann, C., Panchenko, A.: Fighting with the sparsity of synonymy dictionaries for automatic synset induction. In: van der Aalst, W.M.P., Ignatov, D.I., Khachay, M., Kuznetsov, S.O., Lempitsky, V., Lomazova, I.A., Loukachevitch, N., Napoli, A., Panchenko, A., Pardalos, P.M., Savchenko, A.V., Wasserman, S. (eds.) AIST 2017. LNCS, vol. 10716, pp. 94–105. Springer, Cham (2018). https://doi.org/10.1007/978-3-319-73013-4_9
15. Ustalov, D., Teslenko, D., Panchenko, A., Chernoskutov, M.: Mnogoznal: an unsupervised system for word sense disambiguation. In: Proceedings of 2017 International Multi-Conference on Engineering, Computer and Information Sciences (SIBIRCON), pp. 147–150. IEEE, Novosibirsk (2017)
16. Ustalov, D.: Expanding hierarchical contexts for constructing a semantic word network. In: Proceeding of Conference on Computational Linguistics and Intellectual Technologies Dialogue-2017, RSUH, Moscow, pp. 369–381 (2017)
17. Janda, L., Solovyev, V.: What constructional profiles reveal about synonymy: a case study of Russian words for sadness and happiness. Cogn. Linguist. **20**(2), 367–393 (2009)
18. Abramov, N. (ed.): Dictionary of Russian Synonyms and Similar Expressions, 7th edn. Russkiye Slovari, Moscow (1999). (in Russian)
19. Aleksandrova, Z. (ed.): Dictionary of Synonyms of the Russian Language: A Practical Guide, 15th edn. Russkiy yazyk, Moscow (2007). (in Russian)
20. Alektorova, L., Vvedenskaya, L., Zimin, V., et al. (eds.): Dictionary of Synonyms of the Russian Language, 2nd edn. Astrel, AST, Moscow (2002). (in Russian)
21. Gorbachevich, K. (ed.): A Brief Dictionary of Synonyms of the Russian Language. Eksmo, Moscow (2005). (in Russian)
22. Kozhevnikov, A. (ed.): Large Dictionary of Synonyms of the Russian Language: Speech Equivalents: A Practical Guide. Neva, St. Petersburg (2003). (in Russian)
23. Klyueva, V. (ed.): A Brief Dictionary of Synonyms of the Russian Language. Uchpedgiz, Moscow (1961). (in Russian)

24. Apresyan, Y. (ed.): A New Explanatory Dictionary of Synonyms of the Russian Language, 2nd edn. Yazyki russkoy kultury, Moscow (2000). (in Russian)
25. Babenko, L. (ed.): Dictionary of Synonyms of the Russian Language. Astrel, Moscow (2011). (in Russian)
26. Evgenyeva, A. (ed.): Dictionary of Synonyms: Reference Book. Nauka, Leningrad (1975). (in Russian)
27. Dictionary of Synonyms of the Russian Language: Dictionary of Antonyms of the Russian Language. Victoria Plus, St. Petersburg (2007). (in Russian)
28. Efremova, T.: The New Dictionary of the Russian Language. Thesaurus and Word-Building. Russkiy yazyk, Moscow (2000). (in Russian)
29. Ozhegov, S.: Russian thesaurus. https://slovarozhegova.ru/. Accessed 26 April 2019 (in Russian)

Text Mining for Evaluation of Candidates Based on Their CVs

Maria Tikhonova[(✉)]

Doctoral School of Computer Science,
National Research University Higher School of Economics,
Moscow, Russia
m_tikhonova94@mail.ru
https://github.com/MariyaTikhonova

Abstract. The problem of CV (or resume) text mining becomes increasingly relevant nowadays as long as it could simplify the evaluation of future employees and their suitability for the post for which they apply. The paper proposes a procedure for automatic information extraction from text documents, namely from candidate's CVs. The described algorithm is based on Natural Language Processing methods and allows to transform text information into categorical features or classes. These features may further be used as inputs for a machine learning model to predict the suitability of the candidate for the position. Besides the general method, the description of the experiments is given in which the algorithm was used for clusterization of future employees according to their previous position and job spheres they worked in. The obtained classes were used to predict the probability of the candidate's turnover in the first six months. Their addition allowed to raise the model score.

Keywords: Natural Language Processing · Text mining · Clusterization · NLP methods · Machine learning · Word embedding · k-means · Unsupervised learning · Topic modeling

1 Introduction

The problem of automatic CV (or resume) preprocessing and prediction of the candidate's suitability for the position is relevant, especially in big companies where the flow of new employees is high and the staff turnover is big. In this context the task of CV text mining in order to extract information about the candidate and to transform it into features for machine learning algorithms arises. This features could further be used to make a decision about the suitability of the candidate for the position and to predict the probability that he or she will quit the job in the near future.

The paper proposes a procedure based on *Natural Language Processing methods* for effective text mining and information extraction from CV text fields, such as *«Previous Job Sphere»*, *«Previous Position»*, *«About the Candidate»* and others, and their further clusterization. Thus, the algorithm assigns each candidate

© Springer Nature Switzerland AG 2020
W. M. P. van der Aalst et al. (Eds.): AIST 2019, CCIS 1086, pp. 184–189, 2020.
https://doi.org/10.1007/978-3-030-39575-9_19

to one of the classes, which could be regarded as the candidate's «type» and which is quite informative in context of making a decision about the suitability of the candidate.

In the second part of the paper a detailed description of the algorithm is given. The third part describes the usage of the algorithm on the real data and presents the results of the conducted experiments. The algorithm was tested for information extraction for CVs (namely, for text fields *«Previous Job Sphere»* and *«Previous Position»*) of candidates applying in bank branches for the position of *banking product consultant.*

2   Algorithm's Description

Information extraction from a text field of CVs and its clusterization consists of 5 steps:

Step 1. Word embeddings construction;
Step 2. Computation of tf-idf index;
Step 3. Text field embeddings construction;
Step 4. Selection of the optimal cluster number;
Step 5. Final CVs clusterisation.

Below each step is described in more detail.

Step 1. Word Embeddings Construction. On the first step words which are found in the CVs text field are embedded into the linear space $\mathbb{R}_n$. Thus, for every word we obtain an n-dimensional vector. Moreover, we want word embeddings to preserve syntactic and semantic word properties. There exists a number of word embedding algorithms which possess this quality. Among the most popular are Word2Vec [3,4], FastText [1,2] and GloVe [5]. In addition, pretrained word embeddings, built of large collections of documents, are available for free use in many different languages including Russian.

Step 2. Computation of Tf-Idf Index. On the second step for each word in every CV its *tf-idf* index is computed. *Tf-idf (term frequency-inverse document frequency)* is a statistic which represents the importance of a word in a document of the collection. It is proportional to the frequency of a word in a document and inversely proportional its frequency in the collection of documents. Thus, tf-idf discriminates rare words which are frequent in a specific document. That is why in information retrieval it is often used as a weighting factor.

tf of a word t in a document d is defined as:

$$tf(t,d) = \frac{n_t}{\sum\limits_{k \in d} n_k}$$

n_t — a number of times a word t appears in a document d.

idf of a word t in the collection of documents D is defined as:

$$idf(t, D) = \log \frac{|D|}{|\{d_i \in D | t \in d_i\}|}$$

where $|D|$ denotes a number of elements in a set D.

Finally, tf-idf of a word t in document d which belongs to the collection of documents D is computed as the product of $tf(t, d)$ and $idf(t, d, D)$:

$$tf - idf(t, d, D) = tf(t, d) * idf(t, d, D).$$

Step 3. Text Field Embedding Construction. Then, the text field under consideration in every CV is embedded into an n-dimensional vector. For this purpose, vector representations of the words present in the CV are summed with their tf-idf weights:

$$v(d) = \sum_{w \in d} tf - idf(w, d) * v(w)$$

d, w — a text field in the CV and a word in d, respectively,
$v(d)$, $v(w)$ — embeddings for d and w, respectively,
tf- $idf(w, d)$ — tf-idf index of w in d.

Step 4. Selection of Optimal Cluster Number. For the clusterization of the obtained CV embeddings a method known as k-$means$, proposed by Steinhaus [6], is used. In process of work the algorithm minimizes the deviation of points from cluster centres, which are called $centroids$. In other words, for a given K it minimizes $index\ of\ inertia$:

$$J_K(C) = \sum_{k=1}^{K} \sum_{x_i \in C_k} ||x_i - \mu_k||^2 \rightarrow \min$$

K — cluster number,
$J_K(C)$ — index of inertia for the clusterization C with K clusters,
C_k — a set of points which belong to cluster k,
μ_k — a centroid of cluster k.
Thus:

$$C_{opt} = \arg \min_C J_K(C) = \arg \min_C \sum_{k=1}^{K} \sum_{x_i \in C_k} ||x_i - \mu_k||^2$$

C_{opt} — optimal clusterization.

In k-means the number of clusters K is an input parameter which should be selected manually. In order to choose the optimal number of clusters k_{opt} the following heuristic is used. For each number of clusters k under consideration its index of inertia $J_k = \min_C J_k(C)$ is computed and the optimal number of clusters is said to be that, for which the rate of inertia decrease D_k is minimal:

$$k_{opt} = \arg\min_k D_k = \arg\min_k \frac{|J_k - J_{k+1}|}{|J_{k-1} - J_k|}.$$

Step 5. Final CVs Clusterization. Finally, each CV is assigned to one of k_{opt} clusters or classes (where k_{opt} was selected on step (4) according to the information contained in the considered text field.

3 Experiments

The method was tested on real data. It was applied for information extraction for people applying for *banking product consultant* in bank branches in a particular bank. The data was taken from inside bank sources and is not publicly available. The dataset contained 8305 CVs with the labeled information assigned to different categories (*«About the Candidate»*, *«Skills»*, *«Previous Job Sphere»*, *«Previous Position»* and others). For instance, for every candidate we had an enumerated list of his or her previous positions (a sample is shown in Fig. 1). The algorithm was used for text fields *«Previous Job Sphere»* and *«Previous Position»*. All the experiments were implemented on *Python*.

On the first step FastText word vectors pretrained on Russian Wikipedia with 300 components[1] were used. In addition we experimented with word embeddings trained on our data. However, the collection turned out not to be large enough, the vocabulary was limited and the quality of the resulted embeddings was poor.

Then we chose the optimal number of clusters k_{opt} by minimizing the rate of inertia decrease via the procedure described in step 4. For «Previous Job Sphere» the optimal number of clusters k_{opt} was found to be 19 and for «Previous Position» – 17.

	Position 1	Position 2	Position 3
0	Продавец-консультант	Продавец консультант	Охранник
1	Консультант по банковским продуктам	Менеджер по продажам	Сотрудник по эффективности персонала
2	Администратор офисного центра	Менеджер по качеству	Старший консультант по продажам

Fig. 1. A sample for the text field *«Previous Position»*

The results of the clusterization are shown in Tables 1 and 2. It is worth noticing that as long as we conducted our experiments for people applying in bank branches previous job sphere *банк* (which means *bank* in English) separated into an independent class. In addition to this, professions such as *консультант*

---

[1] https://rusvectores.org/ru/models/.

Table 1. Examples of clusters for Previous Job Sphere

Cluster	Examples of Job Spheres
1	банк
2	финансово-кредитное посредничество
	консалтинговые услуги
	услуги по ведению бухгалтерского учета
3	безалкогольные напитки (продвижение)
	алкогольные напитки (продвижение)
	кондитерские изделия (продвижение)
4	строительство жилищное
	архитектура, проектирование
	строительство дорожное и инфраструктурное
5	вуз, ссуз колледж, пту
	школа, детский сад
	лаборатория, исследовательский центр

Table 2. Examples of clusters for Previous Position

Cluster	Examples of Previous Positions
1	консультант по банковским продуктам
	менеджер по финансовым продуктам
	менеджер по продажам
2	кредитный консультант
	финансовый эксперт
	старший кредитный специалист
3	вожатый
	стажер
	курьер
	практикант
4	менеджер call-центра
	специалист контактного центра
	оператор контакт-центра
5	бухгалтер
	главный бухгалтер
	помощник бухгалтера

(*banking product consultant* and *financial product manager* respectively) and similar to them also constitute a separate class.

The obtained classes were further used in the task of prediction candidate's turnover during the first 6 months of his work in order to estimate the performance and usability of the algorithm. The initial ML model for this task used the information about the candidate's age, education, preferred salary and other features describing the candidate which were of numerical or categorical type. The clusterizations of text fields obtained via the algorithm were used as additional input features for the model. This gave a significant information gain and allowed to increase Gini from 0.22 to 0.28 (27% increase) which could be regarded as quite meaningful as long as the task itself containes a high element of uncertainty.

4 Conclusion

In the paper a new method of automatic information extraction from candidates CVs was described. The information obtained via the algorithm could further be used in order to make judgment about the candidate. A series of the experiments was conducted in which the algorithm was tested on real data for two text fields («Previous Job Sphere» and «Previous Position») for candidates applying for the position of banking product consultant. The obtained classes were used as features for a machine learning model to predict the probability of candidates' turnover. In the experiments the use of the new features allowed to improve the score of the model.

In the future we plan to use other word embedding methods and to experiment with embeddings pretrained on the collection of documents with the specific vocabulary in order to improve the method. Besides we plan to apply the algorithm to a larger scope of text fields such as «About the Candidate» or «Skills» and to test it on other tasks.

References

1. Bojanowski, P., Grave, E., Joulin, A., Mikolov, T.: Enriching word vectors with subword information. Trans. Assoc. Comput. Linguist. **5**, 135–146 (2017)
2. Joulin, A., Grave, E., Bojanowski, P., Mikolov, T.: Bag of tricks for efficient text classification. arXiv preprint arXiv:1607.01759 (2016)
3. Mikolov, T., Chen, K., Corrado, G., Dean, J.: Efficient estimation of word representations in vector space. arXiv preprint arXiv:1301.3781 (2013)
4. Mikolov, T., Yih, W.t., Zweig, G.: Linguistic regularities in continuous space word representations. In: Proceedings of the 2013 Conference of the North American Chapter of the Association for Computational Linguistics: Human Language Technologies, pp. 746–751 (2013)
5. Pennington, J., Socher, R., Manning, C.: Glove: global vectors for word representation. In: Proceedings of the 2014 Conference on Empirical Methods in Natural Language Processing (EMNLP), pp. 1532–1543 (2014)
6. Steinhaus, H.: Sur la division des corps materiels en parties. Bull. Acad. Polon Cl. III **IV**, 801–804 (1956)

Vec2graph: A Python Library for Visualizing Word Embeddings as Graphs

Nadezda Katricheva[1]([✉]) [iD], Alyaxey Yaskevich[1] [iD], Anastasiya Lisitsina[1],
Tamara Zhordaniya[1], Andrey Kutuzov[2] [iD], and Elizaveta Kuzmenko[3] [iD]

[1] National Research University Higher School of Economics, Moscow, Russia
n.katricheva@gmail.com, alyaxey.yaskevich@gmail.com, lisitsinan@gmail.com,
ttzhordaniya@edu.hse.ru
[2] University of Oslo, Oslo, Norway
andreku@ifi.uio.no
[3] University of Trento, Trento, Italy
lizaku77@gmail.com

Abstract. Visualization as a means of easy conveyance of ideas plays a key role in communicating linguistic theory through its applications. User-friendly NLP visualization tools allow researchers to get important insights for building, challenging, proving or rejecting their hypotheses. At the same time, visualizations provide general public with some understanding of what computational linguists investigate.

In this paper, we present *vec2graph*: a ready-to-use Python 3 library visualizing vector representations (for example, word embeddings) as dynamic and interactive graphs. It is aimed at users with beginners' knowledge of software development, and can be used to easily produce visualizations suitable for the Web. We describe key ideas behind *vec2graph*, its hyperparameters, and its integration into existing word embedding frameworks.

Keywords: Distributional semantics · Visualization · Word embeddings · Graph representations

1 Introduction

Distributional word embeddings are now arguably the most popular way to computationally handle lexical semantics. In such models, every word is represented with a dense float vector: the number of dimensions is usually in the order of hundreds, and may vary depending on the purpose of the model. The dimensionality of word vectors is too high for them to be visualized directly. Therefore, popular visualizations of embeddings are carried out using methods of dimensionality reduction, the most famous of which are t-SNE and PCA. They transform high-dimensional vectors into points on 2D or 3D planes, making them intelligible for humans. More about these algorithms can be found in Subsect. 3.1.

N. Katricheva and A. Yaskevich—Contributed equally to the paper.

© Springer Nature Switzerland AG 2020
W. M. P. van der Aalst et al. (Eds.): AIST 2019, CCIS 1086, pp. 190–198, 2020.
https://doi.org/10.1007/978-3-030-39575-9_20

In this paper, we show how one can easily and effectively visualize semantic relations between words in the form of graphs with any pre-trained word embedding model and our *vec2graph* library. Here, vertices of a graph correspond to words, while edges indicate semantic similarity between them. Graphs can show not only the proximity of vertices to each other, but also the deeper structure of the connections between words, giving the opportunity to look at the communities of semantically related words. For example, Fig. 1 shows the graph for the English word 'science' and its 5 nearest neighbors by cosine similarity.[1] It reflects common knowledge about various scientific fields. In particular, natural sciences (mathematics, physics, biology, astronomy) are more closely related to each other than to humanities, which constitutes a separate cluster. Below, we describe the process of building and visualizing such graphs in more details.

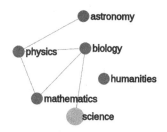

Fig. 1. *vec2graph* visualization for the word 'science' and its nearest neighbors in an embedding model

2   Related Work

The foundations for word embedding models were laid in [2]: in 2003, they presented a neural probabilistic language model, which learned distributed representations of words as a byproduct. In the next seminal paper [11], very efficient Continuous Bag-of-Words and Continuous Skip-gram models were proposed and described. These are shallow neural networks that are used to learn dense word vectors based on the distributional signal from large corpora of natural texts.

Publications directly targeting the visualization of word embeddings are not numerous. It is important to mention [1] in which not only neural language processing analysis methods are reviewed, but also the examples of visualizations for analyzing neural networks in the language domain are given. Authors believe that the availability of open-source tools will encourage users to utilize visualization in their research and development process, and the *vec2graph* library we present is an example of such a tool.

In graph visualizations, variables are encoded and mapped in a straightforward and simple way. In finding the most aesthetic and understandable layouts of our visualizations, we were guided by [6]. Our graphs meet the basic requirements of data visualization; in particular, they are intuitive, informative and easy to read.

---

[1] All English word embedding models used were downloaded from the NLPL Vectors repository [4].

3   Word Embeddings Visualization

3.1   Dimensionality Reduction

There are numerous ways of visualizing word embedding models. An intuitive solution is to use dimensionality reduction: that is, to further embed high-dimensional vector representations into a 2D plane or a 3D space.

A method known as Principal Component Analysis (PCA) emerged at the beginning of the XX century. This is essentially a set of linear transformations aiming to reduce data dimensionality while retaining as much of the original variance across the data as possible [7,14]. To achieve this, the relative center of the data is found, the co-variance (or correlation) matrix is calculated, and the data is rotated and put on a new set of axes, two or three of which are used as the features, dimensions, or the principal components for visualization. Though PCA can be adapted for different kinds of data, due to the transformation and loss of information the data suffers, the final image in most cases is mainly descriptive, not inferential [8]. In other words, PCA produces an approximate simplification of data, which may prevent one from performing rigorous analysis.

Another way to reduce dimensionality is the t-SNE algorithm, introduced in [10]. t-SNE is non-linear, allowing to blow up the most dense vector con-glomerates and shorten the distances between the most remote data regions. Proper application of this technique requires certain skills and craftsmanship in order to tweak its hyperparameters, such as perplexity (vaguely, the number of close neighbors each point has) and the number of iterations, or runs [17]. Additionally, no matter the parameters, t-SNE can yield different results on the same data because of its stochastic nature (even with few words, a desirably expressive image is not what you get on the first run).

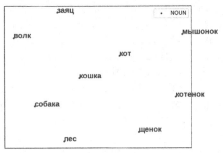

'Semantic maps' can be pro-duced after applying dimension-ality reduction methods to word embeddings, allowing for a partial or entire model overview at one glance. Such visualizations make most sense if they can be zoomed in and out, with word labels showing near the dots while the pointer hov-ers over them, allowing for a more detailed exploration. If only a part of the vector space is shown and if the picture is static, words can

Fig. 2. A 2D t-SNE projection of the embed-dings for Russian words animal names.

actually be used instead of dots for more informativeness (see Fig. 2)[2]. However, as the number of words increases, they inevitably overlap, decreasing readability.

---

[2] All Russian word embeddings used were downloaded from RusVectores service [9].

To try 3D PCA or t-SNE visualization on own data, one can use a TensorFlow sub-module called *TensorBoard Embedding Projector*[3]. Possible options here are the number of dimensions (2 or 3), the method (PCA or t-SNE) and method-specific hyperparameters. Sometimes it can be difficult to immediately capture the relations between words when looking at the resulting visualization. However, *Embedding Projector* is still a powerful tool for word embedding exploration (see Fig. 3 for an example).

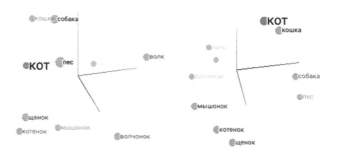

Fig. 3. The Russian word 'кот' (*cat*) and its nearest neighbors projected to 3D using PCA (**left**) and t-SNE with perplexity 5, learning rate 10 and 500 iterations (**right**) in *TensorBoard Embedding Projector*.

3.2   Vectors as Graphs

Graphs as a means of visualizing lexical relations date back to at least XIV century, when Raymundus Lullus conceived of *the first systematic spatial organization of lexical items* [18]. Since then, the idea of representing words as vertices and semantic connections between them as edges has been popular among scholars and researchers. One can recall *WordNet* lexical database [12] or *BabelNet* [13], a multilingual semantic network partially based on *WordNet*, and their graph visualizations. Graphs are also sometimes used to extract sentiment lexicons from word embeddings (SentProp algorithm in [5]), or to create massive semantic map representations (see, for example, *word2vec-graph* library)[4]. Figure 4 shows examples of visualizing lexical relations as graphs. However, the potential of graphs as word embedding schemata seems not to have been fully realized yet.

This is mainly because behind the seemingly simple concept of a graph, rich theoretical heritage lies. It includes such notions as transitivity, centrality, homophily, adjacency matrices and others, which, to our knowledge, have not been extensively explored with regard to word embeddings. Neither our *vec2graph* nor *word2vec-graph* are specifically created for such investigations, but could be used as a basis for further development of the corresponding tools

---

[3] https://projector.tensorflow.org/.

[4] https://github.com/anvaka/word2vec-graph.

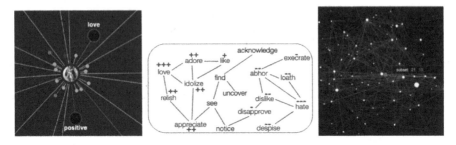

Fig. 4. Word network from BabelNet (**left**), visual summary of the SentProp algorithm (**middle**); example of the *word2vec-graph* library output (**right**).

for analysis of lexical graphs. We also believe that semantic notions like polysemy, synonymy, antonymy, homonymy and the like could be efficiently extracted from word embedding models by means of graph representations.

Both *vec2graph* and *word2vec-graph* allow for exploration of the word meaning and its relations to other words. However, there are significant differences between libraries as well. Our library (*vec2graph*) is aimed at visualizing small groups of 'similar words' (usually the nearest neighbors of one query word by cosine similarity). If the user wants to explore a larger number of words, he or she can set a desired recursion level: in this case, the nearest neighbors of the original query word are visualized into separate graphs with neighbors as new query words. By contrast, *word2vec-graph* only gives the picture of the vector space (embedding model) as a whole. Certainly, large scale visualizations pose great interest, but most of the time scholars try go into detail, and for this reason smaller sets of embeddings need to be explored and demonstrated.

Further on, *vec2graph* solves the important problem of graph visualization availability (with easy parameter tweaking) for people whose coding skills are not advanced. To actually employ *word2vec-graph*, one has to convert vector data into graph data, modify a Python script, deal with binary files, *Node.js*, etc. In contrast, *vec2graph* requires only an embedding model, a query and the path to the directory for writing the output HTML files ready for viewing in any web browser. Besides basic built-in Python modules, *vec2graph* depends only on the well-known *Gensim* library [15] which is used to handle word embedding models.

Technically, we use the cosine similarities between the query word and its nearest neighbors, provided by the *most similar()* method in *Gensim*. With them, *vec2graph* produces small graphs, which are computationally efficient, but still keep all the benefits of graph representations. In the simplest case, the graphs are fully-connected, with cosine similarities between vectors serving as weights on the edges connecting the corresponding word vertices.

Visually, the closer are the words in the vector model, the shorter are the edges between their vertices in the graph (we also use a force-directed graph layout, see Sect. 4). Distances expressed with edge lengths are more demonstrative than cosine similarity values: one can easily understand the relations between

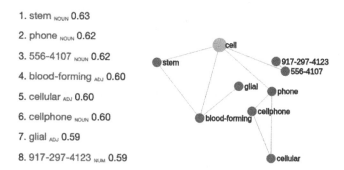

1. stem _NOUN_ 0.63

2. phone _NOUN_ 0.62

3. 556-4107 _NOUN_ 0.62

4. blood-forming _ADJ_ 0.60

5. cellular _ADJ_ 0.60

6. cellphone _NOUN_ 0.60

7. glial _ADJ_ 0.59

8. 917-297-4123 _NUM_ 0.59

Fig. 5. Representations for the word 'cell' embedding. **Left**: a list of nearest neighbors sorted by their similarity to the query word (colors reflect frequency). **Right**: *vec2graph* visualization, lower threshold of cosine similarity set to 0.6.

a set of words at a glance, as can be seen in Fig. 5. Note that the graph additionally uses the information about similarities **between neighbors**, which is completely missing from the nearest neighbors list (it shows only the similarities between the query word and each of the neighbors).

Another feature of graphs representations that sets them apart from lists of nearest neighbors is the **cosine similarity threshold**, set in order to keep edges only for the vertex pairs with cosine similarity equal to or higher than the threshold (which is a hyperparameter). As a result, the visualized graph can stop being fully-connected. This is impossible to implement with the lists of nearest neighbors, since they do not possess the notion of edges between words at all.

This feature serves best if the words in the data are naturally separated into clusters or communities. What happens in case our query word is ambiguous (has multiple senses) is that the vertex which corresponds to the query word is connected to separate disconnected groups of words (vertices). These clusters or communities on the graph represent different senses of the given query word. This can be a way to alleviate the known limitation of distributional word embeddings: their blind eye to homonymy, or ambiguity of natural language words.

An example of this is shown in the right part of Fig. 5, produced by the *vec2graph* module with the threshold set to 0.6. The three visible clusters represent three aspects of meaning of the word '*cell*': the biological term ('*stem*', '*blood-forming*', '*glial*'), the mobile phone related sense ('*phone*', '*cellphone*', '*cellular*') and phone numbers ('*917-297-4123*', '*556-4107*'). In the nearest neighbors list (the left part), these senses are mixed together.

4   Vec2graph Library: Practical Overview

At the implementation level, our *vec2graph* library can be described as follows. The module takes as an input a word embedding model (in any format compatible with the *Gensim* library [15]), a query to the model (word or list of words), and the path to the target directory. It saves interactive graphs formed using *D3.js*-based *JavaScript* [3] and plain *HTML*, being basically a standalone HTML page or pages.

Also, the output can vary in terms of some parameters which optionally could be provided along with the query. Those are:

– the number of nearest neighbors to be produced as vertices;
– the depth to which the algorithm has to drill down into the semantic network (the higher this number is, the deeper the recursion, which means that it produces visualizations for the neighbors of the neighbors of the query word);
– the cosine similarity threshold limiting the number of edges between words;
– the width of edges between nodes in a graph;

The resulting graph is then visualized in the browser using a force-directed graph layout algorithm, as implemented in *D3.js*, employing Verlet integration [16]. The Eq. 1 shows how we convert cosine distances x_{sim} into weights on graph edges:

$$y_{inv} = 1 - x_{sim}$$
$$w_{magn} = y_{inv} \times 100 \qquad (1)$$
$$z_{dist} = w_{magn} \times log(w_{magn})) + 10$$

where x_{sim} is cosine similarity of two words from a word embedding model, y_{inv} is a cosine distance, the 10 constant is the radius of graphic circles representing nodes, and z_{dist} is the final distance value piped to the graph layout algorithm as a weight on the corresponding edge (the distance between the nodes).

Users can choose whether the *D3.js* library will be loaded dynamically from content delivery network (CDN) each time the generated files are viewed in a browser, or downloaded once when building visualizations and then stored locally together with the produced files. This way it is possible to generate interactive visualizations that are completely standalone and can be used for further demonstration or exploration, without requiring Internet access or access to word embedding models anymore (as long as no new words are queried).

If the query consists of several words, the library generates several interlinked HTML files. For example, if the query is '*cat, dog*', and '*dog*' is found within the nearest neighbors of '*cat*', this node will be clickable, and will open the corresponding visualization for '*dog*'.

Figure 6 shows examples of graph visualizations produced by *vec2graph* (8 nearest neighbors). The fully-connected graph (left) for the Russian word 'кот' (*cat*) is generated without setting the threshold of cosine similarity. However, when using the value of cosine similarity 0.8 as a threshold (right), we can see that 'кошак' (colloquial for '*cat*') is not a neighbor of 'кот' any more. We can also see that the words 'мурзик' (*murzik*) and 'барсик' (*barsik*), both common cat names, are connected to each other with an edge, unlike the other neighbors, which are linked to the query word only.

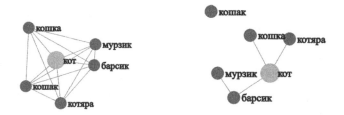

Fig. 6. *vec2graph* visualizations for the Russian word 'кот' (*cat*). **Left**: fully-connected graph. **Right**: threshold of cosine similarity to draw edges set to 0.8.

Vec2graph is extremely easy to use and is distributed via the Python Package Index. In its basic form, it can be run as follows:

```
pip install vec2graph
from vec2graph import visualize
visualize(OUTPUT_DIR, MODEL, WORD),
```

where `OUTPUT_DIR` is the directory to store HTML files with visualizations, MODEL is a pre-trained word embedding model loaded with *Gensim*, WORD is the query word(s). More details and the full source code are available at https://github.com/lizaku/vec2graph.

5   Conclusion

To sum up, we presented *vec2graph*: an open-source Python library to visualize sets of word embeddings as graphs, with words as nodes and edges determined by cosine similarities between these words. Graph representations of continuous vector semantic models can be useful to filter noise relations, which in turn leads to important insights about the structure of human language semantic space. Additionally, it becomes possible to employ the rich inventory of graph theory: random walks, community detection, graph statistics, etc. Last but not least, the library is intentionally made very easy to use, so that meaningful visualizations of word embeddings are available for large audience, without requiring serious programming skills.

In the future, we plan to integrate *vec2graph* directly into *Gensim*. Another direction for future work is automatic community detection (with corresponding coloring of graph nodes) in the resulting graphs, which can be help to handle polysemy in an even more convenient way.

References

1. Belinkov, Y., Glass, J.: Analysis methods in neural language processing: a survey. Trans. Assoc. Comput. Linguist. **7**, 49–72 (2019). https://doi.org/10.1162/tacl_a_00254
2. Bengio, Y., Ducharme, R., Vincent, P., Jauvin, C.: A neural probabilistic language model. J. Mach. Learn. Res. **3**(Feb), 1137–1155 (2003)

3. Bostock, M., Ogievetsky, V., Heer, J.: D-3: data-driven documents. IEEE Trans. Vis. Comput. Graph. **17**, 2301–9 (2011). https://doi.org/10.1109/TVCG.2011.185
4. Fares, M., Kutuzov, A., Oepen, S., Velldal, E.: Word vectors, reuse, and replicability: towards a community repository of large-text resources. In: Proceedings of the 21st Nordic Conference on Computational Linguistics, NoDaLiDa, 22–24 May 2017, Gothenburg, Sweden, pp. 271–276. Linköping University Electronic Press, Linköpings universitet (2017)
5. Hamilton, W., Clark, K., Leskovec, J., Jurafsky, D.: Inducing domain-specific sentiment lexicons from unlabeled corpora. In: Proceedings of the 2016 Conference on Empirical Methods in Natural Language Processing, pp. 595–605. Association for Computational Linguistics, Austin, November 2016. https://doi.org/10.18653/v1/D16-1057. https://www.aclweb.org/anthology/D16-1057
6. Healy, K.: Data Visualization: A Practical Introduction. Princeton University Press, Princeton (2018)
7. Hotelling, H.: Analysis of a complex of statistical variables into principal components. J. Educ. Psychol. **24**(6), 417 (1933)
8. Jolliffe, I.T., Cadima, J.: Principal component analysis: a review and recent developments. Philos. Trans. Royal Soc. A: Math. Phys. Eng. Sci. **374**(2065), 20150202 (2016)
9. Kutuzov, A., Kuzmenko, E.: WebVectors: a toolkit for building web interfaces for vector semantic models. In: Ignatov, D.I., et al. (eds.) AIST 2016. CCIS, vol. 661, pp. 155–161. Springer, Cham (2017). https://doi.org/10.1007/978-3-319-52920-2_15
10. van der Maaten, L., Hinton, G.: Visualizing data using t-SNE. J. Mach. Learn. Res. **9**(Nov), 2579–2605 (2008)
11. Mikolov, T., Chen, K., Corrado, G., Dean, J.: Efficient estimation of word representations in vector space. arXiv preprint arXiv:1301.3781 (2013)
12. Miller, G.A.: WordNet: A lexical database for English. Commun. ACM **38**(11), 39–41 (1995). https://doi.org/10.1145/219717.219748. http://doi.acm.org/10.1145/219717.219748
13. Navigli, R., Paolo Ponzetto, S.: BabelNet: the automatic construction, evaluation and application of a wide-coverage multilingual semantic network. Artif. Intell. **193**, 217–250 (2012). https://doi.org/10.1016/j.artint.2012.07.001
14. Pearson, K.: On lines and planes of closest fit to systems of points in space. London Edinburgh Dublin Philos. Mag. J. Sci. **2**(11), 559–572 (1901)
15. Řehůřek, R., Sojka, P.: Software framework for topic modelling with large corpora. In: Proceedings of the LREC 2010 Workshop on New Challenges for NLP Frameworks, Valletta, Malta, pp. 45–50, May 2010
16. Verlet, L.: Computer experiments on classical fluids. I. Thermodynamical properties of Lennard-Jones molecules. Phys. Rev. **159**, 98–103 (1967). https://doi.org/10.1103/PhysRev.159.98. https://link.aps.org/doi/10.1103/PhysRev.159.98
17. Wattenberg, M., Viégas, F., Johnson, I.: How to use t-SNE effectively. Distill **1**(10), e2 (2016)
18. Wildgen, W.: From Lullus to cognitive semantics: the evolution of a theory of semantic fields. In: Proceedings of the Twentieth World Congress of Philosophy. University of Bremen (1998)

Social Network Analysis

How to Prevent Harmful Information Spreading in Social Networks Using Simulation Tools

Ivan Dmitriev[1] (✉) ⓘ and Elena Zamyatina[2] (✉) ⓘ

[1] Perm State University, Perm, Russian Federation
ivandmitriev5@gmail.com
[2] National Research University Higher School of Economics, Perm, Russian Federation
e_zamyatina@mail.ru

Abstract. The paper discusses the problems of preventing harmful information spreading in social Networks. Social networks are widespread nowadays and are used not only for managers and marketers propagation of advertising, promotion of goods, but also by attackers to spread harmful information. Thus, there is a need to counter the attackers. This paper presents simulation tools and several features that contribute to the successful application for modeling social networks and examine different strategies preventing rumors and harmful information spreading. The authors cite an example of a simulation model for identifying intruders in a social network, software tools and the results of simulation experiments.

Keywords: Social networks · Dynamic modeling · Static modeling · Random graphs · Information dissemination

1 Introduction

Social networks have become an integral part of human life. So we always have the opportunity to quickly connect with friends or other people, find out the news bulletin for today, share our opinion. Social networks are used by large companies, individual entrepreneurs or managers in order to promote products and brands. They are used to spread advertising, technology, knowledge, investment, etc. [1–3]. Social networks are also used by government agencies and various communities to control public opinion. However, rumors [4, 5], and "fake" news [6], and malicious information [7] are successfully spread on the social network. We should not forget that social networks are successfully used by criminals and terrorists for the purpose of criminal conspiracy [8–10].

Large numbers of studies related to monitoring, network analysis and prediction of online networks development and information management have appeared. The work [1] considers several tasks being popular in online social networks: a task of analyses, a task of forecasting, a task of management. So a task of analysis and monitoring implies the collection of statistical data, the identification of changes in the online social networks

The study was carried out with the financial support of the Russian Foundation for Basic Research in the framework of the research project No. 18-01-00359.

© Springer Nature Switzerland AG 2020
W. M. P. van der Aalst et al. (Eds.): AIST 2019, CCIS 1086, pp. 201–213, 2020.
https://doi.org/10.1007/978-3-030-39575-9_21

and an assessment of various indicators. This could be, for example, identifying the initiators of the discussion, monitoring the actual information objects being discussed, assessing the naturalness of the discussions, etc. In forecasting tasks based on data that were obtained during network analysis usually are trying to predict how the network will change in the future. An example is the problem of disseminating information in the network. Management tasks include all the above tasks for the reason that, first of all, it is necessary to perform an analysis of the current situation in the social network, make a forecast of its development, and, depending on the purpose of the research, consider various management strategies. The task of spreading public opinions is an excellent example.

There are various approaches to solving these problems, one of which is the modeling of social networks [2]. This approach, in contrast to the usual network analysis (Social Network Analyzes (SNA)) [1, 11] is more flexible: you can build a social network model that takes into account the characteristics of specific studies, set the required modeling conditions. It is known that the mathematical abstraction of a social network is a graph, and since from the point of view of an outside observer, connections are formed randomly, random graphs are used [2, 12].

The article is structured as follows: static and dynamic approaches to a modeling of social networks are examined. Static approach supposed an analysis of structural characteristics of the online social networks; dynamic approach examines changes in network characteristics depending on time, cause-and-effect relationships. Moreover special software tools are proposed. These tools can be successfully used by both approaches. The following is a description of the task of distributing prohibited information in a network implemented in the simulation system Triad.Net.

2   Online Social Networks Modeling

There are two basic approaches to the analysis of social networks: static and dynamic one (as it was mentioned above) [1].

The first approach involves the study of network structure (topology), its basic properties (contiguity, the degree of centrality, distance and others) [2]. This approach supposes the investigation of the current state of a "snapshot" of a social network. Main attention is paid to the geometric characteristics of the network (structure of network), as well as the different relations between the nodes (members of the social network).

Static (structural) approach allows one to characterize accurately the current state of the system, but does not make it possible to see one to-many pattern that become visible only in the study of the structure of the network in dynamic. Indeed the useful information about social network "can be achieved at points in time through the use of polling and survey data, but the most interesting questions typically lie in the space in between these snapshots in time" [3]. The causal mechanism of the changes in social networks may be obtained due to simulation (in time). The static approach allows to understand such complex adaptive system as society, assists the scientists and managers to take an appropriate decision, but only simulation (discrete event or agent-based) "provides a fully traceable implementation of these concepts that readily accommodates the varying timescales at which events unfold within society" [3].

The study of structural characteristics allows us to fairly accurately characterize the current state of the system, but it does not allow us to understand many of the patterns that are noticeable only when time changes in dynamics. The authors of [1] enumerate tasks in which the structural characteristics of a social network are investigated: (a) searching for exact algorithms for generating recommendations of friends and content based on a social graph [13]; (b) identifying the initiators of the discussion; (c) monitoring of the actual objects under discussion; (d) forecasting the further development of the network, etc.

Today, thanks to the study of the structural characteristics of social networks, a large number of models have been developed that have structural similarities with real networks (for example, the Buckley-Ostgus model [14], the Watts-Strogatz model [15], the copy model [16], etc.

Dynamic modeling studies social networks in dynamics. Over time, the number of nodes in the network, communication between nodes may change. In addition, the processes taking place in the network should be considered: the dissemination of information, opinions, rumors, knowledge, etc.

There are such models of activity in social networks [1].

1. Macro-level models consider the network as a whole, without taking into account connections between nodes;
2. Micro level models view the network, taking into account the links:

   a. models with thresholds are models in which there is a threshold value or a set of threshold values used when a state changes;
   b. models of independent cascades (in which each node gets at a certain step a chance to activate other nodes) [17];
   c. propagation models based on analogies with physics, medicine, and other branches of science [4, 5];
   d. Leakage [18] and contamination [19] models are a popular way of studying the dissemination of information and innovation in social systems.

Dynamic modeling, in contrast to static modeling, allows studying the reasons for why the network is in a certain state, which event is the cause of its transition to another state. Thus, it is possible to solve more complex problems: predict the dissemination of information in networks [20], form public opinion [1], form a discussion topic [13], etc.

3 Combining Dynamic and Static Modeling

So, a lot of research confirms that the combined use of static and dynamic modeling is relevant. Consider some of these researches.

The authors of [21] studied various strategies for disseminating knowledge in the network of employees of the academic center. For this purpose, the authors developed a dynamic model (using the Monte Carlo method). The network structure in this study is static (the number of agents and the connections between them do not change), but the process of knowledge dissemination is dynamic.

The simulation was carried out according to the following scenario: at each moment of time for each pair of related agents, it turned out whether they had been in contact during this period of time or not, after which, if a contact occurred, one of the agents passed some piece of information to another. Upon reaching a certain level of awareness, the agent joined the dissemination of knowledge.

The knowledge dissemination strategy in this study implies choosing of agents who will initially disseminate knowledge. Four strategies were considered: (1) the first five agents selected by degree of centrality, (2) 5 agents with the big number of published work, (3) the first five agents selected by intermediate centrality (4) 5 central agents in clusters.

To assess the effectiveness of strategies, two key indicators were considered: the proportion of aware agents at specific time intervals and the amount of time required to spread knowledge between specific portions of agents.

This model was tested on the network of cooperation of the research center. The model was built on the basis of information about collaboration, the number of publications, etc.

The authors examined the effects of various strategies on the dissemination of knowledge and obtained the following results: the scenario in which agents were selected on the basis of centrality in clusters had the greatest impact on the dissemination of knowledge.

The author of [22] attempted to analyze the distribution of information in an ego-centric network that has a unique node as a source. The study is based on a stochastic multi-agent approach, where each agent is formed according to certain rules using data from the real social network Twitter. The modeling process consists of six phases and is iterative.

The first phase consists of uploading data from a real social network. After that, during the second phase, the topics and moods of the messages (posts) extracted from the sample data are classified. Then, in the third stage, sets of samples for each user are created from the previously classified data. Each set contains user messages and messages from his/her news feed (that is, messages from users who he/she is subscribed to). Each model of user behavior is built from these sets (fourth phase), and the models are used as input for a stochastic simulator (fifth phase). The model is executed. The loop is repeated several times until the most accurate model is found.

Experiments have shown that the proposed approach is promising for modeling user behavior in a social network.

A simulation model for the distribution of harmful information in the network was developed in [19]. An epidemiological model was used as the basis. The classical model of the spread of infection is based on the following cycle of the carrier disease: initially, a person is susceptible to infection. If this person contacts an infected person, he can with some probability become infected. Subsequently, the person over time either recovers, acquiring immunity, or dies; immunity decreases with time, and the person again becomes susceptible to infection.

A similar cycle was implemented in our paper. But we must notice that ordinary users who are susceptible to infection are attackers-agents.

The task of disseminating harmful information can be formulated as follows. The process of distribution of harmful information initiates any attacker agent by sending messages with harmful information to his list of contacts. An attacker can start a single attacker or group of them. They send messages through each time unit.

Subscribers-recipients, having accepted the message, with probability β are included in the attack process (become intruders). It is assumed that the user either read a message, either ignored it or deleted it altogether.

In addition, in each unit of time, the attacked nodes can be protected because of defense mechanisms impact. Thus, they cease to send harmful information and become immune to further attacks.

The simulation results are numerical arrays of data describing the dynamic process of propagation of harmful information (the number of attacking, protected and potentially vulnerable nodes in each time unit).

The simulation experiments were carried out using this model. The next section presents the task of distributing harmful information in the network and illustrates its implementation in the simulation system Triad.Net.

So the popularity of social networks is growing every day. In this regard, there are more and more threats against which you need to protect network users. One of threats: the distribution of harmful information in networks [7, 17]. Recently, a large number of publications related to the study of the process of dissemination of harmful information, rumors, etc. have appeared.

So in [7] an algorithm for large-scale networks monitoring with dynamically changing cascades of harmful information is considered. Several ways to decrease the spread of harmful information through immunization of network nodes are considered in [17]. The search for candidates for immunization is performed dynamically during the process of information dissemination.

Let us consider the task of identifying intruders on the network.

Formulation of the problem:

Given: N - the number of nodes equal to the number of network users; I_0 - the number of malicious subscribers - the primary sources of threat; R_0 is the number of subscribers initially insensitive to attacking influences; β - parameter that reflects the strength of the threat, the likelihood of an attack; γ is a parameter reflecting the degree of resistance to the threat, the probability of subscriber protection (β and γ in this study are defined as constants, but can be expressed as functions depending on the profiles of social network subscribers); t is the process time (in arbitrary units of time).

The process of distribution of harmful information (Z_I) initiates any attacker agent by sending messages with harmful information to his list of contacts. An attacker can be launched by a single attacker or a group of attackers. They send messages through each time unit. Subscribers-recipients, having accepted the message, with probability β are included in the attack process (become intruders). It is assumed that the user either read a message; either ignored it or deleted it altogether.

In addition, in each unit of time, the attacked nodes can be protected due to the impact of defense mechanisms. Thus, they cease to send harmful information and become immune to further attacks.

We implement this task using the Triad.Net modeling system.

The Triad.Net modeling system was developed at Perm State University [23–25] and is intended for modeling computer systems.

The Triad simulation model is divided into three layers: *a structure layer* (a set of objects connected by communication lines, with the help of which objects exchange information with each other), *a routine layer* and a *message layer*. In our case, the set of objects is the set of network users who are in a relationship with each other. The layer of routines - software tools for the implementation of scenarios of behavior of network users. The message layer is intended to describe messages of a complex structure. These messages are shared by network users.

The structure layer is a procedure with parameters. So, the computer network model, which is a dedicated server and several client computing nodes, can be described in Triad language as follows: star (Server, Node [1..n]), where star is a graph constant corresponding to the star network topology. In the Triad language, other graph constants are also defined: cycle (cycle), rectan (lattice), tree (tree), etc.

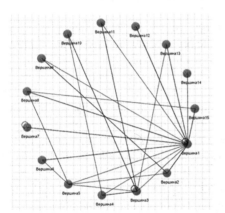

Fig. 1. Bollobás-Riordan model

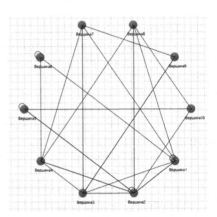

Fig. 2. Buckley-Osthus model

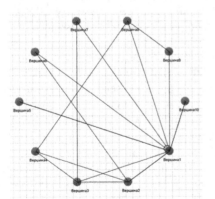

Fig. 3. Copying model

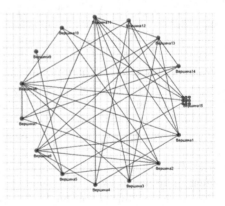

Fig. 4. Erdos- Renyi model

A distributed system consisting of 3 servers (client nodes are connected to each of the servers) can be described as follows: star (Server, Serv [a..c]) + star (Serv [a], Node [1..k]) + star (Serv [b], Node [k + 1..m]) + star (Serv [c], Node [m + 1..n]). Here operations (union of graphs) are used. It should be noted that the following operations with graphs are defined in the structure layer: add/delete vertices, arcs, edges, union, intersection of graphs, etc. The description above defines a whole family of structures. Also there are procedures for constructing random graphs, used as models of social networks (Figs. 1, 2, 3 and 4).

In addition, graph analysis procedures (diameter, number of vertices, edges, etc.), as well as: clustering coefficient, centrality, transitivity, mutual orientation, etc., are implemented in the layer of structures.

Routine is a specific scenario in which agents act. A routine consists of a sequence of events that plan each other. The execution of an event is accompanied by a change in the state of the object. Routine has input and output poles, with their help, agents interact with each other.

A description of routine using the simulation language Triad [25] may be represented as:

```
Routine <Name> { <A section of parameters> | <A defini-
tion of poles> }
[<A section of initialization of routine> ] { <A descrip-
tion of events>}
EndRout
```

The following is the code for the routine used in the task.

```
routine Rout[boolean Defence; boolean bad; real beta; re-
al gama](InOut pol[50])
// initialization
initial
  integer i;
  real Seed:=0;
  if (bad) then
    schedule SendMessage in 0;
  endif;
endi
//Events
// Event send messages
event SendMessage;
  // The command to send a message to all poles
  out " ";
  // With probability gama, an agent turns from an at-
  tacker into a protected one,
  // and stops sending messages.
  if (RandomReal()<gama) then
```

```
  Defence:=true;
    bad:=false;
  endif;
  if (Defence=false) then
    schedule SendMessage in 1;
  endif;
ende

// Input Processing Event
event;
  // With the probability of the beta, the agent joins
  the attack.
  //(Defence=false) check that the agent is not protected
  //(bad=false) check that the agent is not attacking
  if (Defence=false)&(bad=false)&(RandomReal()<beta) then
    schedule SendMessage in 1;
    bad:=true;
    Seed:=Seed+1;
  endif;
ende
endrout
```

Information procedures are used to collect statistical information during simulation experiments. Information procedures in the process of modeling observe the model elements (events, variables, poles). When the state of the observed object changes (i.e. when a variable value changes, when an event is executed, after a message arrives at the input pole or a message is transmitted from the output pole) the information procedure is connected to a specific model element and the data is processed according to the algorithm specified in the information procedure.

A description of procedure using the simulation language Triad [26] may be represented as:

```
Infprocedure<Name>(< A section of parameters >)]
initial< initial conditions >endi
handling// executed when one of the parameters changes
<Information Procedure Operator> { ";" <Information Pro-
cedure Operator> }endh
processing// executed when requesting a result
<Information Procedure Operator> { ";" <Information Pro-
cedure Operator> }endp Endinf
```

Below is the code of the information procedure, which calculated the number of attacking, vulnerable and protected agents.

```
infprocedure ip(
    in array[50] of boolean bad;
    in array[50] of boolean Defence)
    // initialization
    initial
      integer cbad, cdef, cfree, i;
      real prevtime:=0;
      print "time\t"+"bad\t"+"def\t"+"\tfree";
    endi
    handling
      if prevtime!=SystemTime then
        print prevtime+"\t"cbad+"\t"+cdef+"\t"+cfree;
      endif;
      cbad:=0;cdef:=0;cfree:=0;
      for i:=0 to 49 do
        if Defence[i] then
          cdef:=cdef+1;
        endif
        if bad[i] then
          cbad:=cbad+1;
        else
          cfree:=cfree+1;
        endif
      endf;
      prevtime:=SystemTime;
    endh
    endinf
```

During the experiments, several runs of a simulation model with different input parameters were carried out.

According to the results of the experiments, the following conclusions can be drawn:

- the I (t) attack process has an exponential dependence (Fig. 5);
- as β increases, the rate of infection of the nodes increases (attack intensity) (Figs. 5 and 6);
- with an increase in the probability of an attack β and a low probability of protection γ, the time of the defense process increases (Fig. 6);
- the number of protected agents grows slower due to the fact that the agents transition to this state from the attack state (Fig. 6);
- with a low probability of attack β and a high probability of protection γ, the attack process has a non-exponential form (Fig. 7).

The authors conducted a number of other experiments and confirmed the studies [19], which showed that the rate of infection of agents depends not on the network topology, but on the number of connections between agents.

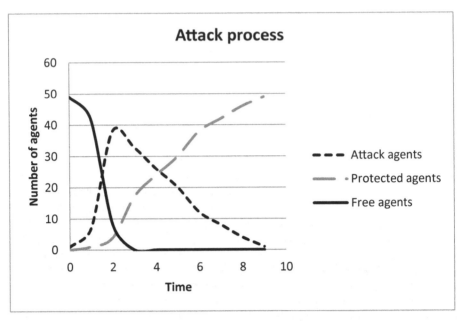

Fig. 5. Attack process, $n = 50$, $\beta = 0.6$, $\gamma = 0.3$

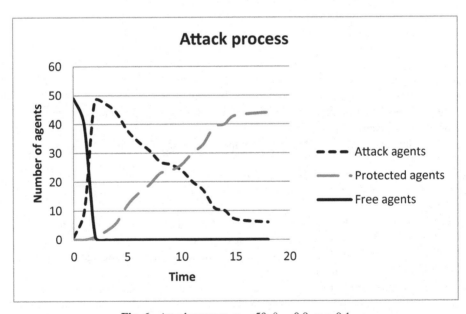

Fig. 6. Attack process, $n = 50$, $\beta = 0.9$, $\gamma = 0.1$

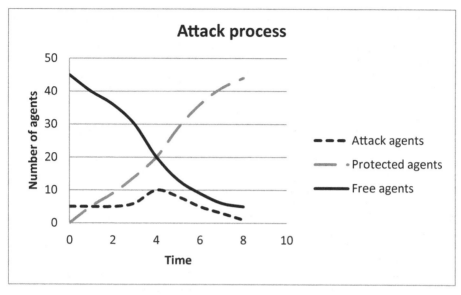

Fig. 7. Attack process, n $= 50$, $\beta = 0.1$, $\gamma = 0.9$

4 Conclusion

The paper considers an example of solving a problem of disseminating malicious information in a social network. To study the dissemination of information in the network, the simulation method and the simulation system Triad.Net were used.

Triad.Net software has a number of characteristics that make it a convenient tool for modeling social networks. Simulation tool Triad.Net includes procedures for constructing graphs, including random ones, used as models of social networks. The number of graph vertices is given as input parameters, so it is possible to construct a graph with a large number of vertices and with a complex structure, while the structure description in the Triad language remains concise. The procedures for analyzing graphs of a layer of structures make it possible to determine the structural characteristics of a graph, which are usually obtained when performing network analysis procedures. In addition, Triad.Net allows to explore and dynamic processes occurring in a social network.

So this paper demonstrated the viability of the Triad.Net for solving problems related to the research of social networks, including the tasks of disseminating of harmful information.

Acknowledgements. The study was carried out with the financial support of the Russian Foundation for Basic Research in the framework of the research project No. 18-01-00359.

References

1. Gubanov, D., Chkhartishvili, A.: A conceptual approach to the analysis of online social networks. Large-Scale Syst. Control (45), 222–236 (2013)

2. Davydenko, V.A., Romashkina, G.F., Chukanov, S.N.: Modelirovanie sotsial'nkh setei, pp. 68–79. Vesntik TSU (2005)
3. Zhao, N., Cheng, X., Guo, X.: Impact of information spread and investment behavior on the diffusion of internet investment products. Phys. A **512**, 427–436 (2018)
4. Zhang, Y., Zhu, J.: Stability analysis of I2S2R rumor spreading model in complex networks. Phys. A **503**, 862–881 (2018)
5. Zan, Y., Wu, J., Li, P., Yu, Q.: SICR rumor spreading model in complex networks: Counterattack and self-resistance. Phys. A **405**, 159–170 (2014)
6. Ilieva, D.: Fake news, telecommunications and information security. Int. J. "Inf. Theor. Appl." **25**(2), 174–181 (2018)
7. Yang, D., Liao, X., Shen, H., Cheng, X., Chen, G.: Dynamic node immunization for restraint of harmful information diffusion in social networks. Phys. A **503**, 640–649 (2018)
8. Bindu, P.V., Thilagam, P.S., Ahuja, D.: Discovering suspicious behavior in multilayer social networks. Comput. Hum. Behav. **73**, 568–582 (2017)
9. Tumbinskaya, M.V.: Protection of information in social networks from social engineering attacks of the attacker. J. Appl. Inform. **12**(3(69)), 88–102 (2017)
10. Filippov, P.B.: Use and implementation of personal data protection in social networks of the Internet. J. Appl. Inform. (2(38)), 71–77 (2012)
11. Dang-Pham, D., Pittayachawan, S., Bruno, V.: Applications of social network analysis in behavioural information security research: concepts and empirical analysis. Comput. Secur. **68**, 1–15 (2017)
12. Raigorodskii, A.M.: Proceedings of Moscow Institute of Physics and Technology (State University). In: Random Graph Models and Their Application, pp. 130–140 (2010)
13. Roth, M., et al.: Suggesting friends using the implicit social graph. In: Proceedings of the 16th ACM SIGKDD International Conference on Knowledge Discovery and Data Mining, Washington, DC, USA, pp. 233–242 (2010)
14. Buckley, P., Osthus, D.: Popularity based random graph models leading to a scale-free degree sequence. Discrete Math. **282**(1–3), 53–68 (2004)
15. Watts, D., Strogatz, S.: Collective dynamics of 'small-world' networks. Nature **393**, 440–442 (1998)
16. Kumar, R., Raghavan, P., Rajagopalan, S., Sivakumar, D., Tomkins, A., Upfal, E.: Stochastic models for the web graph. In: Proceedings of the 41st Symposium on Foundations of Computer Science, p. 57 (2000)
17. Zhou, C., Lu, W.-X., Zhang, J., Li, L., Hu, Y., Guo, L.: Early detection of dynamic harmful cascades in large-scale networks. J. Comput. Sci. **28**, 304–317 (2018)
18. Zhukov, D., Khvatova, T., Lesko, S., Zaltcman, A.: Managing social networks: applying the percolation theory methodology to understand individuals' attitudes and moods. Technol. Forecast. Soc. Chang. **129**, 297–307 (2018)
19. Abramov, K.G.: Modeli ugrozy rasprostraneniya zapreshchennoi informatsii v informatsionno-telekommunikatsionnykh setyakh., Vladimir (2014)
20. Newman, M.E.: A measure of betweenness centrality based on random walks. http://aps.arxiv.org/pdf/cond-mat/0309045.pdf
21. Kang, H., Munoz, D.: A dynamic network analysis approach for evaluating knowledge dissemination in a multi-disciplinary collaboration network in obesity research. In: Proceedings of the 2015 Winter Simulation Conference, Huntington Beach, pp. 1319–1330 (2015)
22. Gatti, M., et al.: Large-scale multi-agent-based modeling and simulation of microblogging-based online social network. In: Alam, S.J., Parunak, H. (eds.) MABS 2013. LNCS (LNAI), vol. 8235, pp. 17–33. Springer, Heidelberg (2014). https://doi.org/10.1007/978-3-642-54783-6_2
23. Zamyatina, E.B., Mikov, A.I., Mikheev, R.A.: TRIADNS computer networks simulator linguistic and intelligent tools. Int. J. "Inf. theor. Appl." (IJ ITA) **19**(4), 355–368 (2012)

24. Zamyatina, E.B., Mikov, A.I.: Programmnye sredstva sistemy imitatsii Triad.Net dlya obespecheniya ee adaptiruemosti i otkrytosti. Informatizatsiya i svyaz (5), 130–133 (2012)
25. Mikov, A.I.: Formal method for design of dynamic objects and its implementation in CAD systems. In: Gero, J.S., Sudweeks F. (eds.) Advances in Formal Design Methods for CAD, Preprints of the IFIP WG 5.2 Workshop on Formal Design Methods for Computer-Aided Design, Mexico, pp. 105–127 (1995)
26. Mikov, A.I.: Avtomatizatsiya sinteza mikroprotsessornykh upravlyayushchikh system. Irkutsk University Publ., Irkutsk (1987)

Effect of Social Graph Structure on the Utilization Rate in a Flat Organization

Rostislav Yavorskiy[1,2], Tamara Voznesenskaya[2], and Ilya Samonenko[2(✉)]

[1] Surgut State University, Surgut 628412, Russia
[2] Higher School of Economics, Moscow 101000, Russia
{ryavorsky,tvoznesenskaya,isamonenko}@hse.ru

Abstract. The goal of our research is to investigate how the communication structure of an organization affects its performance. In the paper, we study a simulation model of a self-organizing team conducting scientific research. The key parameter of the model is the social graph of the organization, which defines the team creation process. For this model, we formally define the average utilization rate of the group. Under some natural condition, the utilization rate is a function of the social graph. Lower and upper bounds of this characteristic are established. The obtained result has evident practical meaning and policy implications for organization management.

Keywords: Social graph · Simulation modeling · Social network analysis · Corporate networks · Team model · Self-organizing teams

1 Introduction

Project teams are important working structures for business continuity due to their high potential, high problem-solving ability and flexibility. Understanding and prediction of human interactions in teams is a key challenge for modern social sciences. Many factors must be considered to describe such processes: social, psychological, economic, etc. In this paper, we study a dynamic model, which describes self-organized collaborative teams in a flat organization.

The paper is organized as follows. A short overview presented in Sect. 2 covers the most relevant publications on mathematical and computational models used to analyze the team creation process and efficiency of self-organizing teams. In Sect. 3 we provide formal definitions, then different restrictions and conditions on the model are considered. The main part of the paper consists of theorems in Sect. 4, which formalize some properties of the model with respect to these conditions.

© Springer Nature Switzerland AG 2020
W. M. P. van der Aalst et al. (Eds.): AIST 2019, CCIS 1086, pp. 214–224, 2020.
https://doi.org/10.1007/978-3-030-39575-9_22

2   Literature Overview

2.1   Team Modeling Approaches

In [1] an analytical model for the *project team selection* problem is proposed by considering several human and nonhuman factors. Because of the imprecise nature of the problem, fuzzy concepts like triangular fuzzy numbers and linguistic variables are used. The skill suitability values of the candidates and the size of each project team are modelled as fuzzy objectives. The proposed algorithm takes into account the time and the budget limitations of each project and interpersonal relations between the team candidates. A simulated annealing algorithm is developed to solve the proposed fuzzy optimization model.

In [11] the *team formation* problem is studied. For a given task T, a pool of individuals $\mathbf{P}$ with different skills and a social graph G that captures the compatibility among these individuals the authors study the problem of finding $X \subset \mathbf{P}$ to perform the task. For a team built from X the effectiveness of teamwork is measured by the communication cost. The authors study two variants of this problem for two different definitions of communication cost and show that both variants are NP-hard. Their approach is quite similar to the one we consider in this paper.

In [2] a meta-model is developed that can be used to model specific *team recommendation* scenarios. The elements of this model are: input space, consisting of team member variables; output space, consisting of output/outcome variables; dependencies that map the input space to the output space; team external variables that influence a team from the outside; constraints that have to be satisfied before a team is composed according to a model.

A general team recommendation framework for finding the best deal teams is presented in [12]. The proposed framework can take into account diverse individual and team features, and accommodate various cost or feature functions. The authors introduce a team quality metric based on a weighted linear combination of these features, the weights of which are learned using a machine learning approach by leveraging historical project outcomes. A combinatorial optimization algorithm is used to search for the possible solution space for the approximate best team.

In [7] a *hybrid simulation* model has been developed to investigate how workers' predisposition to altruistic tendencies, an important personality factor, influences their willingness to help their co-workers on a production task. Model inputs were derived from experimental data, including participants' personalities, perceptions, and decisions regarding whether or not to help team members complete a task.

A mathematical model of teamwork and its organizational structure are presented in [17] in order to characterize a *group of sub-teams* according to two criteria: team size and information technology. The effect of information technology on the performance of team and sub-teams as well as optimum size of those team and sub-teams from a productivity perspective are studied and a quantitative sensitivity analysis is presented in order to analyze the interaction between these two factors through a sharing system.

In [14] the authors develop a *process mining* approach that is able to discover team compositions for collaborative process activities from event logs. The resource perspective (or organisational perspective) deals with the assignment of resources to process activities. Mining in relation to this perspective aims to extract rules on resource assignments for the process activities. The approach is evaluated in terms of computational performance and practical applicability.

Our model presented in Sect. 3 is influenced by all these publications but is not identical to any of them.

2.2   Case Studies and Specific Areas of Application

In [13] the nature of self-managing agile teams and the teamwork challenges that arise when introducing such teams are studied. The authors conducted extensive fieldwork for 9 months in a software development company that introduced Scrum. They describe a project through Dickinson and McIntyre's teamwork model [5], focusing on the interrelations between essential teamwork components. Similarly, in [9] the authors examine the evidence of the application of modelling and simulation techniques to support the management in eXtreme Programming (XP) projects. Sim-Xperience, a simulation model to assist the XP team in the management of their projects is presented. This model has been developed following the agent-based paradigm, well suited to simulate social behaviours. Thus, the model allows one to analyze the effect of different decisions on team management process, observing the evolution of the project development as well as the deviations in comparison with initial estimations.

In [18] three networks were studied, that representing socio-metric interactions in academic teams. The authors described a competence network using a weighted bipartite graph of users and concepts linking the former to the latter. A social network is described using a graph, where vertexes are users and edges are friendship between them. The authors also introduced the idea of a team network as a hypergraph, where the set of hyperlinks describe the joint appearance of users and concepts.

In [8] the results of the study of 23 software organizations in New Zealand and India are presented. The authors identified six informal roles that team members adopt in order to help their teams self-organize: Mentor, Co-ordinator, Translator, Champion, Promoter, and Terminator.

Another application area is considered in [3], where the authors describe the work of operators and a model of incident team formation called the Tailor-made Teams in Operations Centers (T-TOCs). Paper [6] presents a multi-agent model that describes students' participation in a group in a computer-supported collaborative environment. The developed system recognizes their team roles as they work collaboratively, automatically builds their profiles, diagnoses the state of the collaboration considering the balance of team roles as an ideal situation, and proposes corrective actions when the group behaviour is far from this ideal. In [16] the authors developed a teamwork dashboard, founded on learning analytics, learning theory and teamwork models that analyses students' online teamwork discussion data and visualizes the team mood, role distribution and emotional climate.

2.3 Overview of Utilization Rate Analysis

Utilization of IT professionals in software firms is studied in [10]. The authors develop a mathematical model for allocating different categories of IT professionals, according to the requirement of various consultancy projects, as a binary fuzzy multiobjective programming problem. The considered objectives are effort maximization at all phases for executing the projects, and overall cost minimization of the firm. A binary fuzzy goal programming technique is applied to find the solution to the problem.

Article [4] examines the utilization, underutilization, and misutilization of expatriate skills in overseas assignments. Using quantitative data from multinational corporations, the research first examines expatriates' utilization of eight distinct skills and how patterns of skill utilization influence important job attitudes. The study highlights the idea that effective skill utilization depends not only on the selection and training of expatriates themselves but also upon the level of skill and teamwork among host country nationals and the quality of support provided by the MNC as a whole.

In paper [15] the authors propose a formalism to assess parameters that contribute to employee utilization and contribution, such as individual effort, product level experience, defect percentile, project complexity, manager rating. Based on the presented analysis recommendations were suggested made for future resource allocation and areas of improvement.

As it follows from the literature overview above, the topic of team modelling and analysis generates regular interest among researchers. One may conclude that a meaningful modelling approach should take into account the skill profiles of the employees, their roles and communications. Analysis and improvement of employee utilization are widely covered in business administration literature, but papers, which use mathematical modelling for that are quite rare.

3 Formal Definitions

3.1 Simulation Modeling of Team Formation

In this paper, we consider a simplified version of the more general model developed for modelling and analysis of team performance in a research centre [19]. A detailed description of the generalized approach is to appear as a journal publication.

Input parameters of the model to analyze are the following:

- $\mathbf{P} = \{p_1, \ldots, p_n\}$ is a finite set of the group members (employees);
- $\mathbf{S} = \{s_1, \ldots, s_k\}$ is the set of different skills or roles required to perform a team projects;

– π denotes competencies profiles for the team members, that is

$$\forall p \in \mathbf{P} \ (\pi(p) \subseteq \mathbf{S} \wedge \pi(p) \neq \emptyset); \tag{1}$$

– $G \subseteq \mathbf{P} \times \mathbf{P}$ is an undirected social graph of the organization, so $G(p_1, p_2)$ indicates that employees p_1 and p_2 have established relations enough to invite each other to a new project.

The simulation process is executed in the following way:

1. **Project inception.** A free employee p decides to initiate a new project. At this moment the project team consists of the initiator only. If $\pi(p) \neq \mathbf{S}$ then just created team is incomplete and needs to be extended with members who have the missing skills from $\mathbf{S} - \pi(p)$.
2. **Team extension.** Every member of an incomplete team sends an invitation to join to all his contacts in the social graph of the organization. The employee, who received an invitation, may choose to join the team if his skills reduce the gap to the necessary skill set. Any employee can join not more than one project.
3. **Project start and finish.** For starting a new project the gathered team must have at least one member for every skill. So, when the skill set of the joined members reaches or exceeds the project required skills set all the pending invitation are cancelled and the team starts to work on the project. After finishing the project all the team members become free again.
4. **Project idea cancelling.** It may happen that the required skills set is too demanding or all the other employees are busy with their projects. So, the skills of the joined member are not enough and no one else intends to join the project team. Then the project is cancelled and already joined members become free.

The simulation process asynchronously runs the rules above for all the agents (employees). The resulting system behaviour is significantly affected by the following time-related parameters:

τ_{init} — time for a free employee to initiate a new project;
τ_{resp} — response time, time an employee needs before he rejects an invitation, or accepts it and invite his contacts in the social graph to join the team;
τ_{form} — team formation time after which the project initiator cancels the idea;
τ_{exec} — project duration time.

3.2   The Utilization Rate

In business, the utilization rate is an important characteristic that reflects the overall productive use of a firm or an individual. A standard way to calculate the utilization rate is to take the number of productive hours and divide by a total number of working hours. For the model described above we can define the following formal versions of this term:

$\rho_c(p)$ — utilization rate for an employee p in a particular run c of the model;
ρ_c — average utilization rate for the whole group in a particular run c;
ρ_{min}, ρ_{max} — correspondingly minimal and maximal group utilization rate for the given team $\mathbf{P}$ with profiles π and social graph G over all possible runs of the model.

Clearly, the time related parameters τ_{init}, τ_{resp}, and τ_{form} add to the idle time of an employee, and therefore reduce the resulting utilization rate. That is why for the analysis below we assume that

$$\tau_{init} = \tau_{resp} = \tau_{form} = 0. \tag{2}$$

From practical point of view that means that all the communications are fast and decisions are instant in comparison to the duration of projects.

Let c be a particular model run, $p \in \mathbf{P}$, and $\tau^+(p)$, $\tau^-(p)$ denote correspondingly the total productive time of employee p and the time when p was not working on a project. Then the utilization rate of an employee in the model run is defined as follows:

$$\rho_c(p) = \frac{\tau^+(p)}{\tau^+(p) + \tau^-(p)} \tag{3}$$

Let $\mathcal{C}$ be the set of all possible runs of the model. Then we define the *minimal and maximal utilization* rate for given social graph of organization G and profiling function π as follows:

$$\rho_{min}(G, \pi) = \min_{c \in \mathcal{C}} \left\{ \frac{1}{n} \sum_{i=1}^{n} \rho_c(p_i) \right\}, \tag{4}$$

and

$$\rho_{max}(G, \pi) = \max_{c \in \mathcal{C}} \left\{ \frac{1}{n} \sum_{i=1}^{n} \rho_c(p_i) \right\}. \tag{5}$$

Let Δ denote the duration of the modelled period. Then

$$\Delta = \tau^+(p) + \tau^-(p) \text{ for any } p \in \mathbf{P}. \tag{6}$$

So

$$\frac{1}{n} \sum_{i=1}^{n} \rho_c(p_i) = \frac{1}{n} \sum_{i=1}^{n} \frac{\tau^+(p_i)}{\Delta} = \frac{1}{n\Delta} \sum_{i=1}^{n} \tau^+(p_i). \tag{7}$$

It is convenient to calibrate the internal model time in such a way that

$$\Delta = 1 \tag{8}$$

then we get the following formulas for minimal and maximal utilization:

$$\rho_{min} = \frac{1}{n} \min_{c \in \mathcal{C}} \left\{ \sum_{i=1}^{n} \tau^+(p_i) \right\}, \quad \rho_{max} = \frac{1}{n} \max_{c \in \mathcal{C}} \left\{ \sum_{i=1}^{n} \tau^+(p_i) \right\} \tag{9}$$

for fixed G and π.

It is clear that for all the definitions one has

$$0 \leq \rho \leq 1 \tag{10}$$

3.3   Two Skills Case

In this paper, our intention is to provide a more deep analysis of the utilization rate for a special case of the presented model when

$$|\mathbf{S}| = 2 \text{ and } |\pi(p)| = 1 \text{ for any } p \in \mathbf{P}. \tag{11}$$

In practical terms it means that any project needs team members with two types of skills, e. g. a researcher and an engineer, and every member of the group is skilled in something one.

4   Lower and Upper Bounds for the Utilization Rate

In this section, we prove upper and lower bounds for the utilization rate for the two skills version of the model defined above. Let n_1 and n_2 denote the numbers of group members with first and second skill correspondingly.

$$n_i = |\{p \in \mathbf{P} : \pi(p) = s_i\}| \text{ for } i = 1, 2. \tag{12}$$

Theorem 1 (Tight upper bound). *For any social graph of organization* $(\mathbf{P}, G)$ *and skills distribution* π *the following holds:*

$$\rho_{\mathsf{max}}(G, \pi) \leq 2 \frac{\min(n_1, n_2)}{n_1 + n_2} \tag{13}$$

and this upper bound is tight.

Proof. Under assumptions of the two skills model, every project team consists of exactly two members, a researcher and an engineer. So, if their numbers do not coincide, for example, we have more researchers, $n_1 > n_2$, then in any moment of time at least $n_1 - n_2$ researchers are idle due to lack of engineers so at most $2n_2$ team members are working:

$$\rho_{\mathsf{max}}(G, \pi) \leq \frac{2n_2}{n_1 + n_2} \text{ if } n_1 \geq n_2. \tag{14}$$

If G is a complete graph, then the inequality (14) turns into equality, because after finishing one project an engineer can immediately find a mate researcher to start a new project, so n_2 engineers and the same number of researchers are always working.

Similarly

$$\rho_{\mathsf{max}}(G, \pi) \leq \frac{2n_1}{n_1 + n_2} \text{ if } n_1 \leq n_2. \tag{15}$$

Combination of (14) and (15) proves (13).   □

Notice that with the similar argument the upper bound result could be proved for $|\mathbf{S}| > 2$.

Theorem 2 (Tight lower bound). *Assume the following:*

1. *model simulations have zero communication time (2),*
2. *member profile function π satisfies the two skills assumption (11),*
3. *social graph of organization $(\mathbf{P}, G)$ is connected and every employee has a connection to someone with the complementary skill.*

Then the following holds:

$$\rho_{\min}(G, \pi) \geq \frac{2}{n_1 + n_2} \tag{16}$$

and this lower bound is tight.

Proof. Since every employee has a connection to someone with the complementary skill in G at every moment of time at least one project is on the go. Indeed, if everyone is free then instantly any team member may invite a friend with the opposite skill and they will start a new project. So, at least 2 employees are always occupied and lower bound (16) always holds.

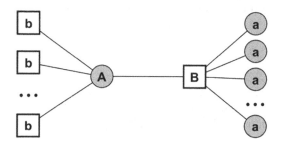

Fig. 1. Example of the social graph of a two-skilled organization. Squares indicate members with one skill, circles with the other one.

In order to prove that this bound is tight considered graph on Fig. 1. With such structure of communications in the organization, it may happen that nodes with labels "A" and "B" form a stable team, which successfully works on a sequence of projects while all the remaining employees are always idle due to lack of contacts with free employees with complementary skills.   □

5   Conclusion

In the paper a simulation model of a flat organization consisting of self-organizing teams has been suggested. The model allows to formally define the utilization rate of the organization on the base of the communication graph. Below are some ideas and possible directions for future research.

5.1 Social Network Analysis

Notice that under the instant time and instant decision assumptions (2) values of $\rho_{\min}$ and $\rho_{\max}$ are functions of the communication graph G labelled by the skill profile function π. So, it would be interesting to study dependencies between the utilization and other graph characteristics, such as diameter, clustering coefficient, centrality, the structure of the core, etc.

5.2 Recommendations for Team Supervisors and Managers

The main results of the paper have a clear interpretation for the organization managers. The worst case example shows, that even if one has to say 10 researchers and 10 engineers, all the people are actively willing to join a project (instant decision assumptions), the social graph is connected and, moreover, every engineer has a connection with a researcher and vice versa, even then it may happen that 18 out of 20 employees are always idle and $\rho = 10\%$ due to the inefficient structure of the communication graph. In practice, it is hardly possible to make the social graph of an organization complete, because of individual specifics and cultural differences between employees. But, adding at least some new communication links may significantly improve the overall performance of the company.

5.3 Enhancements of the Model

One research direction we consider is to enrich the model with more detailed formalization of individual skills, competencies and roles of employees.

Another direction would be to study different individual algorithms of decision-making procedures (here we assume that each team member is focused only on maximizing the overall performance).

References

1. Baykasoglu, A., Dereli, T., Daş, G.: Project team selection using fuzzy optimization approach. Cybern. Syst. **38**, 155–185 (2007). https://doi.org/10.1080/01969720601139041
2. Brocco, M., Groh, G., Forster, F.: A meta model for team recommendations. In: Bolc, L., Makowski, M., Wierzbicki, A. (eds.) SocInfo 2010. LNCS, vol. 6430, pp. 35–50. Springer, Heidelberg (2010). https://doi.org/10.1007/978-3-642-16567-2_3
3. Brown, J.M., Greenspan, S., Biddle, R.: Incident response teams in IT operations centers: the T-TOCs model of team functionality. Cogn. Technol. Work **18**(4), 695–716 (2016). https://doi.org/10.1007/s10111-016-0374-2
4. Feldman, D.C., Bolino, M.: Increasing skill utilization of expatriates. Hum. Resour. Manag. **39**, 367–379 (2000). https://doi.org/10.1002/1099-050X(200024)39:43.0.CO;2-7

5. Dickinson, T.L., McIntyre, R.M.: A conceptual framework for teamwork measurement. In: Brannick, M.T., Salas, E., Prince, C. (eds.) Team Performance Assessment and Measurement, pp. 31–56. Psychology Press, New York (1997)

6. Fares, R., Costaguta, R.: A multi-agent model that promotes team-role balance in computer supported collaborative learning. In: Cipolla-Ficarra, F., Veltman, K., Verber, D., Cipolla-Ficarra, M., Kammüller, F. (eds.) ADNTIIC 2011. LNCS, vol. 7547, pp. 85–91. Springer, Heidelberg (2012). https://doi.org/10.1007/978-3-642-34010-9_8

7. Green, J.J., Krejci, C.C., Cantor, D.E.: A hybrid simulation model of helping behavior. In: 2017 Winter Simulation Conference (WSC), pp. 1619–1630, December 2017. https://doi.org/10.1109/WSC.2017.8247902

8. Hoda, R., Noble, J., Marshall, S.: Supporting self-organizing agile teams. In: Sillitti, A., Hazzan, O., Bache, E., Albaladejo, X. (eds.) XP 2011. LNBIP, vol. 77, pp. 73–87. Springer, Heidelberg (2011). https://doi.org/10.1007/978-3-642-20677-1_6

9. Hurtado, N., Ruiz, M., Capitas, C., Orta, E.: Applying agent-based simulation to the improvement of agile software management. In: Mas, A., Mesquida, A., O'Connor, R.V., Rout, T., Dorling, A. (eds.) SPICE 2017. CCIS, vol. 770, pp. 173–186. Springer, Cham (2017). https://doi.org/10.1007/978-3-319-67383-7_13

10. Jana, R.K., Sanyal, M.K., Chakrabarti, S.: Binary fuzzy goal programming for effective utilization of IT professionals. In: Mandal, J., Satapathy, S., Sanyal, M., Bhateja, V. (eds.) Proceedings of the First International Conference on Intelligent Computing and Communication. AISC, vol. 458, pp. 395–405. Springer, Singapore (2017). https://doi.org/10.1007/978-981-10-2035-3_40

11. Lappas, T., Liu, K., Terzi, E.: Finding a team of experts in social networks. In: Proceedings of the 15th ACM SIGKDD International Conference on Knowledge Discovery and Data Mining, KDD 2009, pp. 467–476. ACM, New York (2009). https://doi.org/10.1145/1557019.1557074

12. Liu, H., et al.: A machine learning approach to combining individual strength and team features for team recommendation. In: 2014 13th International Conference on Machine Learning and Applications, pp. 213–218, December 2014. https://doi.org/10.1109/ICMLA.2014.39

13. Moe, N.B., Dingsøyr, T., Dybå, T.: A teamwork model for understanding anagile team: a case study of a scrum project. Inf. Softw. Technol. 52(5), 480–491 (2010). https://doi.org/10.1016/j.infsof.2009.11.004. http://www.sciencedirect.com/science/article/pii/S0950584909002043

14. Schönig, S., Cabanillas, C., Di Ciccio, C., Jablonski, S., Mendling, J.: Mining team compositions for collaborative work in business processes. Softw. Syst. Model. 17(2), 675–693 (2018). https://doi.org/10.1007/s10270-016-0567-4

15. Ajmani, S., Satya Sai Prakash, K., Prasad, D.S.U.M.: Human resource utilization analysis and recommendations using software analytics. Lect. Notes Softw. Eng. 4(3), 184–188 (2016)

16. Tarmazdi, H., Vivian, R., Szabo, C., Falkner, K., Falkner, N.: Using learning analytics to visualise computer science teamwork. In: Proceedings of the 2015 ACM Conference on Innovation and Technology in Computer Science Education, pp. 165–170. ACM (2015)

17. Tohidi, H., Namdari, A., Keyser, T.K., Drzymalski, J.: Information sharing systems and teamwork between sub-teams: a mathematical modeling perspective. J. Ind. Eng. Int. **13**(4), 513–520 (2017). https://doi.org/10.1007/s40092-017-0199-5
18. Ventresque, A., Yong, J.T.T., Datta, A.: Impact of expertise, social cohesiveness and team repetition for academic team recommendation. In: Datta, A., Shulman, S., Zheng, B., Lin, S.-D., Sun, A., Lim, E.-P. (eds.) SocInfo 2011. LNCS, vol. 6984, pp. 296–299. Springer, Heidelberg (2011). https://doi.org/10.1007/978-3-642-24704-0_33
19. Voznesenskaya, T.V., Krasnov, F.V., Yavorsky, R.E., Chesnokova, P.V.: Modeling self-organizing teams in a research environment. Bus. Inf. **13**, 7–17 (2019). https://doi.org/10.17323/1998-0663.2019.2.7.17

Analysis of Images and Video

Discernment of Smoke in Image Frames

Gurgeet Singh Bhogal[✉] and Anil Kumar Rawat

School of Electrical and Electronics Engineering, Lovely Professional University, Phagwara,
Punjab, India
bhogal237@gmail.com

Abstract. Smoke is a challenging object to detect because of its changing texture, color and shape etc. These features can be extracted with the help of learning algorithms like regression, SVM, decision tree, but they do not provide an optimum result when provided with the large dataset and comparatively, accuracy of the deep learning algorithm increases. The reason of the increase in the accuracy of the algorithm is that, it is trained on the provided dataset and with the increase in the number of input data, the extraction of the dominant features of the desired object will also increase, so that it can able define as well as the detect the similar object. For the detection of the smoke, convolutional neural network is used, which take an image as an input. Transfer learning is used in the algorithm, in which VGG-19 network is used but it does not provide the satisfying results, so it is modified by introducing batch normalization layers in the network. The batch normalization increases the converges rate of the network. The accuracy increases by 3% when the number of epochs increase from 10 to 50.

Keywords: Deep learning · VGG-19 · Convolutional neural network

1 Introduction

Smoke is the product of the combustion material in the presence of the oxygen. Generally, it is considered as the initial product, before the fire which causes much more damage to the property. And also, Smoke coming out of a vehicle causes the pollution in the air. So, to prevent the damage it should be stopped for which it should be detected before causing any conflict. Here the process of Smoke detection comes into play. The traditional detectors which are generally used, they work only in indoor as the smoke generated should pass through them for its detection which is the one of the disadvantages of the detectors. As for open area it is very difficult for these detectors to detect the smoke. So various other methods are used which used the images as an input, to detect smoke in the basis of its properties. Smoke detection mainly involves three steps pre-processing, feature extraction and classification. Pre-processing includes the denoising of the image, background subtraction and contrast matching etc. Denoising of the image can be done using filter or applying some threshold value on the image. In the feature extraction step, the properties of the object to be detected can be used like for example a chair, in this edge of the chair can be used its one of the features. Classification process includes the verification which tells us about the object detected is the desired one or not.

© Springer Nature Switzerland AG 2020
W. M. P. van der Aalst et al. (Eds.): AIST 2019, CCIS 1086, pp. 227–236, 2020.
https://doi.org/10.1007/978-3-030-39575-9_23

Features of the smoke changes with the time for example as the fire increases, the density of the smoke increases and if the oxygen is not regulated sufficiently from the combustion material, smoke is produced. For the better detection of the smoke deep learning can be able to withstand the process, because it itself decides the most suitable features which can be used for the best representation of the desired of object.

1.1 Convolutional Neural Network

Convolutional neural network takes an input of an image and can learn valuable parameters to differentiate that image from the rest of the image. It can detect both spatial and temporal features of an image, it basically depends on the amount of data used for the training of the network. It changes the form of the input image into simpler form, so that it is easy to understand but does not lose its features which are unique to that image. Single layer can extract only the low-level features of an image but with the addition of more layers the level of features extraction can be increased. Pooling layer is used for the reduction of the dimensions to decrease the computation power and also it helps to extract the dominant features. Pooling is of two types max and average. In the max pooling, maximum value of the particular position of the image is taken and in the average pooling layer, the average value of all the values is taken into consideration. Max pooling is more beneficial than the average pooling because it performs the noise reduction along with the dimensionality reduction, but in the other it performs the dimensionality reduction as a noise suppression. For the classification process, convolutional neural network use fully connected layers.

Training the network from scratch required a lot of training data to train the data we require the compatible graphic processing unit which can ease the training part by taking less time. Another way of training is using the pretrained network, they can provide us with better results with less amount of training data. Few of the pretrained network which are known to provide good results are VGG-net, Res-net, inception and Xception. In the VGG-net there are VGG 11,13, 16 and 19, the number describes the depth of the network. The VGG-19 contains 19 deep convolutional layers which is work on the basis of 144 number of parameters. [1]. VGG stands for the Visual geometry group. This network is the improvement of the Alex-net by increasing the kernel size with the addition of the various 7–10 convolutional layers. In the end, for the classification three fully connected layers are used. It achieves the accuracy of 92.3% on the image-net [2]. Objective of the paper:

- Modification of the pretrained network.
- Higher accuracy in a smaller number of epochs.

2 Review of Literature

Kaabi et al. [3] used the machine learning technique which is composed of numerous Restricted Boltz-man layer in which size of the frame can negatively affect the training and testing. Extraction and classification can be performed by the neural network simultaneously which is used by Yin et al. [4], by stacking the 14 layers which has the accuracy rate of 96.37% and false alarm rate of 0.60%. Convolutional neural network is modified in the sense that the individual convolutional layer used is with combination of the normalization layer and the overfitting problem is resisted by using more samples for the training than usual. Filonenko et al. [5] performed the comparative study in which they used the various pretrained network to see which have more accuracy in detecting the smoke and state of the art method outperforms the other network.

Kaabi [6] used deep belief network which is a stacked with Restricted Boltzmann machine(RBM). RBM is one of the machine learning algorithms which is used for dimensionality, reduction, classification and feature learning. Gaussian mixture model is also used to extract the regions containing smoke. Training system consist of two layers one is input layer and one is hidden layer. Loss function is the energy function which is used to define the probability distribution from the parameter model. With the addition of the output layer, it behaves as the multi-layer perceptron which uses stochastic gradient descent method for small sets. Then the classification is done by the logistic regression. Frizzi et al. [3], worked with three layers convolutional, pooling and then fully connected layers. Convolutional layers are the core building blocks which consist of rectangular grids and it is the convolution of the previous layer i.e. the detect the features which are easily recognizable. Pooling layer is the subsample of the input. Fully connected layer is associated with the high-level reasoning. The further the layers go the more the distinctive features are detected. After the layers training is done to detect the smoke from the image. Tao et al. [7] used the deep learning method for the detection of smoke and concluded that for the large dataset, the used traditional algorithms accuracy is decreases in comparison with the author's used algorithm. Smoke detection using deep learning.

3   Smoke Detection Using Deep Learning

In this algorithm, there is the use of transfer learning for the detection of smoke present in the given input. Transfer learning is the process of using the trained model to perform some another related task. The use of transfer learning is advantageous in comparison with the training of the network from the scratch because in it, the judgement is performed on the basis of the knowledge from the previous training of the network. This leads to earlier judgement i.e. the time taken to train the network is very less. Vgg-19 is the pretrained network which is used for the training of the dataset.

It contains 19 deep layers which consist of convolutional layers 3×3 size with number of filters increasing with the factor of 2. After some set of layers, there is max-pooling layer and in the three are three fully connected layers with the classification of 1000 objects. As we have to classify only 2 objects smoke and non-smoke, firstly we modify the last fully connected layer from 1000 to 2 and also modify the last classification layer to classify according to the 2 objects. Accuracy of Vgg-19 trained with the smoke dataset was not sufficient. Problem of internal covariance occur in the deep neural network because of the fact that the internal parameters of the network changes on the proceeding of the result from one layer to another with different distribution of the internal parameters. This results in the different outputs from each layer, this is one of the ill effects of the covariate shift. To reduce this batch normalization, as it increases the representation ability of the individual layers. For example, if the network is trained for the white cat, it does not contain batch normalization layer. It will not be able predict the colored cat, but if we use batch normalization it can somewhat have sufficient accuracy in the detection. Also, it speeds up the training, the higher accuracy can be achieved in a smaller number of the steps. It also decreases the problem of overfitting because of its regularization effect [8, 9] (Table 1).

Table 1. VGG-19 Architecture and Modified VGG-19 architecture

Conv3-64 and Batch- Normalization
Conv3-64 and Batch- Normalization
Max-pool
Conv3-128 and Batch- Normalization
Conv3-128 and Batch- Normalization
Max-pool
Conv3-256 and Batch- Normalization
Conv3-256 and Batch- Normalization
Conv3-256 and Batch- Normalization
Conv3-256 and Batch- Normalization
Max-pool
Conv3-512 and Batch- Normalization
Conv3-512 and Batch- Normalization
Conv3-512 and Batch- Normalization
Conv3-512 and Batch- Normalization
Max-pool
Conv3-512 and Batch- Normalization
Conv3-512 and Batch- Normalization
Conv3-512 and Batch- Normalization
Conv3-512 and Batch- Normalization
Max-pool
Fc-4096
Fc-4096
Fc-2
Soft-Max

4   Discussion

4.1   Dataset Analysis

Dataset is the data, used for the training and the testing of the algorithm. Dataset can be of images, video, audio etc. For the smoke, dataset available on http://staff.ustc.edu.cn/~yfn/vsd.html is used for the testing and training of the algorithm. On the site, there is availability of four set of images and three videos (Fig. 1).

Fig. 1. Images of dataset 1 and 2

Set 1 contain 552 smoke images and 831 non-smoke images. Set 2 contain 668 smoke images and 817 non-smoke images. Set 3 contain 2201 smoke images and 8511 non-smoke images. Set 4 contain 2254 smoke images and 8363 non-smoke images. In the smoke images, only dense smoke is there and in non-smoke images, it is the images of the landscape, buildings, people etc. For the training and testing purpose set 1 and set 2 is used. The size of images is 100×100. Set 1 contain the smoke images, which covered the whole image and they are from different surroundings. The density of the smoke in the image is high and comparatively there is a greater number of images which contain the black color smoke. Set 2 contains the smoke and non-smoke image of same type, but in different conditions.

In video 1, the smoke is of high density and other objects are also present like the vegetation and a person to make the detection difficult. In the video 2, the smoke is of low density, it will be challenging for the network to detect, as it is train on the high-density smoke images. In the background, there is vegetation which is contained in the training set, not the same but somewhat similar. In the video 3, the smoke is of black color and of high density and the network is trained on this type of smoke which will help in its detection. Background of the frame of video is one of the images in the set used for the training of the network.

4.2 Experiments

Transfer learning is the method which is used for the detection of the smoke. Firstly, the Vgg-19 was used for the training of the dataset with the training options of stochastic gradient descent method and initial learning rate of 0.01. After the 1000 iteration the there was no change in the convergence i.e. the network is not able to learn from the dataset. The problem was in the convergence of the graph and batch normalization is the layer which helps in the fast convergence of the network, modification was done. Batch normalization layers were added after convolutional layers in the network and the code was run for 10 epochs which is 1380 iteration. The learning rate is kept 0.001

because other learning rats like 0.1, 0.01 were not providing good results. The converges of the network increase in the tremendous rate i.e. from 0–300 iteration the graph starts converging toward the 100%.

Though the fluctuation is very high, but its accuracy was sufficient to go ahead with changing of the parameter to increase it even further. The accuracy came out to be of 94% with the detection rate of 88.1%, but with high error rate of 11.9%. Augmentation was also performed on the data as the images in the training dataset were only 1383 and training of the network required large amount of data for the network to learn from the images, the traits of the smoke particle. The error rate is more in the smoke because of the factor that the dataset used for the training the network is not sufficient for it, to learn about the smoke. The accuracy can be increased by increasing the number of iteration and dataset (Figs. 2 and 3).

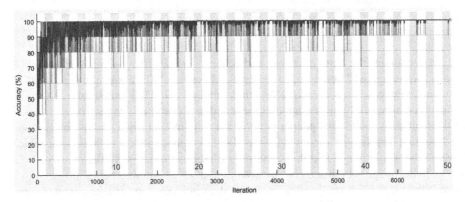

Fig. 2. Between training accuracy and iteration

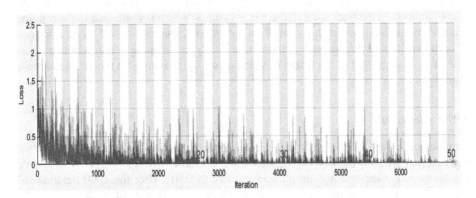

Fig. 3. Graph between training loss and iteration

On this term on the next algorithm the augmentation percentage of the images is increased further and also the number of iterations is increased from 10 to 50 epochs which increase the iteration from 1380 to 6900. The resulted graph by changing the parameters increase the accuracy of the network by 3%.

The convergence of the graph further increases, as in the last epochs the accuracy remains for 100% with the use of the transfer learning the network is learning from the previous knowledge it gained. This helps to achieve good results. The graph between the training loss and iteration shows that loss percentage decrease to approximately 0 value, which means that the provided dataset and augmentation process helps the network to properly define the features which can be used to define smoke.

Overall accuracy of the network came out to be 98.8% with the error rate of 1.2%. when the network is tested on the same set on which it is trained, the accuracy came out to be 100% i.e. the network is completely able to define the smoke present in the set 1 and also the non-smoke images present in it.

4.2.1 Testing On the testing of the algorithm on the videos provided in the dataset.

1. Video 1
    Few problems were faced but also, it gives good results on the data other than the trained data. In the first image, the smoke present is less in density and the network is trained on the dense smoke but the difference in the values can be seen when the frame is cropped to the part where the smoke is present, there is change in the one decimal value i.e. in the normal frame the value is 0.0048 and in cropped frame the value is 0.036 which means that smoke is detected somewhat but the network is being confused by the background in the back of the smoke (Fig. 4).

Fig. 4. Frames of video 1

2. Video 2
    In the first frame of the other video, the smoke is present which is dense in nature but it is only present in some part of the frame and the decimal value is approximately same as the value in the light smoke frame where the focus is on the smoke. In the second image, the part of the smoke is cropped it state the accuracy of detection 99.93% (Fig. 5).
3. Video 3
    Accuracy of the network is 99% in both the frames because of the fact that number of images of black is more in the training data set, the video has high density smoke of black color smoke according to which the algorithm is trained (Fig. 6).

Fig. 5. Frames of video 2

Fig. 6. Frames of video 3

4.3 Comparison

Comparison is done on the basis of detection rate, false alarm rate and accuracy.

$$\text{Detection Rate} = \frac{T_p}{Q_p} \times 100 \tag{1}$$

$$\text{False Alarm Rate} = \frac{F_p}{Q_n} \times 100 \tag{2}$$

$$\text{Accuracy} = \frac{T_p + Q_n - F_p}{Q_n + Q_p} \times 100 \tag{3}$$

In the training details, Tao et al. [4] done the 80,000-iteration using batch size of 128 and in comparison, the proposed performed 6900 iteration using batch size of 12. The author used the Alex-net network for the training of the data and modified its layers for the optimum accuracy (Table 2).

$$\text{Error Rate} = \frac{F_p + F_n}{Q_p + Q_n} \times 100 \tag{4}$$

Accuracy of the proposed algorithm is greater than the papers except two algorithms [10] and [14] but the error rate is less than the other paper algorithm. The accuracy can be increase by increasing the iteration. [4] implements the deep normalization and convolutional neural network, it is having less accuracy in comparison with the proposed algorithm. The epochs performed in this is 100 where as in the proposed it is only for the 50.

Table 2. Comparison between the proposed algorithm and the other algorithm

Paper	Detection rate	False alarm rate	Error rate
[10]	**98.7**	**1.35**	1.33
[11]	97	**1.35**	2.13
[12]	97.8	**1.10**	1.68
[4]	84.5	5.39	10.0
[13]	95.4	**1.67**	3.02
[14]	**98.2**	1.84	1.79
[7]	94.2	**0.86**	–
Proposed algorithm	**97.965**	1.71	**1.196**

The accuracy can be increased, but the algorithm is trained on Intel core i5-4210U Central Processing Unit with 8 GB ram, and Graphical processing unit in the system is less than the required for the training of the deep learning algorithm.

In the comparison, the proposed algorithm performed better than the [7] and in the other comparison, it outperforms the other algorithm in the error rate and four algorithms in detection rate. These algorithms are differently compared because of the formulas used by them to compute the parameters.

5 Conclusion and Future Scope

Deep learning which extract the feature automatically according to the training dataset provided us with good results. For the comparison purpose, only set 1 and set 2 are used for the training and testing. In the first training 1380 iteration, the detection rate came out to be 94% and when we increase the iteration to 6900 the detection rate increased to 97% and it can be further increased. There was a limitation in this algorithm that it was not able to detect the smoke of low density. The dataset used for the detection purpose has smoke which fully cover the image, so the proposed algorithm faces some problems in detecting the smoke in the environment where a greater number of objects were present. For the future, the dataset can be increased to increase the accuracy and also residual convolutional neural networks can be used which can produce better results.

References

1. Tsang, S.: Review: VGGNet—1st Runner-Up (Image Classification), Winner (Localization) in ILSVRC 2014 (2018)
2. KOUSTUBH: ResNet, AlexNet, VGGNet, Inception : understanding various architectures of Convolutional Networks (2018). https://cv-tricks.com/cnn/understand-resnet-alexnet-vgg-inception/

3. Frizzi, S., Kaabi, R., Bouchouicha, M., et al.: Convolutional neural network for video fire and smoke detection. In: IECON Proceedings of Industrial Electronic Conference, pp. 877–882 (2016). https://doi.org/10.1109/IECON.2016.7793196

4. Yin, Z., Wan, B., Yuan, F., et al.: A deep normalization and convolutional neural network for image smoke detection. IEEE Access 5, 18429–18438 (2017). https://doi.org/10.1109/ACCESS.2017.2747399

5. Filonenko, A., Kurnianggoro, L., Jo, K.H.: Comparative study of modern convolutional neural networks for smoke detection on image data. In: Proceedings - 2017 10th International Conference on Human System Interactions, HSI 2017, pp. 64–68 (2017). https://doi.org/10.1109/HSI.2017.8004998

6. Kaabi, R., Sayadi, M., Bouchouicha, M., et al.: Early smoke detection of forest wildfire video using deep belief network. In: 2018 4th International Conference on Advanced Technologies for Signal and Image Processing, ATSIP 2018, pp. 1–6. https://doi.org/10.1109/ATSIP.2018.8364446

7. Tao, C., Zhang, J., Wang, P.: Smoke detection based on deep convolutional neural networks. In: Proceedings - 2016 International Conference on Industrial Informatics - Computing Technology, Intelligent Technology, Industrial Information Integration, ICIICII 2016, pp. 150–153 (2017). https://doi.org/10.1109/ICIICII.2016.0045

8. Doukkali, F.: Batch normalization in Neural Networks – Towards Data Science. Web (2017)

9. Ioffe, S., Szegedy, C.: Batch Normalization: Accelerating Deep Network Training by Reducing Internal Covariate Shift (2015). https://doi.org/10.1007/s13398-014-0173-7.2

10. Yuan, F., Shi, J., Xia, X., et al.: Co-occurrence matching of local binary patterns for improving visual adaption and its application to smoke recognition. IET Comput. Vis. 13, 178–187 (2018). https://doi.org/10.1049/iet-cvi.2018.5164

11. Yuan, F., Shi, J., Xia, X., et al.: Sub oriented histograms of local binary patterns for smoke detection and texture classification. KSII Trans. Internet Inf. Syst. 10, 1807–1823 (2016). https://doi.org/10.3837/tiis.2016.04.019

12. Yuan, F., Shi, J., Xia, X., et al.: High-order local ternary patterns with locality preserving projection for smoke detection and image classification. Inf. Sci. (Ny) 372, 225–240 (2016). https://doi.org/10.1016/j.ins.2016.08.040

13. Mehta, R., Eguiazarian, K.E.: Texture classification using dense micro-block difference. IEEE Trans. Image Process. 25, 1604–1616 (2016). https://doi.org/10.1109/TIP.2016.2526898

14. Qi, X., Xiao, R., Li, C.G., et al.: Pairwise rotation invariant co-occurrence local binary pattern. IEEE Trans. Pattern Anal. Mach. Intell. 36, 2199–2213 (2014). https://doi.org/10.1109/TPAMI.2014.2316826

Face Recognition Using DELF Feature Descriptors on RGB-D Data

Andrey Kuznetsov[⊠] [iD]

Samara National Research University, Samara, Russia
kuznetsoff.andrey@gmail.com

Abstract. Face recognition is a very challenging and important task in many areas. This sort of algorithms is used in security systems, for person authorization tasks, person reidentification, etc. In face recognition speed and quality are very important, but in this research, we concentrate on quality improvement. In this paper we propose a new solution for face recognition using RGB-D data obtained with Kinect sensor. The proposed solution is based on a new type of feature descriptors – Deep Learning Features (DELF), which showed high classification results on Google landmarks dataset. We compared DELF descriptors with HOG, MSER and SURF descriptors using two classification schemes: error correcting output codes (ECOC) and decision trees. Conducted experiments showed an improvement in classification quality using F1 score metric when face recognition is based on DELF descriptors and ECOC classifier.

Keywords: Face recognition · Classification · Feature descriptor · DELF · HOG · MSER · SURF · ECOC · Decision tree

1 Introduction

Face recognition is performed in many areas: person identification, authorization, mimics recognition, etc. However, facial images taken in an uncontrolled environment may contain different head positions, facial expressions, lighting, and clothing. Since the variation type is unknown for such images, the development of an algorithm capable to processing all these factors at the same time becomes critical.

A sparse representation classifier (SRC) allows you to process face images with occlusion (e.g., sunglasses), which removes or adjusts the pixels of the masking elements. However, these pixels may have a similar texture with a human face and therefore cannot be identified. Some researchers also tried to solve the problem of head position invariance using 2D images. For example, Gross et al. [1] built Eigen-light fields, which are two-dimensional models of faces from all points of possible perspectives. This method requires a large training set containing images with different head positions. Moreover, images must be taken under fixed shooting conditions. Head position variations are non-linear and cannot be modeled by linear methods. That is why the performance of these methods rapidly decreases with extreme head positions.

Face images in various lighting conditions can be generated using a 3D model [2]. In addition, a face image can be used to correct the position of the head or to create a

© Springer Nature Switzerland AG 2020
W. M. P. van der Aalst et al. (Eds.): AIST 2019, CCIS 1086, pp. 237–243, 2020.
https://doi.org/10.1007/978-3-030-39575-9_24

new one. For example, the iterative closest point algorithm (ICP) [3] detects the optimal transformation to minimize the nearest point distance between two point clouds. Some approaches [4–6] use the final ICP error for face recognition. However, this point-to-point error is sensitive to changes in facial expressions.

To cope with the problem of facial expression invariance, Bronstein et al. [5] proposed a representation of facial surfaces based on isometric deformations. Mian et al. [4] proposed a multi-modal partial processing method that uses texture information and focuses on hard parts of the face. Kakadiaris et al. [7] suggested using the annotated face model (AFM) to transform the input three-dimensional model into a model invariant to facial expression. Passalis et al. [8] proposed an advanced AFM method that uses face symmetry to process missing data obtained by self-trapping in un frontal head positions. An extensive review of methods for recognizing 3D facial images is given in [9]. Existing 3D face recognition methods are not designed to handle cases with masking elements. Moreover, they all suggest the presence of high-resolution 3D scanners. Such scanners are expensive, cumbersome and have a slow speed of work, which limits their use (Table 1).

Table 1. Comparison of 3D scanners

Device	Processing speed (sec.)	Price ($)	Precision (mm)
3dMD	0.002	50000	0.2
Minolta	2.5	50000	0.1
Kinect	0.033	200	1.5–50

On the other hand, some 3D scanners process the scene relatively quickly and at the same time have low cost. For example, the Kinect sensor is not expensive, has a high scanning speed and compact size. These properties are most attractive for everyday recognition. However, the data provided by Kinect contain a lot of noise and have low depth resolution. Figure 1 shows a comparison of Kinect data with Minolta, however the 3D face model obtained using Kinect is far from perfect.

Fig. 1. 3D face model comparison for Kinect and Minolta sensors

In Sect. 2 we briefly describe the proposed algorithm steps including preprocessing and decision making. Section 3 is devoted to DELF description. In Sect. 4 classification algorithms used in this research are briefly described. Section 5 contains the dataset

[10] used and the results of conducted experiments. In this paper, we do research on the applicability of the Kinect depth sensor data used separately or in conjunction with the RGB image, when recognizing faces in different positions of the head and with different facial expressions. To show the advantage of the proposed solution results, we examined several descriptors and classifiers.

2 The Proposed Algorithm

The general scheme of face recognition systems is well known, and the proposed app-roach is based on existing pipeline. It can be divided into two main parts: face detection and face classification.

The first one part is implemented as a two-step procedure. The first one step obtains RGB and depth data from Kinect sensors and combines them to similar resolution. The RGB data on Kinect sensor has resolution 1280×960 pixels, whereas the depth sensor provides the data with resolution 640×480. To use these data sources together they are adjusted to a common resolution - 1280×960. The second step is face detection on an image. This step is based on Viola-Jones algorithm.

The second part of the proposed algorithm is based on descriptors calculation for the adjusted RGB-D data and further classification. At this step we select several well-known face descriptors: HOG, MSER and SURF and the new one approach - Deep Learning Features (DELF) [11].

3 Delf

Well known methods like SIFT and SURF provide good results in keypoints detec-tion but have limitations on applied transforms that can lead to incorrect detection or missed detection. This is why we decided to analyze the state-of-art Deep Local Fea-tures (DELF), proposed by Noh et al. [11]. These features are based on CNN trained with image-level annotations on a large landmark image dataset. The CNN structure described in the initial paper contains 2 main blocks (Fig. 2): dense localized feature extraction and keypoint selection (attention mechanism). Both blocks are coupled tightly in one CNN architecture. Both blocks are trained on the Google Landmark dataset which contains 1060709 images from 12894 landmarks and 111036 additional query images. This approach detects and describes semantic local features of an image, which can be geometrically verified between images showing the same object instance.

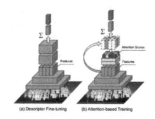

Fig. 2. DELF CNN architecture

The first block is based on a fully convolutional network (FCN) taken from the ResNet50 model, using the output of the conv4 x convolutional block. To handle different scales an image pyramid is constructed, and FCN is applied to all the scales. As a result, a dense grid of features is constructed. Finally, the features are localized on the basis of their receptive fields. The network is trained with a cross-entropy loss for image classification. During this stage no attention layers are used. Instead of them average pooling is applied. The input images are initially center cropped and rescaled to 250×250. Random 224×224 crops are then used for training. The second block is trained to assign high scores to relevant features and select the features with the highest scores. Attention mechanism leads to selection of the most valuable features from the dense field. In fact, this block represents a sophisticated pooling layer to reduce the number of features (features dimensionality reduction is another important step of the algorithm, but it does not concern training). This block is trained in the same way as the first one using image-level labels. However, the attention score function is not trained jointly with the feature extraction stage but is learned given the fixed descriptors after their fine-tuning. Random image rescaling is done during attention block training.

4   Classification Schemes

4.1   Error Correcting Output Codes (ECOC)

In a multi-class classification, the challenging part is to select a label from more than 2 possible options. For example, in the digit recognition task, we need to map each digit to one of 10 classes.

Some classifiers, such as a decision tree or a Bayes classifier, can solve a problem with several classes directly. Error Correcting Output Codes (ECOC) classifier consists of a set of binary classifiers (classifiers ensemble). ECOC is used in multi-class classification tasks. The ECOC algorithm consists of two stages:

1. a cascade of binary classifiers is constructed, each of which is responsible for eliminating some uncertainty regarding the correctness of the input data class;
2. a decision is made using a voting scheme that considers the decision of each classifier.

4.2   Decision Tree

Decision tree [12] is a hierarchical structure, which has non-terminal nodes defining feature space partition, and terminal nodes defining elementary classification function (in the simplest case – the class number). While constructing a decision function, the domain, which is a K-dimensional hypercube, is divided by axes. As a result, we receive a tree structure. There are different strategies for decision trees constructing. In this research we use C 4.5 decision tree [13].

5 Experiments

To carry out research we used a standard PC (Intel Core i5-3470 3.2 GHz, 8 GB RAM). As a source of data for research we used a free dataset [10]. The dataset consists of images for 31 persons. For every person there are 51 images made for 17 different head positions and 3 different facial expressions. The examples are presented on Fig. 3.

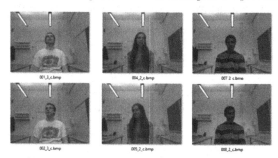

Fig. 3. Examples of images from the dataset [10]

As the images from the dataset contain not only face information but also background part, we use the first step of our approach to detect faces using Viola-Jones algorithm. For some head positions Viola-Jones was not able to detect any face, so we manually extracted it. The example of a face detection error is presented on Fig. 4.

Fig. 4. Face detection step with several mistakes that were corrected manually

When faces are extracted, we do the adjustment step for RGB and depth data. An example of this procedure is presented on Fig. 5. As we can see RGB and depth data have the same size.

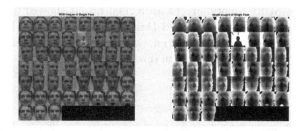

Fig. 5. RGB and depth data adjustment step result

After the first step of data preprocessing, we move to the second step of face recognition. As it was stated above, we used 4 types of descriptors with two classification schemes. Dividing the whole preprocessed dataset on train and test sets (1271 images for the train and 310 images for test) we could estimate the quality of the proposed solution.

To evaluate the performance of the proposed algorithm $F1\_Score$ ($F1$) measure is used, which is calculated using true positive (TP), false positive (FP) and false negative (FN) values:

$$p = \frac{TP}{TP + FP}, \ r = \frac{TP}{TP + FN}, \ F1 = \frac{2 \cdot p \cdot r}{p + r}.$$

The results of different classification algorithms comparison using four types of descriptors is presented in Table 2. The average feature extraction time for all descriptors is 0.5 s.

Table 2. F1 score values for different combinations of descriptors and classification algorithms

Classifier	DELF	HOG	MSER	SURF
RGB data				
ECOC	**0.95**	0.92	0.85	0.76
Decision tree	0.84	**0.86**	0.76	0.78
RGB-D data				
ECOC	**0.97**	0.82	0.85	0.91
Decision tree	**0.88**	0.86	0.74	0.76

The results show that DELF descriptors use lead to a slight increase in face recognition task for RGB and RGB-D input data.

6   Conclusion

In this paper, we proposed an algorithm for face recognition on RGB-D data using new deep descriptors DELF. We compared several classification schemes and 3 more descriptors calculation methods to show the advantage of DELF over the well-known HOG, MSER and SURF algorithms. Experimental results showed that DELF-based solution leads to quality increase for RGB-D input data as well as for RGB input data. Further we plan to fine tune DELF descriptors and compare classification results with deep CNN solutions for face classification (FaceNet, etc.).

References

1. Gross, R., Matthews, I., Baker, S.: Appearance-based face recognition and light-fields. IEEE Trans. Pattern Anal. Mach. Intell. **26**(4), 449–465 (2004)

2. Toderici, G., et al.: Bidirectional relighting for 3D-aided 2D face recognition. In: IEEE International Conference on CVPR, pp. 2721–2728 (2010)
3. Besl, P.J., McKay, N.D.: A method for registration of 3-D shapes. IEEE Trans. Pattern Anal. Mach. Intell. **14**(2), 239–256 (1992)
4. Mian, A., Bennamoun, M., Owens, R.: An efficient multimodal 2D-3D hybrid approach to automatic face recognition. IEEE Trans. Pattern Anal. Mach. Intell. **29**(11), 1927–1943 (2007)
5. Bronstein, A., Bronstein, M., Kimmel, R.: Expressioninvariant representations of faces. IEEE Trans. Image Process. **16**(1), 188–197 (2007)
6. Georghiades, A., Belhumeur, P., Kriegman, D.: From few to many: illumination cone models for face recognition under variable lighting and pose. IEEE Trans. Pattern Anal. Machine Intell. **23**(6), 643–660 (2001)
7. Kakadiaris, I., et al.: Three-dimensional face recognition in the presence of facial expressions: an annotated deformable model approach. IEEE Trans. Pattern Anal. Mach. Intel. **29**(4), 640–649 (2007)
8. Passalis, G., Perakis, P., Theoharis, T., Kakadiaris, I.: Using facial symmetry to handle pose variations in real-world 3D face recognition. IEEE Trans. Pattern Anal. Mach. Intel. **33**(10), 1938–1951 (2011)
9. Bowyer, K.W., Chang, K., Flynn, P.: A survey of approaches and challenges in 3D and multi-modal 3D+2D face recognition. Comput. Vis. Image Underst. **101**(1), 1–15 (2006)
10. Høg, R.I., Jasek, P., Rofidal, C., Nasrollahi, K., Moeslund, T.B.: An RGB-D database using microsoft's kinect for windows for face detection. In: The IEEE 8th International Conference on Signal Image Technology and Internet Based Systems, Italy (2012)
11. Noh, H., Araujo, A., Sim, J., Weyand, T., Han, B.: Large-scale image retrieval with attentive deep local features, pp. 3476–3485 (2017)
12. Quinlan, J.R.: Induction of decision trees. Mach. Learn. **1**(1), 81–106 (1986)
13. Wang, M., Gao, K., Wang, L., Miu, X.: A novel hyperspectral classification method based on C5.0 decision tree of multiple combined classifiers. In: Proceedings of the 2012 4-th International Conference on Computational and Information Sciences (ICCIS 2012), pp. 373–376 (2012)

Efficient Information Support of the Automatic Process and Production Control System

Oksana Logunova[1(✉)], Ivan Bagaev[1], Daria Arefeva[2(✉)], and Evgeniy Garbar[3]

[1] Nosov Magnitogorsk State Technical University, Magnitogorsk, Russia
logunova66@mail.ru, inprofess@yandex.ru
[2] LLC United Service Company, Magnitogorsk, Russia
arefewa.daria@rambler.ru
[3] LLC ITC "Fact", Magnitogorsk, Russia

Abstract. The paper describes a method of efficient specialized information support for the automatic process control system by saving only information about key elements of an image with a complex structure. When carrying out studies, the authors introduced a concept of the image with a complex structure, including an object under study and a set of elements affecting its uniformity. To describe the image, the authors introduced a structural unit of information with regard to the object under study, including a brightness class and brightness histogram, an initial point and description of the border; for elements inside the object under study: an initial point of the area of every element and a description of an element border. To implement a structural unit of information, the authors developed a functional diagram of software, consisting of two blocks: a block forming a mathematical description of the image to be stored in the corporate data warehouse; and a block using a mathematical description for an expert evaluation subject to requirements for a recovery of an initial image.

Keywords: Image · Image structure · Structural unit of information · Mathematical description · Data warehouse

1 Introduction

A rapid growth of industrial enterprise information environment fosters development of methods of efficient organization and management of specialized information support and software for the automatic control system for processes, production and technology process planning. Modern technologies of information support for process and production control systems involve designing and developing a structure of a data base and data banks, and alternative ways of data representation and storage, including graphic data [1].

At large enterprises a number of records in databases for storing process parameters amount to millions of records with over 500 fields and a size of every

© Springer Nature Switzerland AG 2020
W. M. P. van der Aalst et al. (Eds.): AIST 2019, CCIS 1086, pp. 244–255, 2020.
https://doi.org/10.1007/978-3-030-39575-9_25

record up to 15,000 MB. For example, at one of the largest Russian steelmaking companies, for a continuous casting division for full calendar 2017 and six months of 2018 only, data warehouse (DW) contained 6,211,239 records, and every record had 306 fields with a total size of 9180 MB. Regarding the rolling division, the maximum number for the same period totaled 18,998,408 records with 408 fields and 9760 MB of every record. A total size of records for all divisions in DW amounts to 1 EB. It should be noted that DW not only stores current data, but also backs them up for further recovery.

Methods of efficient organization and specialized information support of process and production control systems include [2–5]:

- optimizing a database structure, taking into account functional features of search requests for parameters, characterizing a production process: revision of query plans, redistribution of query weight classes, query refactoring, changes in a structure of indices, fine tuning of query plans;
- changing a composition and type of fields of a process parameter data base to increase reliability and completeness of data, characterizing the process: database denormalization and normalization, adding, removing and changing fields.

If it is possible to store original information in a smaller size (at least 30% less), data are to be optimized. Theoretical and practical studies on the design and development of methods of efficient organization and management of specialized information support and software for the automatic control system for processes, production and technology process planning include papers aiming at:

- offering methods of building and using control systems with a distributed structure, which should match control parameters for every section [6,7];
- analyzing features of building information support of PCS for potentially dangerous facilities [8,9].

At present, a graphic presentation of information in DW becomes more and more important. Graphic information gives a full presentation of a current or dynamic state of the facility. Now both conventional and new algorithms of graphic information processing are under active development [10]. However, to store graphic information in the DW operational space, we have to provide a lot of space, and, as a rule, such information is excluded from a set of stored information.

To use graphic information, our task is to find a compromise, when choosing a set and form of presentation of information, which is supposed to be stored in DW. The most popular approach is to save images using compression algorithms [11–15]. One of improved lossy compression algorithms JPEG is given in studies [11] and lies in dividing into an object and background of the image and applying different compression ratios for each taken separately. This contributes to increasing a total graphic data compression ratio and saving the image object better in quality than the background quality. The algorithm given in the paper is based on marking the most informative areas of the image, not having a uniform distribution of brightness inside the image and borders, defining the object. Less popular lossy compression methods include a method described in [12], which is based on a decomposition of quasi-cyclic components of the image into vectors of relevant sub-band matrices by applying a variance sub-interval frequency response analysis/synthesis. In simulation experiments, conducted by the authors of the paper, the compression method showed a high-level compression of images, whose energy was in a few frequency domains. A rather easy-to-understand lossy compression algorithm is described in paper [13]. Every pixel of the image has a one-to-one correspondence to X and Y from a random sequence of numbers, acquired by a Henon discrete mapping. Thanks to cancelling a fragment of the acquired, randomly transferred image at one of the steps of the compression algorithms, disk space taken up by graphic data decreases. The authors state that up to 70% of the image area, depending on its texture, is successfully subject to cancelling without significant losses. Papers [14,15] describe an image compression algorithm adapted to a graphic data structure. The adaptation is deemed to be a process of dividing the image into non-overlapping fragments, taking into account their information value or morphological structure. To solve a task of the adaptation, the authors introduce a concept of the mask, which show one or several fields of interest. It is assumed to mark the mask by statistical characteristics defined by a brightness histogram based on values of average entropy, characterizing changes in image brightness. An adaptive compression method described by the authors makes it possible to save information about the area of higher interest in the image close to its original accuracy and keep the file size small. The described algorithms are mainly an improved JPEG format and may be successfully used subject to limitations stated by the authors. Due to a variety of reasons attributed to a field of application, a small file size and relevant quality of graphic information is not always acceptable, when it is important to keep an original quality of marked objects in the image only, disregarding other ones.

Regarding current warehouses of industrial data about a process and equipment status, there is a contradiction: on the one part, graphic information about the object makes it possible to eliminate a human factor, when performing an expert evaluation, and, consequently, improve the quality of decisions taken in view of full information, applying modern information automated or automatic processing methods; on the other part, directly placed graphic information fills DW quickly, entailing high expenses for purchasing and maintaining hardware.

To eliminate contradictions, we developed a method of efficient organization and management of specialized information support and software for the automatic control system for processes, production and technology process planning.

2   Research Methods

Let us assume that when conducting a study on a production process, we acquire an image, showing an object under research in an arbitrary shape and its non-uniformity in darker irregular shaped elements (Fig. 1a). Such type of the image shall be called an image with a complex structure.

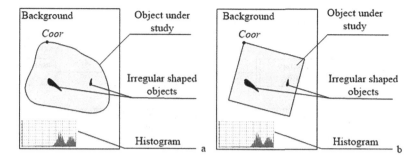

Fig. 1. Image structure: a - general view of the image with a complex structure; b - particular case of the image with a complex structure, using a sulfur print as an example.

Features of images with a complex structure are as follows:

- the object under study is limited and its borders are described by the continuous function $\Gamma = \Gamma(x; y)$;
- the object under study and elements inside have an irregular shape;
- the object under study in the image and elements inside are randomly arranged.

A simplified case of the image with a complex structure is considered to be a scanned image of a continuous cast square billet sulfur print, where the non-uniformity means formation and development of internal defects [16,17]. In this case, the object under study has a square shape, requiring the identification of one of its vertices *Coor* and a tilting angle of one of its sides with respect to borders of a complete image (Fig. 1b). The borders of the object under study are described by four equations of straight lines, whose positions are set online.

Considering the fact that a ratio between average background brightness and a main object under study is different for a set of images, it is required to introduce a parameter or a set of parameters to classify the difference between the background and the object. Paper [17] presents an image cascade classification algorithm, solving this task by assessing shape-generating parameters of the brightness histogram. In our further research we believe that a classification task has been already solved, and a route of marking and segmenting the object under study has been chosen.

A size of the file for saving the image with its full information capacity amounts to 7 MB. When designing a database for DW, conventional methods of image placement (routes 1 and 2) are considered and a new method (route 3) is offered, as given in Fig. 2.

The following symbols are introduced in Fig. 2: PP is a production process; DW is the data warehouse; I0 is an original image; I_1 is the image after its compression; I_2 is the image after its preliminary processing; Ar_{Pr} is an array of finite elements of the image with a complex structure; Sr_{MD} is a structural unit

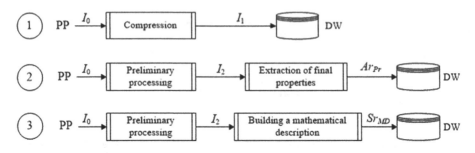

Fig. 2. Routes of a graphic data layout in DW.

of information built on a mathematical description of the image with a complex structure.

Route 1 describes a process of placing the image in DW after its compression by conventional algorithms [12–15].

Route 2 involves two stages:

- preliminary image processing, including algorithms of a classical theory of an image processing [10], aims at improving an original image for its further segmentation;
- extracting final properties, which results in creating an array of evaluations for internal elements of the object under study (for example, a point-based expert evaluation of development of internal defects).

New route 3 involves two stages:

- preliminary image processing, which is similar to the stage in route 2;
- creating a mathematical description of the image by a preset structure for key objects of the image.

Following routes 2 and 3, at stage 1 images are classified by algorithms described in studies [17], and an individual route of conventional preliminary processing algorithms is built.

A concept of a key object for every type of the image is additionally defined for every task.

The above routes are characterized by properties given in Table 1.

Having generalized properties, characterizing routes of placing graphic data in DW, we put forward a hypothesis about the advantages of representing the image as a mathematical description. To use a new technique, we developed a method of building a mathematical description of the image with a complex structure for the efficient organization of specialized information support for the process control system.

To describe the method, let us consider the image with a complex structure and introduce a system of coordinates on its surface (Fig. 3). According to [17], every image has a corresponding class with regard to a ratio between background brightness and the object under study. Let us set the value of the class $K_m, m =$

Table 1. List and evaluation of properties of routes for a graphic data layout in DW

No	Route properties	Route number		
		1	2	3
1	Providing the smallest size of the file with the image	+	+	−
2	Saving information about properties of elements in the image	−	+	+
3	Saving information about the image structure	+	−	+
4	Recovery of an original image	−	−	+
5	No need for saving the file with the image	−	−	+

$1...M$, where m – class number, M – number of image classes of the task under consideration. Image brightness is characterized by a histogram, containing 256 color shades. To get a mathematical description of the image histogram, it is enough to present a structure as the linear array $\boldsymbol{G} = (g_0, g_1, ..., g_{225})$.

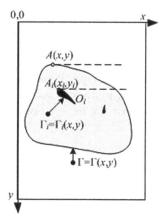

Fig. 3. Scheme of a structured image.

In the field of the image we mark the object under study, whose mathematical description contains coordinates of the border initial point $A = A(x, y)$ and the object border as analytic or table function $\Gamma = \Gamma(x, y)$. Due to the border of the object under study, we can segment the area for a detailed view of elements, defining a deviation from uniformity. The above parameters of the image form the first part of the structure of an image mathematical description and contain the fields: the brightness class, the image brightness histogram, the initial point of the object under study, and a description of the border of the object under study.

By segmenting the image for the object under study, we mark an array of elements to describe the deviation from the uniformity $\boldsymbol{E} = \boldsymbol{E}(E_1, E_2, ..., E_n)$

inside its border, and every element is a structure to save coordinates of the initial point of entry into the element area and the analytic or table function of the border of every element $A_q = A_q(x_q, y_q), q = 1..n$, where n - number of elements of the object under study. Thus, we formed the second part of the structure for a mathematical description of the image – parameters of the object.

Considering that during the production process a set of images for every unit of products is formed or selectively, for products of a critical application, a structural unit of information is formed, containing a structural field of a mathematical description of the image:

$$\boldsymbol{Sr}_{MDj} = \boldsymbol{Sr}_{MDj}(FI_j\{K_{jm}, G_j, A_j, H_j\}, OI_j\{|A_{jq}, H_{jq}|\}), \tag{1}$$

where j - record ID in DW; FI - a structural unit of information to describe the image in general; OI - a structural unit of information to describe sets of elements inside the border of the object under study; q - element number in the field of the object under study, $q = 1..n$; n - number of elements of the deviation from object uniformity. A graphic representation of a structural unit of information for the field of a mathematical description of the image in DW is given in (Fig. 4).

A structure for a mathematical description of the image involves development of a relevant algorithmic environment, enabling the automatic expert evaluation of elements in the image in compliance with state or industry-specific standards and recovery of a similar original image, if necessary. Figure 5 shows a flow diagram of software for creating a mathematical description of the image with a complex structure and its use for the expert evaluation of the deviation from uniformity of the object under study.

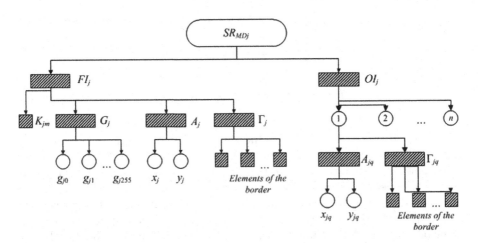

Fig. 4. Graphic representation of a structural unit of information built on a mathematical description of the image with a complex structure.

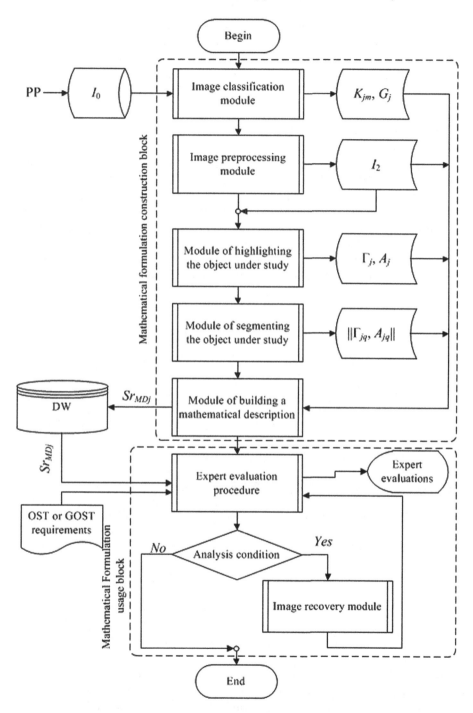

Fig. 5. A flow diagram of software for creating a mathematical description and its use for the expert evaluation of the deviation from uniformity of the object under study in the image.

In Fig. 5 a condition of the evaluation determines a need for recovery of the image to compare expert evaluations acquired in the automatic mode with a direct visual evaluation of the expert. Following the diagram given in Fig. 5, software was developed to carry out tests and evaluate efficiency of saving only a mathematical description of the image in corporate DW.

3   Results of the Research

One of tasks solved with a theory of segmentation of the image with a complex structure is saving space occupied by the image in corporate DW with information about the quality of continuous cast billets, when designing a continuous casting process feedback control system [18–20].

Sulfur print images of continuous cast billet transverse templates are chosen as initial data. Figure 6 shows typical images, corresponding to different classes. A total amount of the tested base is 110 images for a period of 2011–2018. Every image contains an object under study in a shape similar to a square one.

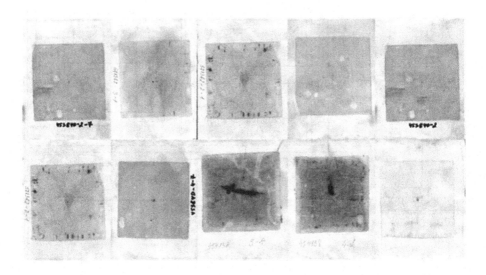

Fig. 6. A set of images for software testing.

Using software, we can open image I_0, process it and get image I_2. A list of segmented elements and a mathematical description of the image are built for image I_2 to be stored in DW (Fig. 7).

We calculated the dispersion with a recovered image and a ratio of compression by conventional methods with regard to every image. Results of the simulation experiment with averaged parameters are given in Table 2.

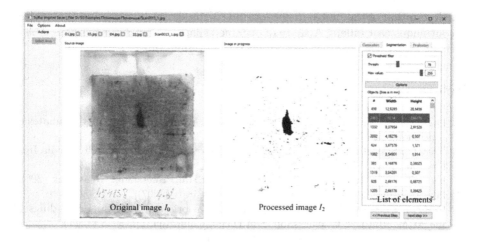

Fig. 7. A software window.

Table 2. A final share of the image after its compression with regard to image formats and dispersion

Compression ratio	Conventional compression methods				Amethod of saving key parameters	
	JPEG	Dispersion	PNG	Dispersion	SISImg	Dispersion
Minimum	24.67%	14.6175	51.24%	94.0750	0.53%	0.1218
Average	8.24%	3.0487	43.83%	64.4573		
Maximum	2.98%	0.3743	43.31%	64.8741		

4   Conclusions

The analysis of theoretical and practical developments of using, compressing, and storing images showed that the use of images contributed to an increase in completeness and reliability of information, when taking decisions following the expert evaluation of the object under study. The object under study may correspond with many branches, including the evaluation of quality of steel products, scheduling maintenance and repairs of equipment, evaluation of uniformity of the object during a medical examination, assessment of a flat surface, etc. The authors offered a method of saving key parameters of images with a complex structure, differing from conventional ones by the fact that DW stores a mathematical description of an image structure, including information about a structure of a main object under study and a structure of elements inside the object. A structure of the image characterizes a brightness class, brightness histogram, an initial point of the object under study and all its elements, a description of a border of the object and a border of elements inside the object. The stated elements of the structure are sufficient for an expert evaluation in compliance with relevant OST and GOST standards. Following the functional diagram, the authors developed software, which proved efficiency of saving only a mathematical description of the image in corporate DW, as shown by assessing the quality of continuous cast billets. A share of occupied space of a final image size is 0.53% without losing the information content of an original image.

References

1. Logunova, O.S., Devyatov, D.K., Nurov, K.K.: Computerized quality estimates of continuous-cast billet. Steel Transl. **35**(9), 36–42 (2005)
2. Tow, D.: SQL Tuning: Generating Optimal Execution Plans. O'Reilly Media Inc., Newton (2003)
3. Kyte, T.: Effective Oracle by Design. McGraw Hill Professional, New York (2003)
4. Lewis, J.: Oracle. Cost-Based Oracle Fundamentals (2007)
5. Codd, E.F.: Normalized data base structure: a brief tutorial. In: Proceedings of the ACM SIGFIDET, Workshop, San Diego, California, pp. 1–18 (1971)
6. Isaev, Yu.Yu.: Methods of efficient organization and information support and software of the APCS with a distributed structure. Ph.D. (Eng.) thesis: 05.13.06, 94 p. A place of thesis defense: Moscow Automobile and Road Construction Institute (Technical University), Moscow (2009). il. RSL TD, 61 09-5/3168
7. Smolin, P.A.: Methods of efficient organization and specialized information support and software of the generative computer-aided process planning system with a distributed structure. Ph.D. (Eng.) thesis: 05.13.06, 107 p. A place of thesis defense: Moscow Automobile and Road Construction Technical University (MADI), Moscow (2011). il. RSL TD, 61 12-5/491
8. Sineshchuk, M.Yu.: Features of ensuring information security of PCS for potentially dangerous facilities. Mod. Technol. Ensuring Civ. Defense Emerg. Response Recov. **2**(1(6)), 49–51 (2015)
9. Korneeva, M.D., Trutsina, Yu.Yu.: Information security of automatic process control systems. Educ. Sci. Borders: Soc. Hum. Sci. (3), 211–214 (2016)
10. Gonzalez, R., Woods, R.: Digital Image Processing. Tekhnosfera, Moscow (2005)
11. Foody, G.: A relative evaluation of multi-class image classification by support vector machines. IEEE Trans. Geosci. Remote Sens. **42**(6), 1335–1343 (2004)
12. Ivanov, V.G., Lyubarsky, M.G., Lomonosov, Yu.V.: Reducing content-related redundancy of images by classifying objects and a background. J. Autom. Inf. Sci. (3), 93–102 (2007)
13. Zhilyakov, E.G., Chernomorets, A.A., Goloshchapova, V.A.: An image compression method based on the expansion of image quasi-cyclic components in eigenvectors of relevant subband matrixes. Belgorod State Univ. Sci. Bull. Ser.: Hist. Polit. Sci. 13-1(108), 191–195 (2011)
14. Gora, S.Yu., Dovgal, V.M.: A method and tools to complete a task of compressing images applying chaotic dynamics. Sci. Notes Electron. Sci. J. Kursk State Univ. 4-2(24), 25–28 (2012)
15. Samira, E.K., Sulema, E.S.: An adaptive method of image compression. Herald Khmelnytskyi Natl. Univ. Ser.: Techn. Sci. (2), 125–131 (2010)
16. Posokhov, I.A.: Visualizing and processing information about the quality of continuous cast billets. Electrotechn. Syst. Complexes 2(31), 35–43 (2016)
17. Logunova, O.S., Shakshin, V.V., Logunov, S.M.: A mathematical description of objects of an irregular shape in colored images. In the Collection of Papers: Automation of Process and Production Processes in Metallurgy, pp. 54–62 (2009)

18. Logunova, O.S., Matsko, I.I., Posokhov, I.A., Lukyanov, S.I.: Automatic system for intelligent support of continuous cast billet production control processes. Int. J. Adv. Manuf. Technol. **74**(9–12), 1407–1418 (2014)
19. Logunova, O.S., Devyatov, D.K., Yachikov, I.M., Kirpichev, A.A.: A mathematical simulation of macroscopic parameters of solidification of continuous cast billets. Izv. Ferrous Metall. (2), 49–51 (1997)
20. Logunova, O.S., Matsko, I.I., Safonov, D.S.: Simulation of the thermal state of the infinite body with dynamically changing boundary conditions of the third kind. Bull. S. Ural State Univ. Ser.: Math. Simul. Program. (27), 74–85 (2012)

An Ensemble of Learning Machine Models for Plant Recognition

Vladimir Mokeev$^{(\boxtimes)}$ (ID)

South Ural State University, Chelyabinsk, Russia
mokeyev@mail.ru
https://ite.susu.ru/ite-staff/staff.html

Abstract. Plant recognition is an important problem and can be performed manually by specialists, but in this case, a lot of time is required and therefore is low-efficiency. Thus, automatic plant recognition is an important area of research. In this paper, we propose an ensemble of models to increase the performance of plant recognition. The ensemble of models presents a composite model which has two-level architecture. At the first level of the stacking model, convolutional neural networks are used, which demonstrate high performance in solving problems of object recognition. At the second level, gradient boosting methods are used. The model is taught using an open database of plant images containing 12 different species. Experiments with a plant dataset show that the proposed model is significantly better than other classification methods. High classification accuracy makes the model very useful for supporting the plant recognition system for working in real conditions.

Keywords: Plant · Recognition · Convolutional neural networks · Stacking · Gradient boosting · Random forest

1 Introduction

In this paper, we consider the problem of recognizing a plant image. Plant recognition is a critical problem especially for biologists and chemists and can be performed by human experts manually, but it is time-consuming and low-efficiency. There are many methods and approaches to solve this problem, which can be divided into a few groups. The first group of plant recognition methods uses the shape of the leaves, the texture, and the color characteristics of the plant images. In Ref. [1] the shape parameters of a leaf (length, width, area, perimeter) are used for recognition plant. In Refs. [2] and [3] the shape based approaches using leaves contours are described. A method of recognizing plant leaves by combining the texture and shape features is proposed in Ref. [4]. In Ref. [5] the shape and texture parameters are used to differentiate between lobed and unlobed leaves. Color characteristics with shape parameters are used for weed recognition in fields [6], the shape, color and texture characteristics are jointly used [7,8]. To extract parameters of plant the different modeling techniques are applied: fuzzy

© Springer Nature Switzerland AG 2020
W. M. P. van der Aalst et al. (Eds.): AIST 2019, CCIS 1086, pp. 256–262, 2020.
https://doi.org/10.1007/978-3-030-39575-9_26

logic [9], fractal dimensions [10], Fourier analysis [11], wavelets [4,12], curvelets [13] and Zernike moments [14] are used. The different classifiers are used: neural networks [4], genetic algorithm [5], support vector machines [12,13].

The second group of methods presents reduction methods, such as Principal Component Analysis (PCA), Linear Discriminant Analysis (LDA), Locality Preserving Projections (LPP) and Locally Linear Embedding (LLE). They are often applied to reduce the dimensionality of the extracted features. PCA and LDA are used to detect legumes in Ref. [15]. The approach for feature subset selection to classify the leaf images based on Genetic Algorithm (GA) and Kernel PCA (KPCA) is described in Ref. [16].

The third group of methods includes Convolutional Neural Networks (CNN). CNN's are seen as a breakthrough for image classification and seem to be very promising for this purpose CNN's have been very successful in many machine learning areas, including image recognition [17,18,25], speech recognition [19], object detection [20,21], image superresolution [22–24]. CNN is a special architecture of artificial neural networks, proposed by Yann LeCun in 1988. Despite the success of CNN application, the mechanism behind them is still not fully understood.

Although several effective plant recognition methods have been proposed there is a potential to improve the performance of plant recognition. In this paper, we study an ensemble method, which presents a technique that combines a few machine learning methods into one predictive model. There are different techniques: bagging, boosting and stacking. We focus on stacking technique which is introduced by Wolpert in the year 1992 [26]. We study two level stacking models for plant recognition. At the first level, different CNN models are used. At the second level, the results are refined using meta-classifier (random forest, gradient boosting and etc).

2   Stacking Method

Stacking method uses several machine learning technique to obtain better predictive performance than could be obtained from any of the constituent machine learning methods alone. The stacking procedure has two levels. The first level classifiers are called base learners and second level classifier is named a meta-learner. At the first level, CNN's are applied. To CNN learn to make predictions from training the dataset, we perform k-fold cross-validation procedure. The cross-validation predictions of different CNN's are combined to form of a new matrix, which is called the "level-one" data. CNN consists of four types of neural layers, which are convolutional layers, pooling layers, dropout layers, and fully connected layers. Different kinds of layers play different roles.

The second level input data uses the output of the first level models ("level-one" data) and the second level model learns to make predictions from this data. Thus, we train the second level classifier on the level-one data. The following methods can be used as second-level classifiers: Random Forests, gradient boosting, Neural Network, etc.

Random forest is one of the popular classification methods. The reason for this was not only the high accuracy of Random forest. Firstly, the Random forest guarantees protection from overfitting, even in the case when the number of features significantly exceeds the number of observations. Secondly, a training sample with the use of a random forest may contain features measured in different scales: numerical, categorical, and nominal, which is unacceptable for many other classification methods. Gradient boosting is a technique, which generates a prediction model in the form of an ensemble of weak prediction models. The gradient boosting is based on the gradient descent algorithm, which minimizes prediction loss when new models are added. Extreme gradient boosting (XGB) follows the idea of gradient boosting, which transforms many weak predictive models into a strong model.

The image dataset is divided into two datasets. The first dataset is used to train and validate the model. The second dataset (test dataset) is used to assess the performance of the trained model. We use a 90/10 division, i.e. 90% of the images are used for training and validation, and 10% is left for the test dataset. When models of first and second levels are trained, we use the procedure of 10 fold cross-validation. The predicted values of the validation dataset are used to assess the model performance. Since the training set has a small size, the procedure of data augmentation is used. As a way of data augmentation, image rotation is used. As a result, the size of the training dataset increases, although the resulting training examples are, of course, highly interdependent.

3   Experiments and Results

Dataset and Performance Metrics. To build the stacked model for plant recognition the dataset of plant images is used. The dataset has been recorded at Aarhus University Flakkebjerg Research station in collaboration between the University of Southern Denmark and Aarhus University. The dataset includes 4750 plant images from 12 species at few growth stages. The number of samples in each species is varied from 221 to 654. The stacked model is built on an RGB plant image subset of 12 kinds of plants, which are Black-grass, Charlock, Cleavers, Common Chickweed, Common Wheat, Fat Hen, Loose Silky-bent, Maize, Scentless Mayweed, Shepherds Purse, Small-flowered Cranesbill, Sugar Beet. We perform experiments with a closed set experimental protocol. This means that for each image in the dataset the stacking model needs to predict to which of the 12 classes the input image corresponds to. Based on that we report the following metrics for the experiments: precision, recall and F1 score. In the multi-class classification case, the micro and macro average of metrics are used. A macro-average of metrics defines the metric for each class and then computes an average of class metrics, whereas a micro-average of metrics aggregates the contributions of all classes to compute the average metric. In a multi-class classification setup, micro-average is preferable if there is a class imbalance.

The Architecture of the Model. We design CNN using the best values of tuning parameters and evaluate the performance of CNN after training. The image samples from each category of the plant were randomly split into ten batches and 10-fold cross-validation was used to produce the final result. CNN trained on the image dataset lead to an overall accuracy of 84.6%. We also calculate the confusion matrices from the evaluation results on the validation dataset. A confusion matrix is a table that is often used to describe the performance of a classification model on a validation or testing dataset for which the true values are known. It allows easy identification of confusion between classes e.g. one class is commonly mislabeled as the other. A confusion matrix gives us insight not only into the errors being made by a classifier but more importantly the types of errors that are being made. Analysis of the confusion matrix shows that most Black-grass samples were classified as positive Loose Silky-bent ones, and only 38.2% were classified correctly. Contrariwise 38 samples of Loose Silky-bent species were classified as positive Black-grass ones.

An analysis of the plant recognition result obtained above shows that the Black-grass and Loose Silky-bent species are very similar to each other. As such, at the first level, two CNN models are created. The first CNN model (CNN-A) is used to recognize 12 species. The second CNN model (CNN-B) recognizes only two species (Black-grass, Loose Silky-bent).

CNN-A model was trained using the tuning parameters shown above. CNN-B model was trained using next parameters: resolution 64×64, number of convolution layers is equal to 5, number of max-pooling layers is equal to 4, number of dropout layers is equal to 5. After appropriate experimentation, these values gave the best results during training. The learning rate defines a drop-based schedule which drops the learning rate by a factor every twelve epochs.

Results. When training CNN-A and CNN-B models, data augmentation is used to create additional artificial images and, therefore, to prevent overfitting that may occur. The three variants of training dataset augmentation are considered. At first variant, the training dataset augmentation is not used. At second variant, the images obtained by rotation original images on angles in the range from 0 to 360 with step 15 are added to the training dataset. At third variant, the images obtained by rotation original images on angles in the range from 0 to 360 with step 10 are added to the training dataset. At second variant and third variant, the size of the training dataset is increased by a factor of 24 and 36 respectively.

Comparison of the CNN-A models with different augmentation sizes is presented in Table 1. In Table 1 the micro avarage of performance metrics are defined on the validation and the testing datasets respectively.

As can be seen from Table 1, the performance metrics obtained on the validation dataset grow with increasing the augmentation size. However, the performance metrics obtained on the testing dataset reach maximum values in the second variant and not increase in the third variant. These facts can be explained by the overfitting the CNN-A model. Thus, the best results are achieved by increasing the training dataset by 24 times (second variant).

Table 1. Performance metrics of CNN-A model on validation and validation and testing dataset.

Dataset	First variant			Second variant			Third variant		
	Precision	Recall	F1 score	Precision	Recall	F1 score	Precision	Recall	F1 score
Validation	0.841	0.841	0.841	0.948	0.948	0.948	0.983	0.983	0.983
Testing	0.814	0.814	0.814	0.938	0.938	0.938	0.937	0.937	0.937

To training CNN-B model, the input data of the CNN-B is formed by extraction from training and validation datasets images labeled as Black-grass or Loose Silky-bent species. To prediction on a testing dataset by CNN-B, the input data of the CNN-B model is formed only from those images of testing dataset that were recognized by the CNN-A model, as the images of Black-grass or Loose Silky-bent species. Table 2 presents the performance metrics of the CNN-B model trained in a training dataset with different degrees of augmentation. Micro-average of performance metrics are obtained on the predicted values of the validation dataset and the testing dataset.

Table 2. Performance metrics of CNN-B model on the validation dataset.

Dataset	First variant			Second variant			Third variant		
	Precision	Recall	F1 score	Precision	Recall	F1 score	Precision	Recall	F1 score
Validation	0.755	0.755	0.755	0.799	0.799	0.799	0.837	0.837	0.837
Testing	0.676	0.676	0.676	0.725	0.725	0.725	0.802	0.802	0.802

As can be seen from the tables, the best performance of model training is achieved with an increase in the training dataset of 36 times (third variant).

In the second level, the "level-one" data is formed from the predicted values obtained using the CNN-A and CNN-B models. As a result, two datasets are formed. The first dataset (training and validation datasets) is used for training and validation of the model at the second level. The second dataset (level-one testing dataset) is used to predicate and evaluate the performance of the second-level model and ensemble of models in general. The next methods are used to build second-level models: random forest (RF), gradient boosting (GB), extreme gradient boosting (XGB). When GB is used, we obtain the next best values of the tuning parameters: number of trees to fit is 80, maximum depth of a tree is 20, learning rate is 0.05. When XGB is used, we define the next best values of the tuning parameters: number of trees to fit is 94, maximum depth of a tree is 2, learning rate is 0.04. When RF is applied, we define the next best values of the tuning parameters: number of trees of forest is 325, maximum depth of a tree is 10, the number of features to consider when looking for the best split is 0.4. Using the best values of tuning parameters we fit models by RF, GB, CGB. Table 3 shows the micro average of performance metrics obtained using different second-level models.

Table 3. Performance metrics of the second level model on the validation and testing dataset.

Dataset	RF			GB			XGB		
	Precision	Recall	F1 score	Precision	Recall	F1 score	Precision	Recall	F1 score
Validation	0.949	0.949	0.949	0.952	0.952	0.952	0.948	0.948	0.948
Testing	0.948	0.948	0.948	0.953	0.953	0.953	0.946	0.946	0.946

As can be seen from Tables, the best results are shown by a model using a gradient boosting at the second level as a classifier.

4   Conclusion

This paper considers the problem of plant recognition with the help of a model ensemble. We studied the stacking technique which has two levels. In the first level uses convolutional neural networks. In the second level gradient boosting shows best performance as a meta-classifier. The most successful convolutional neural network of the first level achieved an F1 score of 0.938 on the testing dataset. While the use of a model ensemble allowed us to increase the F1 score to 0.953 on the testing dataset. Based on the high level of performance of the results obtained, it becomes evident that the stacking technique is a good way of improving the model performance for plant recognition.

References

1. Sakai, N., Yonekawa, S., Matsuzaki, A.: Two-dimensional image analysis of the shape of rice and its applications to separating varieties. J. Food Eng. **27**, 397–407 (1996)
2. Wang, Z., Chi, Z., Feng, D., Wang, Q.: Leaf image retrieval with shape feature. In: International Conference on Advances in Visual Information Systems (ACVIS), pp. 477–487 (2000)
3. Du, J., Wang, X., Zhang, G.: Leaf shape based plant species recognition. Appl. Math. Comput. **185**, 883–893 (2007)
4. Chaki, J., Parekh, R., Bhattacharya, S.: Plant leaf recognition using texture and shape features with neural classifiers. Pattern Recognit. Lett. **58**(1), 61–68 (2015)
5. Beghin, T., Cope, J.S., Remagnino, P., Barman, S.: Shape and texture based plant leaf classification. In: International Conference on Advanced Concepts for Intelligent Vision Systems (ACVIS), pp. 345–353 (2010)
6. Perez, A.J., Lopez, F., Benlloch, J.V., Christensen, S.: Color and shape analysis techniques for weed detection in cereal fields. Comput. Electron. Agric. **25**, 197–212 (2000)
7. Kebapci, H., et al.: Plant image retrieval using color, shape and texture features. Comput. J. **54**(9), 1–16 (2010)
8. Bama, B.S., Valli, S.M., Raju, S., Kumar, V.A.: Content-based leaf image retrieval using shape, color and texture features. Indian J. Comput. Sci. Eng. **2**, 202–211 (2011)

9. Wang, Z., Chi, Z., Feng, D.: Fuzzy integral for leaf image retrieval. In: IEEE International Conference on Fuzzy Systems, pp. 372–377 (2002)

10. Du, J.-X., Zhai, C.-M., Wang, Q.-P.: Recognition of plant leaf image based on fractal dimension feature. Neurocomputing **116**, 150–156 (2013)

11. Yang, L.-W., Wang, X.-F.: Leaf image recognition using fourier transform based on ordered sequence. In: Huang, D.-S., Jiang, C., Bevilacqua, V., Figueroa, J.C. (eds.) ICIC 2012. LNCS, vol. 7389, pp. 393–400. Springer, Heidelberg (2012). https://doi.org/10.1007/978-3-642-31588-6_51

12. Wang, Q.-P., Du, J.-X., Zhai, C.-M.: Recognition of leaf image based on ring projection wavelet fractal feature. In: Huang, D.-S., Zhang, X., Reyes García, C.A., Zhang, L. (eds.) ICIC 2010. LNCS (LNAI), vol. 6216, pp. 240–246. Springer, Heidelberg (2010). https://doi.org/10.1007/978-3-642-14932-0_30

13. Prasad, S., Kumar, P., Tripathi, R.C.: Plant leaf species identification using curvelet transform. In: IEEE International Conference on Computer and Communication Technology (ICCCT), pp. 646–652 (2011)

14. Kadir, A., Nogroho, L.E., Susanto, A., Santosa, P.I.: Neural network application on foliage plant identification. Int. J. Comput. Appl. **29**, 15–22 (2011)

15. Valliammal, N., Geethalakshmi, S.N.: An optimal feature subset selection for leaf analysis. Int. J. Comput. Commun. Eng. **6**, 152–157 (2012)

16. Visnevschi-Necrasov, T., Barreira, J.C.M., Cunha, S.C., Pereira, G., Oliveira, M.B.P.P.: Phylogenetic insights on the isoflavone profile variations in Fabaceae spp.: assessment through PCA and LDA. Food Res. Int. **76**(1), 51–57 (2015)

17. Steve, L., Lee Giles, C., Tsoi, A.C., Back, A.D.: Face recognition: a convolutional neural network approach. IEEE Trans. Neural Netw. **8**(1), 98–113 (1997)

18. Hubel, D.H., Wiesel, T.N.: Receptive fields, binocular interaction, and functional architecture in the cat's visual cortex. J. Physiol. (Lond.) **160**, 106–154 (1962)

19. Lecun, Y., Bengio, Y.: Convolutional networks for images, speech, and time-series. In: Arbib, M.A. (ed.) The Handbook of Brain Theory and Neural Networks. MIT Press (1995)

20. Redmon, J., Divvala, S., Girshick, R., Farhadi, A.: You only look once: unified, real-time object detection. In: Proceedings of the IEEE Conference on Computer Vision and Pattern Recognition, pp. 779–788 (2016)

21. Erhan, D., Szegedy, C., Toshev, A., Anguelov, D.: Scalable object detection using deep neural networks. In: Proceedings of the IEEE Conference on Computer Vision and Pattern Recognition, pp. 2147–2154 (2014)

22. Dong, C., Loy, C.C., He, K., Tang, X.: Learning a deep convolutional network for image super-resolution. In: Fleet, D., Pajdla, T., Schiele, B., Tuytelaars, T. (eds.) ECCV 2014. LNCS, vol. 8692, pp. 184–199. Springer, Cham (2014). https://doi.org/10.1007/978-3-319-10593-2_13

23. Cui, Z., Chang, H., Shan, S., Zhong, B., Chen, X.: Deep network cascade for image super-resolution. In: Fleet, D., Pajdla, T., Schiele, B., Tuytelaars, T. (eds.) ECCV 2014. LNCS, vol. 8693, pp. 49–64. Springer, Cham (2014). https://doi.org/10.1007/978-3-319-10602-1_4

24. Dong, C., Loy, C.C., He, K., Tang, X.: Image super-resolution using deep convolutional networks. IEEE Trans. Pattern Anal. Mach. Intell. **38**(2), 295–307 (2016)

25. Mokeev, V.V.: On application of convolutional neural network for classification of plant images. In: Proceedings - 2018 Global Smart Industry Conference GloSIC 2018, p. 8570141 (2018)

26. Wolpert, D.H.: Stacked generalization. Neural Netw. **5**, 241–259 (1992)

Hand Gestures Detection, Tracking and Classification Using Convolutional Neural Network

Oleg Potkin$^{(\boxtimes)}$ and Andrey Philippovich

Moscow Polytechnic University, Moscow 107023, Russian Federation
olpotkin@gmail.com
https://mospolytech.ru/

Abstract. The article describes a software pipeline for detecting, tracking and classification of static hand gestures of the Russian Sign Language in a video stream using computer vision and deep learning techniques. The dataset used for this task is original, includes 10 classes and consists of more than 2000 unique images. The solution includes a hand detection module that uses a color mask, a gesture tracking module, a static gestures classification module in the detected region of the image based on convolutional neural network, as well as an auxiliary image preprocessing module and dataset augmentation module.

Keywords: Deep learning · Convolutional neural networks · Computer vision · Detection · Tracking · Classification · Hand gestures · Russian Sign Language

1 Introduction

Robotic systems have become key components in various industries. Recently, the concept of Human-Robot Collaboration has attracted the attention of a wide range of researchers. Many examples suggest that a human possesses incomparable problem-solving skills, largely due to advanced sensory-motor abilities, but has limited strength and accuracy [1]. However, robotic systems have resistance to fatigue, a higher speed, accuracy, and performance, but significant limitations in flexibility. Well defined and implemented Human-Robot Collaboration concept can free a human from heavy tasks through an intuitive and reliable interface.

Gestures are one of the ways to exchange information, communicate. The information underlying facial expressions and gestures are at the basis of an efficient communication channel of human interaction [2].

Hand gesture recognition refers to the mathematical interpretation of human palm movements by a computing device. In order to interact with humans, robotic systems must correctly understand human gestures and execute appropriate commands with a sufficient degree of accuracy.

© Springer Nature Switzerland AG 2020
W. M. P. van der Aalst et al. (Eds.): AIST 2019, CCIS 1086, pp. 263–269, 2020.
https://doi.org/10.1007/978-3-030-39575-9_27

Interfaces based on the custom gestures recognition are various on its application, implementation techniques, and readiness for the market. So, for example, system DICE (Dynamic and Intuitive Control Experience) from Mercedes-Benz provides a gesture-based user interface for the multimedia system of a car [3]. Google LLC patented the system for driving pattern recognition and safety control that offers to track the driver's hands movements and transform certain gestures into commands to control onboard electronics [4]. Russian researches also demonstrated a concept of protection against spam bots based on the gestural CAPTCHA mechanism [5].

Moreover, technological giants such as Apple, Kuka Robotics, BMW, Facebook, Netflix, and others are actively developing the direction of promising human-machine interaction interfaces, where gesture interaction is one of the most popular direction. And the task of precise and confident recognition of gesture commands is in the high priority.

The present study demonstrates a system for detecting, tracking, and classification of static hand gestures of the Russian Sign Language in a video stream using computer vision and deep learning techniques. The topic is relevant and represents a basis for a scalable and robust control system based on the gesture commands, which can be used as an interface for human-machine interaction.

2   Hand Detection Using Color Mask

Human hands and body have unique visual features. In the task of recognizing gestures based on images, gestures consist of fragments of hands and/or body images. Therefore, the usage of such visual signs in the identification of gestures is quite reasonable.

Color is a simple visual function for identifying gestures from background information. However, color-based gesture recognition systems are strongly influenced by lighting and shadows [6]. Another common problem in body parts detecting based on skin color is that the skin color is very different among human races. (Fig. 1).

Fig. 1. Skin color palette

Another way to identify gestures is to use the unique shape and contour of the human body. A significant contribution to the formalization of this method was made by Serge Belongie and his colleagues [7].

However, the use of color masks for such tasks is justified in view of the relative simplicity of the method itself, especially at the prototyping stage [8–10].

Finding the ROI (region of interest) - hand region in this case, divided into the following steps (Fig. 2):

- initial blur the image (Gaussian blur was applied with kernel size = 3) in order to avoid noises;
- convert the image from RGB to HSV color space [11];
- define the upper and lower boundaries of the HSV pixel intensities to be considered as a skin;
- in order to remove binary noises, apply a series of erosions and dilations to the color mask using an elliptical kernel (OpenCV library features);
- detect and draw a hand's ROI. In order to simplify the solution, ROI, in this case, is the area with the highest number of neighbor white pixels;
- ROI is ready for the classification task.

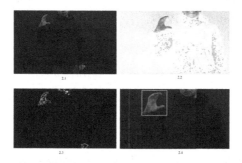

Fig. 2. ROI detection using skin color mask: 2.1 - original, 2.2 - HSV filter, 2.3 - color mask, 2.4 - ROI

3   Dataset and Data Preprocessing

In order to train, cross-validate and test of the classifier, original dataset was developed[1] Dataset includes 2071 images of hands in a certain gesture configuration. Each hand configuration is a class itself and corresponds to a certain static gesture of the Russian Sign Language. In total, 10 classes are represented (Fig. 3). Images were made with different people, light conditions and backgrounds.

Before sending the data to the classifier, it is necessary to perform image preprocessing - transformations that will help to get rid of minor characteristics for the classifier, which in turn will increase the performance of the classifier. Image preprocessing reduces the number of unimportant details (e.g. color information from RGB space) [12].

Table 1 represents the image parameters from the dataset before and after preprocessing.

In case of small datasets more synthetic data is needed. In order to do so, data augmentation techniques are used. It is common knowledge that the more

---

[1] Source code and dataset (gestureset) are available on the GitHub: https://github.com/olpotkin/DNN-Gesture-Classifier.

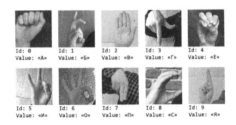

Fig. 3. Random samples from the dataset

Table 1. Parameters of images from dataset: before and after preprocessing

Parameter	Original image	Preprocessed image
Format	.PNG	.PNG
Width (px)	128	64
Height (px)	128	64
Color space	RGB	Grayscale
Color Depth	3	1

data a deep learning algorithm has access to, the more effective it can be. Even when the data is of lower quality, algorithms can actually perform better, as long as useful data can be extracted by the model from the original data set [13].

Following operation were applied in order to get more synthetic images (Fig. 4):

– random rotation (all directions, $+/-10°$);
– random resized crop.

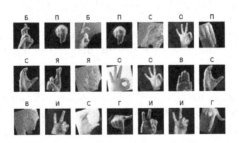

Fig. 4. Random samples from the dataset

In order to minimize overfitting of the classifier, the dataset was divided into 3 parts:

– training set contains about 80% of the dataset (1671 images);

– test set, contains about 20% of the data set (400 images) and used for evaluation of the model (these samples never used during training and cross-
validation processes);
– cross-validation set, contains about 20% of the training set (300 images).

4 Neural Network Architecture for the Classification Task and Performance Review

In this section described CNN architecture applied for classification task on the
abstract level. As a benchmark were taken two models: LeNet-5 CNN architecture [14] and static gesture classifier from [15].

The main disadvantage of LeNet-5 is overfitting in some cases and no built-in
mechanism to avoid this. So, benchmark architecture was improved by adding
dropout layers. Improved LeNet-5 architecture for gesture classification task was
represented in [15]. The main disadvantage of this model is a comparably low
performance on the test set (91.38%).

New improved architecture is more complex, includes more convolutions and
units in dense layers (Fig. 5).

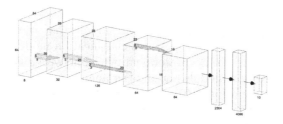

Fig. 5. Improved CNN architecture for the gestures classification task

Training and test procedures were performed on the benchmark classifiers.
Results of benchmark classifiers and improved classifier are shown in Table 2.

Table 2. Classification results: benchmarks vs current implementation

CNN	Training loss	Training accuracy	Test loss	Test accuracy
LeNet-5	0.0681	0.9985	0.5134	0.8708
Static gesture classifier from [15]	0.1411	0.9369	0.2385	0.9138
Imroved classifier	0.1310	0.9433	0.2119	0.9360

The neural network is developed using the PyTorch framework with the following hyper-parameters after the optimization: learning rate is 0.001, batch size
is 20, the number of epochs is 24, the metrics – accuracy, the optimizer type –
Adam.

5   Conclusion/Future Work

As a result of the research, a unique dataset with 10 classes and more than 2000 unique images was developed and published as well as software pipeline for detecting, tracking and classification of static hand gestures of the Russian Sign Language in a video stream using computer vision and deep learning techniques (Fig. 6).

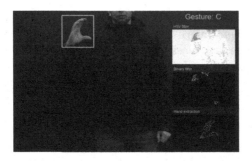

Fig. 6. Gesture control interface

The solution includes a hand detection module that uses a color mask, a gesture tracking module, a static gestures classification module in the detected region of the image based on convolutional neural network, as well as an auxiliary image preprocessing module and dataset augmentation module. For neural network design development PyTorch framework was used.

The classifier demonstrates an accuracy of classification in 93.6% on the test dataset that is better than the result of the previous version – 91.38% and LeNet-5 classifier – 87.08 %. The represented results of accuracy are sufficient to be a strong basis for the initial industrial prototype of gesture control interface and further research in this direction.

In the next generation of the project, the semantic segmentation approach for classification task will be used with Fully Convolutional Network [16]. It assumes another way for dataset labeling and requires more data. In order to solve this problem new images will be created using new synthetic data techniques (random perspective transformation, random noise, Generative Adversarial Nets [17], etc.)

References

1. Krüger, J., Lien, T., Verl, A.: Cooperation of human and machines in assembly lines. CIRP Ann. Manuf. Technol. **58**, 628–646 (2009)
2. Bauer, A., Wollherr, D., Buss, M.: Human-robot collaboration: a survey. Int. J. Humanoid Rob. **5**, 47–66 (2008)
3. DailyTechInfo: DICE - gesture-based control HMI from Mersedes-Benz (2012). [https://dailytechinfo.org/auto/3291-dice-sistema-zhestovogo-upravleniya-avtomobilem-ot-mersedes-benz.html]

4. Urmson, C., Dolgov, D., Nemec, P.: Driving pattern recognition and safety control (2011). [https://patents.google.com/patent/US8634980]
5. Shumilov, A., Philippovich, A.: Gesture-based animated CAPTCHA. Inf. Comput. Secur. **24**(3), 242–254 (2015)
6. Letessier, J., Bérard, F.: Visual tracking of bare fingers for interactive surfaces. In: Proceedings of the 17th Annual ACM symposium on User interface software and technology, pp. 119–122 (1970)
7. Belongie, S., Malik, J., Puzicha, J.: Shape matching and object recognition using shape contexts. Pattern Anal. Mach. Intell. **24**, 509–522 (2002)
8. Doliotis, P., Stefan, A., McMurrough, C., Eckhard, D., Athitsos, V.: Comparing gesture recognition accuracy using color and depth information. In: Proceedings of the 4th International Conference on PErvasive Technologies Related to Assistive Environments - PETRA 2011 (2011)
9. Nalepa, J., Grzejszczak, T., Kawulok, M.: Wrist Localization in Color Images for Hand Gesture Recognition. In: Gruca, D.A., Czachórski, T., Kozielski, S. (eds.) Man-Machine Interactions 3. AISC, vol. 242, pp. 79–86. Springer, Cham (2014). https://doi.org/10.1007/978-3-319-02309-0_8
10. Habili, N., Lim, C., Moini, A.: Segmentation of the face and hands in sign language video sequences using color and motion cues. IEEE Trans. Circuits Syst. Video Technol. **14**(8), 1086–1097 (2004)
11. Shaik, K., Ganesan, P., Kalist, V., Sathish, B.: Comparative study of skin color detection and segmentation in HSV and YCbCr color space. Procedia Comput. Sci. **57**, 41–48 (2016)
12. Förstner, W.: Image preprocessing for feature extraction in digital intensity, color and range images. In: Dermanis, A., Grün, A., Sansó, F. (eds.) Geomatic Method for the Analysis of Data in the Earth Sciences. LNEARTH, vol. 95, pp. 165–189. Springer, Heidelberg (2000). https://doi.org/10.1007/3-540-45597-3_4
13. Wang, J., Perez, L.: The effectiveness of data augmentation in image classification using deep learning. Stanford University (2017)
14. LeCun, Y., Jackel, L., Bottou, L.: Learning algorithms for classification: a comparison on handwritten digit recognition. AT&T Bell Laboratories (1995)
15. Potkin, O., Philippovich, A.: Static gestures classification using convolutional neural networks on the example of the Russian sign language. In: Supplementary Proceedings of the Seventh International Conference on Analysis of Images, Social Networks and Texts (AIST 2018), pp. 229–234 (2018)
16. Long, J., Shelhamer, E., Darrell, T.: Fully convolutional networks for semantic segmentation. In: The IEEE Conference on Computer Vision and Pattern Recognition (CVPR), pp. 3431–3440 (2015)
17. Goodfellow, I., Pouget-Abadie, J., Mirza, M., Xu, B.: Generative adversarial nets. Advances in Neural Information Processing Systems 27 (NIPS 2014) (2014)

Synthesizing Data Using Variational Autoencoders for Handling Class Imbalanced Deep Learning

Taimoor Shakeel Sheikh[1]([✉])[iD], Adil Khan[1][iD], Muhammad Fahim[2][iD], and Muhammad Ahmad[3,4][iD]

[1] Institute of Artificial Intelligence and Data Science, Innopolis University, Innopolis, Russia
{t.sheikh,a.khan}@innopolis.ru
[2] Institute of Secure Cyber-Physical Systems, Innopolis University, Innopolis, Russia
m.fahim@innopolis.ru
[3] Dipartimento di Matematica e Informatica - MIFT, University of Messina, Messina, Italy
[4] Department of Computer Engineering, Khwaja Fareed University of Engineering and Information Technology, Rahim Yar Khan, Pakistan
mahmad00@gmail.com

Abstract. This paper addresses the complex problem of learning from unbalanced datasets due to which traditional algorithms may perform poorly. Classification algorithms used for learning tend to favor the larger, less important classes in such problems. In this work, to handle unbalanced data problem, we synthesize data using variational autoencoders (VAE) on raw training samples and then, use various input sources (raw, combination of raw and synthetic) to train different models. We evaluate our method using multiple criteria on SVHN dataset which consists of complex images, and perform a comprehensive comparative analysis of popular CNN architectures when there is balanced and unbalanced data and determine which operates best in class imbalance problem. We found that data synthesis via VAE is reliable and robust, and can help to classify real data with higher accuracy than traditional (unbalanced) data. Our results demonstrate the strength of using VAE to solve the class imbalance problem.

Keywords: Imbalanced data · Convolutional Neural Network (CNN) · Variational autoencoder (VAE)

1 Introduction

During the last few years classification has become an important task in field of machine learning and pattern recognition. A wide range of classification algorithms i.e. decision trees, neural networks, rule induction, bayesian network,

© Springer Nature Switzerland AG 2020
W. M. P. van der Aalst et al. (Eds.): AIST 2019, CCIS 1086, pp. 270–281, 2020.
https://doi.org/10.1007/978-3-030-39575-9_28

nearest neighbor, support vector machines, and deep learning convolution methods have been well developed and successfully applied to many different application domains.

Datasets in which one or more classes are particularly rare or don't have enough samples, but are more important, are termed as unbalanced or imbalanced datasets. In an unbalanced data set, the majority class has more samples as compared to minority class which contains only a small percentage in a complete set due to which the prediction of small classes in classification models tend to be rare, infrequent or ignored. Consequently, during testing samples belonging to the minority classes are misclassified more often than those belonging to the majority (frequent) classes as shown in Fig. 1a. In many multi-class classification problem, there are many minority classes. Even in some applications, these minorities classes are important i.e. (for information extraction, feature reconstruction).

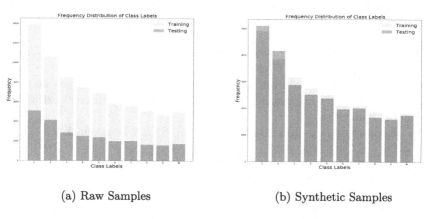

(a) Raw Samples   (b) Synthetic Samples

Fig. 1. Number of cropped samples (32×32) of each digit from the SVHN dataset used for the classification task, where the green and blue color bar represents training and testing samples in (a) raw and (b) synthetically generated samples w.r.t. VAE Best model. *(Best viewed in color)* (Color figure online)

Thus, in imbalanced datasets, classification methods rely on ill defined or not fully characterized set of data for each class. Consequently, these methods suffer from unbalance data problems, and result in learning ineffective weights used for classification tasks. Some methods solve this problem by assuming that the training instances are evenly distributed among different classes [2,3,7,10,20,29] or by using traditional sampling methods [1,4,5,16,17,27,32] which are used to support class imbalance. Hence, the classifications coming from this setup are unreliable.

An alternate solution to handle such problems (less or incomplete representation of raw data) is to generate synthetic samples from the raw (training) samples, and use combination of raw and synthetic samples to train the classification models at the training or learning phase, and similar samples that appear during testing should be accepted successfully.

Accordingly, in this paper, we implement and evaluate the idea of dealing with the class imbalance problem for classification tasks, by training CNN with additional representation of the synthetic data. The aim is to successfully distinguish between classes during testing, despite having limited training samples for minority classes. To implement this idea, we employ an unsupervised deep learning method to learn latent space and synthesize samples of minority classes using variational autoencoders (VAE). We report the detailed experimental setup with comparative analysis of results on SVHN dataset using three popular CNN models and with multiple selection criteria of training samples. We demonstrate that our method can successfully discriminate between classes, and performs significantly better than just using the raw (training) samples. To conclude, our main contributions are:

1. A simple and efficient framework for analyzing and dealing with class imbalance problem by generating training data instead of collecting and manual annotation of new data.
2. Feature learning and synthesis of data using the latent space of VAE and further training different CNN models.
3. A detailed experimental evaluation which shows the effectiveness and diversity of our data generation method in combination with three classical methods of classification.
4. We demonstrate our experimental results on SVHN benchmark.

The rest of the paper is organized as follows. Section 2 describes related work to present the effort and success of research community. Section 3 explains the methodology for the selection of optimal classification model when there is unbalanced data, and by using additional synthetic samples generated by VAE. Section 4 explains the experimental details, discussion about the achieved results and implementation details of our methodology. Section 5 concludes our paper.

2   Related Work

2.1   Unbalanced or Imbalance Class Scenarios

There exists several methods in literature for solving (handling) the class unbalance or imbalance problem. These methods are commonly divided into the following three categories.

Class Imbalance at Data Level. These techniques utilize the distribution of a dataset to solve problem by synthetic re-sampling of the data. It can be achieved through: under-sampling the majority class, oversampling the minority class, or by combining both (under and over sampling) techniques.

Under Sampling: This method tries to balance the distribution of class by randomly removing samples from the majority class samples (i.e. noisy, borderline or redundant) which may degrades the performance of system [14,21]. However, in this way valuable information could be compromised. Liu reused the same data without adding new information [24].

Over Sampling: The alternate to under sampling is to over sample the number of minority class samples. This can be done by generating new synthetic data of minority class [23,28]. On the other hand, [6,34] interpolates or duplicates the minority class to reduce the data imbalance but with the assumptions that naturally occurring class distribution is not best for learning, and substantially better performance can be obtained by using a different class distribution which directly depends on the complexity of the dataset (different variants of images or signals i.e. illuminations, cropped or rotated, magnified etc.).

Class Imbalance at Classifier Level. It is another way to handle class imbalance, where one optimizes the performance of learning algorithm on unseen data. A mine classification algorithm was introduced by William [35] which employs cost-sensitive learners by favoring the minority class. Farquad introduced an approach which not only effectively balance the data but also provides more number of samples for minority class [9] by training SVM as a preprocessor and the actual target values of training data are then replaced by the predictions of trained SVM which are later used to train different methods i.e. Multilayer Perceptron (MLP), Logistic Regression (LR), and Random Forest (RF). Usually one-class classifiers accept the sample which belongs to one class by rejecting others but [33] Wasikowski introduced a method which achieved better performance on multi-dimensional data set. To use boosting methods in combination with SVMs is also an effective method to solve the class imbalanced problem [7].

Class-Imbalance at Feature Level. Above-mentioned methods are not able to deal with high dimensional class imbalance problems [33]. To deal with class imbalance problem in high dimensional spaces, feature selection is a key technique [15]. A detailed performance study of feature selection from nine different metrics in classifying text data drawn from the Yahoo Web hierarchy is done by Ertekin [8]. Maldonado [25] introduced a method based on a backward elimination approach for feature ranking and embedded classification using SVM, by selecting those attributes that are relevant to discriminate between classes under imbalanced data conditions.

3 Methodology

Our goal is to solve class imbalance problem, so we synthesize samples using learned latent distribution of the raw training data and then, use various input sources (raw, combination of raw and synthetic) data to train different classification models.

3.1   Learned Latent Representation

Let's consider the raw training samples $x \in \mathcal{X}$. To effectively synthesize samples from x, we must first convert these samples into more a suitable representation. There are several methods proposed in literature to learn such representations and simultaneously generate data samples i.e. VAE [18] and GAN (Generative Adversarial Network) [13]. We used VAE because they are easy to train (due to limited resources) as compared to GAN model and have an important generative property.

More specifically, our model uses hierarchical latent space. First, we transform x into a latent space through a family of VAE as shown in Fig. 2. (We detail the VAE's architecture and parameters in Sect. 4.5).

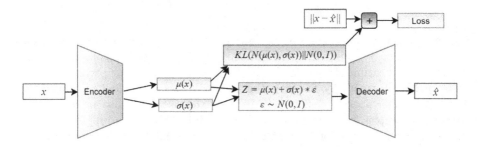

Fig. 2. Architecture of VAE (variational autoencoder)

The latent space representation from the VAE is a learned distribution represented by two parameters (μ, σ). During the training phase, our goal is to reconstruct the input space, while imposing a prior distribution into the latent space. That is, if Θ is a set of parameters for the VAE model, we use back-propagation to learn these parameters, and to find the optimal model that lets us to reconstruct the digits. To achieve this, we select those parameters of the VAE models that minimize the loss function with Eq. (1).

$$\Theta^* = \arg\min_{\Theta} Loss = \|x - \hat{x}\|^2 + \mathrm{KL}(N(\mu(x), \sigma(x)) \parallel N(0, I)) \qquad (1)$$

where KL is the Kullback-Leibler divergence to measure the distance between the prior and the learned distribution [11,12,18]. We assumed that our learned latent distribution $N(\mu(x), \sigma(x))$ is distributed as a Gaussian for the models, and try to minimize its distance towards a zero-centered unitary Gaussian, $N(0, I)$.

Consequently, given training samples of any class, we can map them to a distribution space which has a high level representation with a useful property: i.e. to draw samples of different variants that are close to the original values with some randomness. This particular behaviour is important to synthesize new samples. However, if we carefully select which samples to use, we can synthesize less likely samples from the given class.

3.2   Synthetic Data Sampling

Given any learned latent distributions that we named $\mathcal{V}$, we can synthesize training data using Eq. (2).

$$D_G = \mu + \sigma\epsilon : (\mu, \sigma) \in \mathcal{V}, \epsilon \in \mathcal{E} \tag{2}$$

where each new sample is drawn from the distributions defined by our method Sect. 3.1, ϵ is generated from a noise distribution (normal) $\mathcal{E}$.

3.3   Classification Models

Once we have raw and synthesized samples of training sets $\mathcal{X}$, we trained three different deep classification models by split training data in two subsets (training and validation) and with different percentages (%) and combination of training samples as shown in Table 1. The three models are:

LeNet. LeCunLeCun [22] was a pioneer to introduce the 7-level convolutional network which classifies digits and it was applied by several banks to recognize hand-written numbers on cheques digitized in 32 × 32 pixel gray-scale input images. The ability to process higher resolution images requires larger and more convolutional layers, so this technique is constrained by the availability of computing resources.

AlexNet. A very similar architecture to LeNet was proposed by Alex [19] in which a network was deeper, with additional filters per layer, and stacked convolutional layers. But this architecture design significantly outperformed all the prior competitors and won the challenge by reducing the top-5 error from 26% to 15.3%. It was trained for 6 days simultaneously on two Nvidia Geforce GTX 580 GPUs which is the reason for why their network is split into two pipelines.

VGG-16 Net. The runner-up at the ILSVRC 2014 competition is VGG-16 Net which was developed by Simonyan and Zisserman [30]. It is a similar architecture to AlexNet but with 16 convolutional layers of size 3 × 3 convolutions with alot of filters. They trained this model on 4 GPUs for more than 2 weeks. Currently these trained VGG-16 weights have been used in many research applications and challenges as a baseline feature extractor from the images.

4   Results and Evaluation

4.1   Dataset

To validate our method we used SVHN dataset[1] [26], which was collected using a combination of automated algorithms and the Amazon Mechanical Turk (AMT)

---

[1] Publicly available at http://ufldl.stanford.edu/housenumbers/.

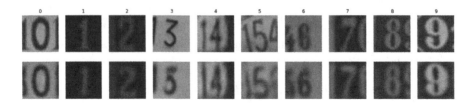

Fig. 3. Digits from SVHN dataset generated using VAE. The first row are raw (training) digits, while the second tow corresponds to the synthetic (generated) samples. Each sample is assigned a single digit label (0 to 9) which corresponds to the center digit.

framework [31]. The dataset is available in two standard formats but we used cropped digits having character level ground truth with a fixed resolution of 32 × 32 pixels as shown in Fig. 3. Further, this dataset is subdivided into three subsets: 73 257, 26 032, 531 131 digits for training, testing and extra (training). The extra samples are the less difficult ones and easier than the other two subsets.

4.2 Experimental Setup

For our experiments, we evaluated our method by synthesizing samples from the raw(training) data of the SVHN dataset [26] using learned latent distribution and further we trained three different classification models (i.e. AlexNet [19], VGG-16 Net [30] and Lenet [22]). For CNN models training, we used two different ways first with the raw (training) samples by varying proportion from 100% to 25% with a step size of 25% and then as a combination of raw and synthesized training samples from the VAE using multiple criteria as reported in Table 1. During these experiments the training samples is sub-divided into two parts (train and validation part). The training data was used to learn the features and the models performance is evaluated with the validation part. During CNN testing, the model was evaluated only with the test portion of the dataset (which were not part of the training data). Some of the other important details of the system are as follows.

4.3 Model Training

We performed a train, validation, and test data split to find the best hyper-parameters of the model (using training and validation split), and then evaluated on the test partition of the dataset. For the classification task we used three different models. We train them using raw (training) samples and with the combination of raw and synthesized samples, as previously described using different percentages of data samples and then evaluated, during testing. Our evaluation metrics is the accuracy, where the higher the score represents the best model for this particular task. The results reported in this section are from our implementation using Python 3.6.5 executed on an NVIDIA GeForce GTX 1080-Ti with TensorFlow and keras framework.

4.4   Comparative Analysis of Results

We report the accuracy performance of training three models with and without synthetic data in Table 1. It is apparent that, in the case of unbalanced data, training using raw and synthetic samples (which are generated using our VAE) is superior to when training is done without such samples. This trend is true for all models. Thus the empirical evidence favors the idea of combating class imbalance by means of synthesizing artificial data. The results also demonstrate that such synthesis can be done by employing VAE, which are easy to train as compared to more complex generative models like GAN's.

Table 1. Comparisons of accuracy metric on SVHN dataset with different models and percentages (%) of raw and synthetic samples used in the experiments. Raw training samples (RS) are randomly selected with 100%, 75% ,50% and 25%. Further we select a different percentage of synthetic samples (SS) (i.e. 95%, 75% ,50% and 25%) generated from the 25% of raw training samples (RS) using VAE. Blanks cells denotes the cases with no synthesis samples.

RS (%)	SS (%)	Test Accuracy (%)		
		AlexNet	VGG-16 Net	LeNet
100		0.9225	0.9276	0.8843
75		0.9189	0.9113	0.8703
50		0.9073	0.8728	0.8502
25		0.8472	0.7972	0.7833
25	95	**0.8643**	**0.8701**	**0.8488**
25	75	0.8201	**0.8412**	**0.8072**
25	50	0.8050	0.7813	0.7591
25	25	0.723	0.7150	0.7047

Furthermore, we also report the result of our experiments when using different percentage of raw training samples (without synthetic data) in Table 1. The highest accuracy results were obtained with 100% samples in comparison to other percentages (%) for all models because in other cases the samples are randomly selected with some % of training set which includes several variants of complex images (rotate, crop, mixed etc.) per digit and some amount of perspective distortion which affects the features learning of CNN. First row of Fig. 3 shows the few instances of each digit and second row corresponds to the synthetic data generated using VAE.

In Fig. 4 we showed the accuracy and loss curves of three classification models which performs best (with and without synthetic data). In all scenarios, our goal was to find the optimal models when trained with raw and synthetically generated data to achieve smooth and stable results with high accuracy and low loss. Subsequently, if we compare each model individually they are approximately close to each other whereas, in case of only raw samples there is variation which effects the model performance to classify the digits accurately. We observed that the method's performance depends on two factors the way we synthesis data via learned latent space of VAE (generate random samples) and other is the features learned by CNN.

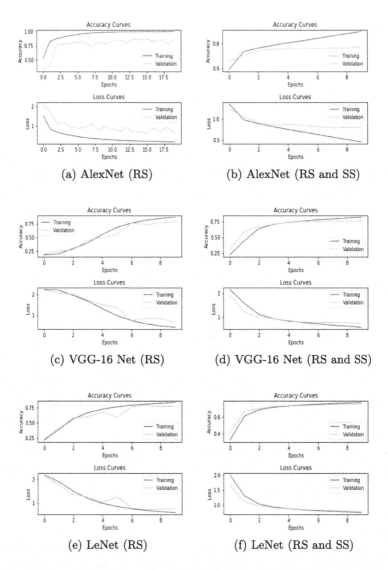

Fig. 4. Accuracy and Loss curves w.r.t. Best Model on SVHN dataset for the method when 25% of raw samples (RS) are selected with no synthetic samples (a,c,e) and 25% of raw samples (RS) concatenated with 95% synthetic samples (SS) (b,d,f) in our experiments.

4.5   Implementation Details

Our proposed framework comprises of two main modules: CNN [19,22,30] and VAE as shown in Fig. 2. For synthetic data generation, we used a CNN based VAE [18] with a Gaussian distribution. There are five hyper parameters that we tuned for our VAE model: amount of layers, number of filters, learning rate, batch size, and dimensionality of latent space. One more parameter which was common for all settings is the reconstruction loss. We evaluated our proposed model with amount of layers from 2, 3. The number of filters, we used: 64, 128 and for learning rate: 10^{-2}, 10^{-3} and 10^{-4}. The batch size was also used: 32, 64, 128. And the most important is the sizes of the Gaussians vary from 80 to 140 with a step size of 20, we selected these dimensions due to resource constraints and for executing all of these experiments. The decoder is a mirror image of the encoder in terms of layer structure. For the reconstruction loss function in Eq. (1), we used mean squared error (MSE) or binary cross entropy. If the input values are in $[0, 1]$, then we use cross entropy, otherwise we use MSE.

These are the following hyper-parameters that we tune for each CNN model: learning rate, batch size and epoch. We evaluated model with five values of learning rate: 10^{-1}, 10^{-2}, 10^{-3}, 10^{-4}, and 10^{-5}. For the batch size, we used 32, 64, 128, 256 value. And for the epochs we used 10, 20, 30, and 40. For the loss function, we used categorical cross entropy because our input are in range $[0, 1]$.

5   Conclusion

In this work, we showed an effective way to handle class imbalanced learning by using VAE in combination with different literature classification models. For our experimental evaluations, we used variational inference to generate synthetic samples using different percentage of raw (training) samples by learning a latent distribution. Our method does not require any additional data except the training data, we trained different classification models which can be easily distinguished between classes during testing and we also used multiple criteria's to select training data (with and without synthetic data) samples. We evaluated and showed the robustness of our method against the traditional (unbalanced or raw training samples) method on SVHN dataset. Our results demonstrate the strength of using VAE (generative property) to solve the class unbalance or imbalance problem.

References

1. Abdi, L., Hashemi, S.: To combat multi-class imbalanced problems by means of over-sampling techniques. IEEE Trans. Knowl. Data Eng. **28**(1), 238–251 (2016)
2. Ahmad, M., Protasov, S., Khan, A.M., Hussian, R., Khattak, A.M., Khan, W.A.: Fuzziness-based active learning framework to enhance hyperspectral image classification performance for discriminative and generative classifiers. PLoS One **13**(1:e0188996) (2018)

3. Ahmad, M., Khan, A., Khan, A.M., Mazzara, M., Distefano, S., Sohaib, A., Nibouche, O.: Spatial prior fuzziness pool-based interactive classification of hyperspectral images. Remote Sens. **11**(9), 1136 (2019)
4. Ando, S., Huang, C.Y.: Deep over-sampling framework for classifying imbalanced data. In: Ceci, M., Hollmén, J., Todorovski, L., Vens, C., Džeroski, S. (eds.) ECML PKDD 2017. LNCS (LNAI), vol. 10534, pp. 770–785. Springer, Cham (2017). https://doi.org/10.1007/978-3-319-71249-9_46
5. Batista, G.E., Prati, R.C., Monard, M.C.: A study of the behavior of several methods for balancing machine learning training data. ACM SIGKDD Explor. Newslett. **6**(1), 20–29 (2004)
6. Chawla, N.V., Bowyer, K.W., Hall, L.O., Kegelmeyer, W.P.: Smote: synthetic minority over-sampling technique. J. Artif. Intell. Res. **16**, 321–357 (2002)
7. Chawla, N.V., Japkowicz, N., Kotcz, A.: Special issue on learning from imbalanced data sets. ACM Sigkdd Explor. Newslett. **6**(1), 1–6 (2004)
8. Ertekin, S., Huang, J., Bottou, L., Giles, L.: Learning on the border: active learning in imbalanced data classification. In: Proceedings of the Sixteenth ACM Conference on Information and Knowledge Management, pp. 127–136. ACM (2007)
9. Farquad, M., Bose, I.: Preprocessing unbalanced data using support vector machine. Decis. Support Syst. **53**(1), 226–233 (2012)
10. Fawcett, T., Provost, F.: Adaptive fraud detection. Data mining Knowl. Discov. **1**(3), 291–316 (1997)
11. Figueroa, J.A., Rivera, A.R.: Is simple better?: Revisiting simple generative models for unsupervised clustering
12. Figueroa, J.A., Rivera, A.R.: Learning to cluster with auxiliary tasks: a semi-supervised approach. In: 2017 30th SIBGRAPI Conference on Graphics, Patterns and Images (SIBGRAPI), pp. 141–148. IEEE (2017)
13. Goodfellow, I., et al.: Generative adversarial nets. In: Advances in Neural Information Processing Systems, pp. 2672–2680 (2014)
14. Guo, X., Yin, Y., Dong, C., Yang, G., Zhou, G.: On the class imbalance problem. In: 2008 Fourth International Conference on Natural Computation, vol. 4, pp. 192–201. IEEE (2008)
15. Hukerikar, S., Tumma, A., Nikam, A., Attar, V.: Skewboost: an algorithm for classifying imbalanced datasets. In: 2011 2nd International Conference on Computer and Communication Technology (ICCCT-2011), pp. 46–52. IEEE (2011)
16. Japkowicz, N.: The class imbalance problem: significance and strategies. In: Proceedings of the International Conference on Artificial Intelligence (2000)
17. Katharopoulos, A., Fleuret, F.: Not all samples are created equal: Deep learning with importance sampling. arXiv preprint arXiv:1803.00942 (2018)
18. Kingma, D.P., Welling, M.: Auto-encoding variational bayes. arXiv preprint arXiv:1312.6114 (2013)
19. Krizhevsky, A., Sutskever, I., Hinton, G.E.: Imagenet classification with deep convolutional neural networks. In: Advances in Neural Information Processing Systems, pp. 1097–1105 (2012)
20. Kubat, M., Holte, R.C., Matwin, S.: Machine learning for the detection of oil spills in satellite radar images. Mach. Learn. **30**(2–3), 195–215 (1998)
21. Kubat, M., Matwin, S., et al.: Addressing the curse of imbalanced training sets: one-sided selection. In: ICML, Nashville, USA, vol. 97, pp. 179–186 (1997)
22. LeCun, Y., Bottou, L., Bengio, Y., Haffner, P., et al.: Gradient-based learning applied to document recognition. Proc. IEEE **86**(11), 2278–2324 (1998)
23. Ling, C.X., Li, C.: Data mining for direct marketing: problems and solutions. In: KDD, vol. 98, pp. 73–79 (1998)

24. Liu, P., Wang, Y., Cai, L., Zhang, L.: Classifying skewed data streams based on reusing data. In: 2010 International Conference on Computer Application and System Modeling (ICCASM 2010), vol. 4, pp. V4–90. IEEE (2010)
25. Maldonado, S., Weber, R., Famili, F.: Feature selection for high-dimensional class-imbalanced data sets using support vector machines. Inf. Sci. **286**, 228–246 (2014)
26. Netzer, Y., Wang, T., Coates, A., Bissacco, A., Wu, B., Ng, A.Y.: Reading digits in natural images with unsupervised feature learning (2011)
27. Pouyanfar, S., et al.: Dynamic sampling in convolutional neural networks for imbalanced data classification. In: 2018 IEEE Conference on Multimedia Information Processing and Retrieval (MIPR), pp. 112–117. IEEE (2018)
28. Qazi, N., Raza, K.: Effect of feature selection, smote and under sampling on class imbalance classification. In: 2012 UKSim 14th International Conference on Computer Modelling and Simulation, pp. 145–150. IEEE (2012)
29. Riddle, P., Segal, R., Etzioni, O.: Representation design and brute-force induction in a boeing manufacturing domain. Appl. Artif. Intell. Int. J. **8**(1), 125–147 (1994)
30. Simonyan, K., Zisserman, A.: Very deep convolutional networks for large-scale image recognition. arXiv preprint arXiv:1409.1556 (2014)
31. Turk, A.M.: https://www.mturk.com/mturk/welcome (2010)
32. Van Hulse, J., Khoshgoftaar, T.M., Napolitano, A.: Experimental perspectives on learning from imbalanced data. In: Proceedings of the 24th International Conference on Machine Learning, pp. 935–942. ACM (2007)
33. Wasikowski, M., Chen, X.W.: Combating the small sample class imbalance problem using feature selection. IEEE Trans. Knowl. Data Eng. **22**(10), 1388–1400 (2010)
34. Weiss, G.M., Provost, F.: The effect of class distribution on classifier learning: an empirical study (2001)
35. Williams, D.P., Myers, V., Silvious, M.S.: Mine classification with imbalanced data. IEEE Geosci. Remote Sens. Lett. **6**(3), 528–532 (2009)

New Approach for Fast Residual Strain Estimation Through Rational 2D Diffraction Pattern Processing

Eugene S. Statnik[1,3](✉) , Fatih Uzun[4], Alexei I. Salimon[2,3],
and Alexander M. Korsunsky[2,4]

[1] Skoltech, Center for Design, Manufacturing and Materials, 121205 Moscow, Russia
Eugene.Statnik@skoltech.ru
[2] Skoltech, Center for Energy Science and Technology, 121205 Moscow, Russia
[3] NUST "MISiS", Center for Composite Materials, 119049 Moscow, Russia
[4] Department of Engineering Science, University of
Oxford, Parks Road, Oxford OX1 3PJ, England, UK

Abstract. Highly brilliant synchrotron X-ray beams enable fast non-destructive evaluation of residual strains even in large and thick engineering objects under *operando* conditions. Large datasets of 2D patterns are acquired at high rate, meaning that near real time evaluation demands fast data processing and mapping of residual strain. The somewhat time-consuming traditional data processing algorithms involve the conversion of each 2D X-ray diffraction pattern within a substantial stack to 1D intensity plots by radial-azimuth sectoral binning ("caking"), followed by peak fitting to determine peak center positions. A new approach was realized and programmed as the open-source code to perform residual strain evaluation by direct 'polar transformation' of 2D X-ray diffraction patterns. We compare residual strain calculations by 'polar transformation' and 'caking' respectively for an Aluminum alloy bar containing a Friction Stir Weld and subjected to *ex-situ* three-point bending. 'Polar transformation' method shows good agreement with the traditional 'caking' technique, but also exhibits excellent speed, and robustness.

Keywords: 'Caking' · 'Polar transformation' · Residual stress/strain · Synchrotron X-ray diffraction

1 Introduction

In the last two decades, various approaches were developed connecting the conventional mechanical methods with modern microscopy techniques for experimental evaluation of residual stress and strain at the micro-scale. Diffraction of micro-focused high-energy synchrotron X-ray beams (μSXRD) proved to be one of the fastest and most convenient techniques for residual strain estimation [1]. For the empirical evaluation of residual stress, the μSXRD method offers unique benefits [2]. Firstly, μSXRD is a non-destructive technique which preserves material integrity during the test, unlike other

© Springer Nature Switzerland AG 2020
W. M. P. van der Aalst et al. (Eds.): AIST 2019, CCIS 1086, pp. 282–288, 2020.
https://doi.org/10.1007/978-3-030-39575-9_29

destructive or semi-destructive methods which cannot be directly checked by repeat measurements. Secondly, different approaches require stress-free reference samples, which are usually challenging to construct and prepare. Furthermore, their spatial resolution and depth penetration are typically orders of magnitude worse than those of μSXRD [3]. An additional benefit of the synchrotron-based method is that it can provide information about the average bulk strain over the depth of the sample, and not just from the surface as for other techniques [4, 5].

The term residual stress refers to the presence of internal forces in a body in the absence of external fields [6]. They play a crucial role in the deformation behavior of processed engineering materials [7]. The knowledge of residual stress distribution allows optimizing the durability of engineering components and assemblies [8]. The stress is a derived extrinsic property that is defined as force per unit cross-sectional area, and is not directly measurable. Consequently, the approaches to stress evaluation rely on the measurement of some intrinsic property, such as strain, or indirect deduction of force.

The use of high energy synchrotron X-ray diffraction provides the possibility of measuring residual elastic lattice strain and then deducing stress by using the material's elastic properties in the generalized Hooke's law expression.

Thus, the emphasis of synchrotron diffraction data interpretation falls on strain determination from μSXRD scattering patterns. To streamline and speed up the analysis of many data files, a 'polar transformation' approach has been proposed that obviates the need for the traditional radial binning technique.

2 Methods and Materials

2.1 Sample Preparation

A 4 mm-thick rolled plate of Aluminum alloy AA6082-T6 plate was used for the manufacturing of a Friction Stir Welded joint. The welded plate was subjected to wire Electro Discharge Machining to cut out a sample with parallel sides. The sample was then subjected to four-point bending to induce permanent plastic deformation associated with specific bending strain profile. The resulting residual stress state within the sample represents a combination (superposition) of welding-induced and bending-induced effects.

2.2 SXRD Experiment

The experiment was conducted at Diamond Light Source on the Test Beamline B16. The beam energy was defined at 18 keV using multi-layer monochromator. The variation of strain across the bent bar was investigated by scanning the sample across the beam (collimated to 0.1 mm square beam spot) as shown in Fig. 1.

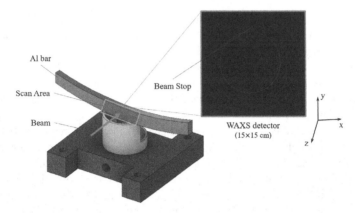

Fig. 1. Illustration of the investigated Aluminium alloy bar with the scan area.

2.3   Center Estimation of Calibration Pattern

In this study calibration test was performed using silicon powder that enables to define the exact center position of the diffraction pattern and use this for all subsequent analyses. The analysis of the 2D diffraction pattern of the reference specimen is demonstrated in Fig. 2.

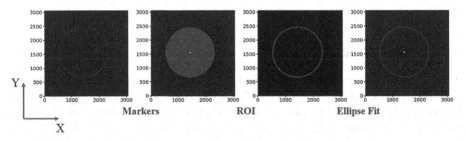

Fig. 2. The scheme of the center estimation for calibration pattern (3056 pixel = 15 cm).

The correct center of the calibration diffraction pattern was found by several simple operations, as follows:

(a)   select the specific circle and make markers in four directions (0°, 90°, 180° and 360°, respectively) for the initial center estimation;

(b)   set indent from the circle and create a region of the interest (ROI);

(c)   choose a circle and perform a numerically stable direct Least Squares fit [9] using an ellipse, to account for various sources of distortion, namely, the combination of detector misalignment (tilt of the detector normal direction with respect to the incident beam), and non-equibiaxial residual strain.

2.4 'Caking'

Defining precise residual strain values requires careful interpretation of 2D μSXRD patterns so that the radial peak position reaches the accuracy approaching 10^{-4} or even better. Currently, there is a reliable and efficient approach for 2D μSXRD data interpretation in use based on radial-azimuthal binning (colloquially referred to as 'caking') which allows obtaining strains with high accuracy. However, this method has some drawbacks in terms of the large processing time and a large number of steps required. 'Caking' is a multi-stage process that involves the following stages: calibration, conversion of the 2D pattern into a 1D profile, and Gaussian fitting for peak center determination to calculate strain values.

The first experimental step is the assembly of ancillary equipment for positioning (and possibly loading) the sample, placing the detector and beam stop, calibrating the sample-detector distance, obtaining exact information regarding the image center of diffraction patterns, and calibrating the geometrical distortion related to detector's orientation angle. Next, it is essential to collect scattering patterns from reference sample(s) and processing them via 'caking' to reduce 2D diffraction pattern to 1D radial functions of the azimuthal angle φ that can be visualized as line plots of scattered beam intensity against radial position R as shown in Fig. 3.

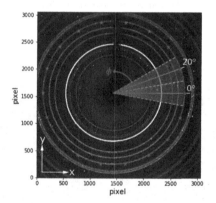

Fig. 3. 2D diffraction pattern of calibration with selected azimuthal angle φ in 20° increments to produce 1D intensity distributions (3056 pixel = 15 cm).

These operations were then applied to each 1D profile for azimuthal angles around values of 0°, 45°, 90°, 135°, 180°, 225°, 270°, 315°, and 360°.

Step three is fitting of most intense peak positions (R_i) as a function of the azimuthal angle (φ_i) by a sine function, namely,

$$R_i = a + b \sin(2(\phi_i + c))$$ (1)

where a, b and c are offset, amplitude and phase shift, respectively.

The relationship between residual strain and d-spacing is defined as

$$\varepsilon = \frac{d - d_0}{d_0} = \frac{d}{d_0} - 1 = \frac{a + b}{a_0 + b_0} - 1$$ (2)

where ε is an elastic residual strain, $d = a + b$ is d-spacing of a strained sample, and $d_0 = a_0 + b_0$ is the value of d-spacing when the sample is strain free (from calibration image).

Nevertheless, conventional 'caking' approach has drawbacks, as it requires defining regions, selecting directions, splitting into sectors and averaging them that are imprecise for a textured sample whose ring intensity is nonuniform or spread on ring region. This should be considered in the context of the processing time of this analysis increasing manifold for large amounts of images.

2.5 'Polar Transformation'

In this investigation, we consider a new technique of determining residual strain based on the geometric transformation [10] of the 2D diffraction pattern from a Cartesian to a polar coordinate system with respect to the pattern center as shown in Fig. 4, namely,

$$R = \frac{X_i - X_0}{height\,of\,image} \cdot magnitude(I);$$

$$\phi = \frac{(Y_i - Y_0)}{360°} \cdot angle(I)$$

(3)

Following this cartesian-to-polar transformation, Debye–Scherrer rings can be displayed as lines on the radial-azimuthal contour plot. Following this, the entire transformed intensity pattern can be fitted with a sine function (1) to determine the residual strain variation (2).

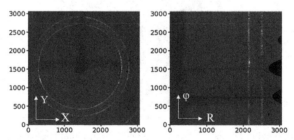

Fig. 4. The typical 2D diffraction pattern (a) and its polar transformation (b) (3056 pixel = 15 cm).

3   Results and Discussions

In this study, the accuracy of a new technique based on the polar transformation of 2D μSXRD patterns was validated by comparison with the conventional 'caking' method. The programming language Python 3.7 with set standard libraries (os, numpy, scipy, matplotlib, lmfit, opencv and skimage) was used for obtaining 1D profiles from 2D patterns, Gaussian and polar transformation fitting respectively.

The approaches described above were applied to the series of summed diffraction patterns as a function of vertical position. Figure 5 shows the dependence of peak center positions on the azimuthal angle. The set of experimental data of 'caking' significantly less than in case 'polar transformation'. It influences essential difficulties related with comparison two methods: the lack of experimental points for 'caking' approach is connected with a particular small angle range by contrast with 'polar transformation' where fitting the entire 360°. At the result, the errors are reduced in determining the fitting parameters, and hence strain.

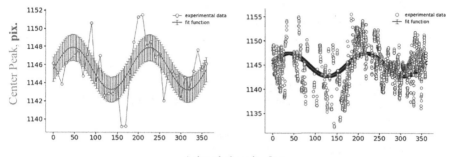

Fig. 5. The dependence of peak center position against azimuth angle range on (0°, 360°, step 15°) for 'caking' (a) and the complete revolution for 'polar transformation' (b): the blue 'o' markers are experimental data, the red solid lines are sine fitting (with black error bars). (Color figure online)

Figure 6 demonstrates the correlation between data of residual elastic strains at the azimuth angles from 0° to 360° with a step equals 15° respectively.

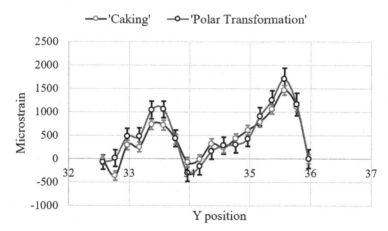

Fig. 6. Comparison of residual elastic strains obtained from 'caking' (red line) and 'polar transformation' (blue line) techniques. (Color figure online)

The presented Cartesian-to-polar transformation technique allows determining the strain and subsequently stress values for various in plane directions. The results of this approach show good agreement with 'caking' microstrain results.

In this study, a technique for the processing of 2D X-ray diffraction patterns from synchrotron experiments is presented that is referred to as the 'polar transformation' interpretation approach. The advantage of 'polar transformation' lies in the ability to perform fast processing of large numbers of 2D diffraction patterns without repeated binning conversions, with the specific purpose of extracting strain information.

The key benefits of the new method are:

(a) reduction of computational effort and processing time (the library time was imported from Python for time process measurement; the results showed 41 vs 1011 s for 'polar transformation' and 'caking', respectively);
(b) simplification of the analysis by omitting the conversion of 2D patterns to 1D profiles followed by Gaussian peak fitting;
(c) minimizing possible error sources associated with radial-azimuthal binning.

The new technique for the processing of 2D X-ray diffraction patterns collected at synchrotrons in large volumes conforms to these requirements. Furthermore, the efficiency of the original method may additionally facilitate online strain mapping in engineering objects during data collection. Cartesian-to-polar transformation of 2D synchrotron X-ray diffraction patterns allows separating Debye-Scherrer rings distortions into the contributions from detector misalignment and strain, respectively.

References

1. Korsunsky, A.M.: Variational eigenstrain analysis of synchrotron diffraction measurements of residual elastic strain in a bent titanium alloy bar. J. Mech. Mater. Struct. **1**, 259–277 (2016)
2. Lord, J., Cox, D., Ratzke, A., Sebastiani, M., et al.: A good practice guide for measuring residual stresses using FIB-DIC. In: Measurement Good Practice Guide, p. 143 (2018)
3. Reimers, W.: Analysis of residual stress states using diffraction methods. Acta Phys. Pol., A **2**(96), 229–239 (1999)
4. Zhang, H.J., Sui, T., Salvati, E., et al.: Digital image correlation of 2D X-ray powder diffraction data for lattice strain evaluation. J. Mater. **11**, 427–440 (2018)
5. Paul, S.: X-ray diffraction residual stress techniques. Met. Handb. **10**, 380–392 (1986)
6. Korsunsky, A.M., Wells, K.E., Withers, P.J.: Mapping two-dimensional state of strain using synchrotron X-ray diffraction. Scr. Mater. **39**, 1705–1712 (1998)
7. Yaowu, X.U., Rui, B.A.O.: Residual stress determination in friction stir butt welded joints using a digital image correlation-aided slitting technique. Chin. J. Aeronaut. **30**, 1258–1269 (2017)
8. Lunt, A.J.G., Korsunsky, A.M.: A review of micro-scale focused ion beam milling and digital image correlation analysis for residual stress evaluation and error estimation. Surf. Coat. Technol. **283**, 373–388 (2015)
9. GitHub. https://github.com/bdhammel/least-squares-ellipse-fitting. Accessed 18 July 2019
10. OpenCV documentation. https://docs.opencv.org/2.4/modules/imgproc/doc/geometric_trans formations.html. Accessed 18 July 2019

Fast Identification of Fingerprint

Vladimir Gudkov[✉]

Chelyabinsk State University, 129, Bratiev Kashirinykh, Chelyabinsk 454001, Russia
`diana@sonda.ru`

Abstract. The paper discusses a method of fast identification of fingerprints based on templates. The method is based on the properties of local neighborhoods of minutiae and consists of two stages. At the first stage, the neighborhoods of minutiae are compared independently from the request and reference templates. For each pair of minutiae, their similarity is estimated. To improve performance rating classes are introduced. They filter out unlikely pairs of minutiae. Some of the best ratings for each pair of minutiae of request and reference templates are accumulated in the histogram. As a result of the first stage, a histogram is constructed for the entire database. Based on the histogram, some of the most similar pairs of templates are selected. At the second stage, the already selected pairs of templates are compared additionally, taking into account the consolidation of minutiae. A significant increase in identification performance, in contrast to other publications, is achieved by eliminating false candidates in the first stage.

Keywords: Fingerprint · Fast identification · Minutiae · Histogram

1 Introduction

Currently, fingerprints are widely used to identify a person's identity [1]. To do this, use Automated Fingerprint Identification Systems, access control systems, etc. They are essentially a pattern recognition system that recognize a person by determining the authenticity of a specific physiological characteristic [2]. Personal identification involves the interaction of computer systems (CSs) with large databases [1]. To improve the performance of CSs, identification based on fingerprint image templates [3], that is, on their models. Templates can provide acceptable image recognition quality [4, 5], but the performance of CSs is limited [6]. Increasing the performance without loss of recognition quality is possible due to the increase in hardware cost. However, this method of increasing productivity is not acceptable. Therefore, a new challenge arises [7]. How can, by means of a slight decrease in the quality of recognition, do raise the productivity of CSs by tens and hundreds of times without a noticeable increase in their cost? This is the task, which is currently actual. It can be solved by applying new mathematical models of images, new recognition algorithms and software building paradigms [8, 9]. In this paper, such questions of identification of the person are considered.

The work is structured as follows. The introduction briefly describes the current state of the identification task, the formulation of the problem and its relevance. Section 2

© Springer Nature Switzerland AG 2020
W. M. P. van der Aalst et al. (Eds.): AIST 2019, CCIS 1086, pp. 289–300, 2020.
https://doi.org/10.1007/978-3-030-39575-9_30

outlines the main known methods for improving the quality and performance of identification. Section 3 reveals the structure of classifiers based on topology and ridge counting. It also outlines the algorithm to build a histogram of similarity of the request and reference templates. Section 4 proposes the new fingerprint matching itself for the database share. It also reveals the essence of the fast identification algorithm. Section 5 is devoted to presenting the results of experiments. In conclusion, the findings of the study and the further direction of work are discussed.

2 Related Works

Methods that improve quality and increase identification performance are well known [2]. Primary selection rate of pairs of minutiae has a significant impact on performance [5, 6]. At the same time, minutiae, as well as the relations between them, can be stored in a template in various ways. In this case, the ridge count [7, 8] or Euclidean distance [9, 10] between pair of minutiae is most often used. Usually ridge count is determine by the number of ridges that are intersected by the segment which connecting pair of minutiae [11].

Local minutiae topological structures (LMTS), noise and distortion resistant, has been proposed by Jiang at the end 2000 [12]. Later, Feng proposed a method of combining LMTS into pairs [13], and Cao - into star structures [14]. In [15], generalized methods of combining LMTS into structures of two, three elements and k-points of a structure are considered.

In paper [16], the special k-plet structure builds for each minutia. For verification or identification, the Coupled Breadth First Search (CFBS) method was proposed. Algorithm based on CFBS method performs a matching of two graphs that based on fingerprint image skeleton. Tracing the skeleton lines form k-plets. Later this idea have developed in [17]. In this case, the fingerprint has performed as oriented graph.

In [17], the authors describe Minutiae Adjacency Graph (MAG). Each minutia supplies MAG construct. When it is built, the minutiae closest to the given minutia are selected on the basis of a certain threshold value. Each edge in graph is represented in the form of $\{i, j, d_{ij}, r_{ij}, \varphi_{ij}\}$, where i, j – are the numbers of minutiae, d_{ij} – is an Euclidean distance between i and j minutiae, r_{ij} – is the value of ridge count between them, φ_{ij} – is an orientation of the edge.

In [18], the authors extend the MAG graph method. They build MAG graph not only for minutiae, but also for loops, deltas and whorl of pattern.

The above-mentioned algorithms are designed to identify images of medium or high quality value. Quality is assessed by the NIST criterion [19]. Minutiae detected on the fingerprints define the graphs. During identification, the edges of the graphs are comparing and summarizing their similarity score. From all minutiae, select those pairs of minutiae for which the total estimate of similarity score is maximal.

In general, these algorithms show good performance, but their acceleration is problematic. The restriction arises because the division of the formed structures into classes is not determined at the stage of image processing. The implementation of the necessary functions for this will lead to an increase in the number of calculations and a decrease in identification performance. Obviously, the preliminary classification of structures and their ordering can further improve the performance of algorithms.

Thus, the formulation of the problem is not to speed up the selection of pairs of minutiae that are potentially identical. This is not enough. It is necessary to build a histogram based on the selected pairs of minutiae and their similarity estimates, which will allow weeding out unlike patterns in advance. A detailed analysis of the templates left only for a small share of database. This will further enhance the performance.

The complexity of the problem lies in the variability of images, and in the content of the templates. This affects the classification of structures. To compensate for this, it is necessary to develop a stable classifier, as well as a method for quickly assessing the similarity of pairs of minutiae. In addition, it is necessary to select the quantile degree of similarity of templates on the histogram carefully, that is unknown from analogs.

3 Classificatory

Consider the fingerprint identification algorithm. The fingerprint template consists of the following data: a list of minutiae and a list of structures (nests) for these minutiae.

Each minutia m_i consists of its coordinates and orientation [3]. For convenience, minutiae are supplied with the type $t_i \in \{0, 1\}$: ending or bifurcation, with $t_i = 1$ for ending and $t_i = 0$ for bifurcation.. The description of minutia using the formula

$$m_i = \{(x_i, y_i), \alpha_i, t_i\}, \quad i \in 1..n, \tag{1}$$

where (x_i, y_i) are coordinates of m_i; α_i is a direction of m_i in the range $[0; 2\pi]$, t_i is a type of m_i, n is a number of minutiae.

Consider the cross sections for two types of minutiae: the ending (Fig. 1) and the bifurcation (Fig. 2). The direction vector of every minutia goes upwards in the number of lines tangential to the line [20]. Perform a sequence of operations.

Let us carry out projections from each minutia to the right and to the left on the adjacent lines perpendicular to the direction vector of each point. Fix all projections.

Select one minutia and draw a section to the right and to the left. To ensure stability, we will cross section perpendicular to the intersected lines. The section cuts each line into two parts, which we call links. Let us enumerate the links clockwise.

For ending number 19, the enumeration begins with the line on which the ending is located (link number zero in Fig. 1). For bifurcation number 19, the enumeration begins with a line dividing into two other lines (link number zero in Fig. 2). In the figures, the section for the ending generates seventeen links (0...16), for the bifurcation – seventeen links (0...16). As a rule, we chose the depth of the section arbitrarily (5–17 lines).

This method of enumeration of links allows to enumerate the links in a spiral way and arbitrarily change the depth of the section. However, the method of numbering is not essential.

When constructing a nest for a minutia, we trace the course of each link, starting from the section. We trace the link to a meeting with another minutia located on the line, or to a projection from another minutia (minutiae number 20...27 in Figs. 1 and 2). A minutia located on the left or right of the link can form this projection.

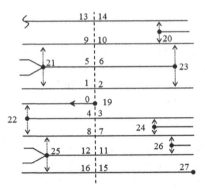

Fig. 1. A section for ending as a nest.

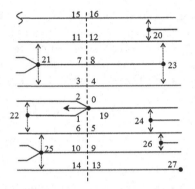

Fig. 2. A section for bifurcation as a nest.

There are many links. Their number calculates using the formula

$$w_i = 4x + 2 + (-1)^{t_i}, \qquad (2)$$

where t_i – is a type of minutia m_i, x – number of intersected lines.

While building a nest, for every link minutiae and projections form events (Fig. 3). Every event corresponds to the number [7, 21]. The last bit depends on the direction of the minutia, which formed the event. The link number and the number of the minutia that generated event supply each event. Events 1100 and 0000 do not take into account. The line either closes or breaks off in the background. There are no other events on line.

For each minutia m_i, all generated links are form the nest. In the nest, the ridge count r_{ik} is measured from the main minutia m_i (center) to the minutia m_k, which belong to this nest. So, the value of ridge count r_{ik} equal to the number of lines intersected by the section connecting the minutiae m_i and m_k.

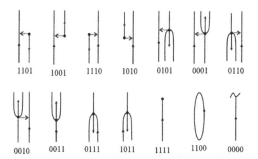

| 1101 | 1001 | 1110 | 1010 | 0101 | 0001 | 0110 |
| 0010 | 0011 | 0111 | 1011 | 1111 | 1100 | 0000 |

Fig. 3. Events detected on the links.

The set of links in the nest for minutia m_i has the next form

$$N_i = \{m_k, d_{ik}, r_{ik}, \gamma_{ik}, v_{ik}\}, \quad m_k = \{x_k, y_k, \alpha_k\}, \quad k \in K, \tag{3}$$

where d_{ik}, is the distance between minutiae m_i and m_k, r_{ik} is the ridge count between the same minutiae, value γ_{ik} – azimuth, which equal to the angle of the turn of α_i to the ray from minutia m_i to minutia m_k, α_i from formula (1), v_{ik} is an event for minutia m_k [7, 20] in the nest of minutiae m_i (see Fig. 3), K is the number of links for this nest. Note that the parameters of links are very stable. They are not depend on the image orientation, choice of the coordinate center and image scale except of d_{ik} value.

Two minutiae m_i and $m_k, k \in K$, form a dipole in the nest (Fig. 4). We supply each dipole with its length d_{ik}, equal to the distance between these minutiae, the ridge count r_{ik}, and the angle φ_{ik} as the difference of the vectors α_i and α_k of minutiae m_i and m_k, see formula (1) and (3). Thus, the dipole has the form

$$D_{ik} = \{\gamma_{ik}, d_{ik}, r_{ik}, \varphi_{ik}\}. \tag{4}$$

There are many dipoles, one of whose ends coincides with the central minutia m_i. For this nest, we form a Multiplicity of Nest Dipoles (MND). We present two dipoles for the request and reference fingerprints at the Fig. 4.

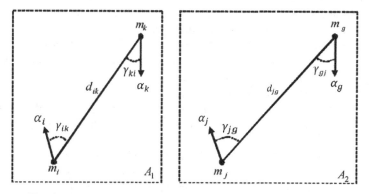

Fig. 4. Parameters of dipoles for request and reference templates.

Template stores all MND of fingerprint. Compare each MND of the request template with each MND of the reference template. In this case, a list of the nearest pairs of MND is formed and their similarity is estimated. For evaluation, the geometric and topological characteristics are applied [22]. The following formula applies:

$$\psi\big(\omega(m_i, m_k), \omega(m_j, m_g)\big) = f(\Delta d, \Delta \gamma_1, \Delta \gamma_2, \Delta \gamma_3). \tag{5}$$

Here $\Delta d = \|d_{ik} - d_{jg}\|$ is the norm of the difference of the lengths of dipoles; $\Delta \gamma_1$ is the norm of the difference of azimuth angles γ_{jg} and γ_{ik}; $\Delta \gamma_2$ is the norm of the difference of the azimuth angles of γ_{gj} and γ_{ki}; $\Delta \gamma_3$ is the norm of the difference between the angles φ_{ik} and φ_{jg} (4); function f is a first degree polynomial. Without loss of generality, we consider the value $0 \le \psi \le 1$. The closer this value of ψ to one, the more similar the dipoles.

The topology similarity of dipoles depends on the event v_{ik} of request template and the event v_{jg} of the reference fingerprint. It depends on the first bit of events as shown on the Fig. 3. Event as a number define the link state. It is zero if the angle of minutia direction is oriented along the course of link; otherwise, it is one [22, 23]. We use a logical operation "exclusive or":

$$\lambda\big(\omega(m_i, m_k), \omega(m_j, m_g)\big) = v_{ik} \oplus v_{jg}. \tag{6}$$

It is clear that $\lambda \in \{0, 1\}$.

The final similarity of two dipoles is a combination of formulas (5) and (6) in the form of

$$s_{ik,jg} = \psi\big((m_i, m_k), (m_j, m_g)\big) * \lambda\big((m_i, m_k), (m_j, m_g)\big), \tag{7}$$

where

$$s_{ik,jg} = s\big((m_i, m_k), (m_j, m_g)\big).$$

The formulas (5)–(7) estimate the similarity of two dipoles. For quick calculation of the normalized differences using formula, it is necessary to reduce lengths of dipoles and angles value to the integer range $0 \ldots 127$. It is easy with the help of quantization. These values allow us to use the methods of multibyte add, subtract, absolute value [24].

Thereafter, it is possible to select the most similar dipoles, associated with minutiae, but it requires brute force dipoles of the request and reference templates.

For each minutia, we select at least two closest minutiae from its nests. For example, for minutia m_i two minutiae m_l and m_r are selected (the more minutiae are selected, the higher accuracy and the lower performance). Therefore, minutia m_i forms two dipoles (see Fig. 5). The ridge count equal r_{il} and r_{ir}, respectively for selected minutiae.

For minutiae m_l and m_r, the azimuth angle is defined as $\gamma_{ik} \in \{0; 359\}, k \in \{l, r\}$. With the help of quantization we reduce this angle to $\gamma_{ik} \in \{0; 63\}$. This set gives us 64 possible classes.

The number of classes is not enough; there is high probability of erroneous comparison of two different dipoles. The additional bit in the dipole description allow us to double the number of classes. This bit is the same as the value of v_{ik} by formula (3), where $k \in \{l, r\}$, $v_{ik} \in \{0, 1\}$ (Fig. 5). Thereby, the number of classes increases to 128.

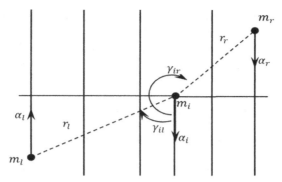

Fig. 5. Two dipoles and their parameters for central minutia m_i.

Let us add the reverse bit v_{ki} that the nest of minutia m_k forms, $k \in \{l, r\}$ (see Fig. 5). The method of its calculation is the same as for the nest of minutia m_i. Therefore, we define the class c_{ik} like a byte shown in Fig. 6.

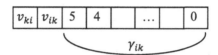

Fig. 6. Structure of the class c_{ik}.

Thus, the number of classes expended to 256. Now rewrite the structure of MNDs according to formula (4) in the form

$$D_{ik} = \{c_{ik}, d_{ik}, r_{ik}, \varphi_{ik}\}, \ k \in \{l, r\}. \tag{8}$$

Before matching dipoles ordered by classes. This order depends on the value of classes. Then the intersection of the list of classes for request and reference templates is calculated [23]. If the classes are equal, the angles φ_{ik} and φ_{jg} difference, the distances d_{ik} and d_{jg} difference, the ridge count r_{ik} and r_{jg} difference are estimated (see Fig. 4). If the values of these differences within the specified tolerances, then minutiae m_i and m_k saved in the list L_0. So minutiae m_i and m_k are considered as possible candidates of identical minutiae pairs in the list

$$L_0 = \{(m_i, m_k)\}, k \in \{l, r\}.$$

Tolerance values are determined at the training stage. Comparison of dipoles based on the classes as fast as comparison of integers. Any other estimation performs after comparison of classes. It increases productivity. Therefore, the list L_0 is built quickly.

Then take a pair of minutiae m_i and m_k from the L_0 list. Consider them as minutiae m_i and m_j (see Fig. 4). The nest of each minutia forms multitude of events $\{v_{ik1}, v_{ik2}, \ldots\}$ and $\{v_{jg1}, v_{jg2} \ldots\}$ according to formula (3) and (6). Two multitudes of events define two 32-bit words. The number of link on which the event is detected determines the bit index in the word. Thus, the word reflects the topological properties of the whole nest. The word counting for each nest can be done before template matching. We compare these two 32-bit words using a logical operation of the exclusive "or" ("XOR"). The number of matching bits is counted [24]. After that, the list L_1 is forming as selecting the best matching score b_{ij} for pair of minutiae m_i and m_j:

$$L_1 = \{(m_i, m_j, b_{ij})\}, i \in \{1, n_1\}, j \in \{1, n_2\}, \tag{9}$$

where n_1 is the number of minutiae in the request template and n_2 – in the reference template.

The list L_1 is ordered by matching score b_{ij}. Then from the list L_1 we select the M best pairs or less if the templates too small and estimate the similarity of two templates in the form of

$$s = \sum_{m=1}^{M} b_{i_m, j_m}, \tag{10}$$

where indices i_m, j_m are unique and not duplicated.

Similarity estimation (9) of two templates is performed very quickly. Comparing request template with each reference template from database, we construct the histogram of similarities $H(s)$, where s determined by the formula (10). Obviously, this is an align-free method.

Value M is determined at the training stage and depends on the type of impression: rolled or plain fingerprint. Increasing the value M leads to better selection of genuine pairs.

4 Matching

Performance improvement is not only in the use of dipole classes, but also in the application of the properties of histogram estimates.

Let us define a quantile q of the best estimations for the histogram $H(s)$ using (10). This will determine a small proportion of all referenced templates from database. We denote this part of the reference templates as $H_q(s)$. Compare request template with every reference template from $H_q(s)$ using formulas (1)–(9).

For a single pair of request and reference templates now we make a qualitative matching.

From the list L_1 we select $K = \sqrt{n_1 n_2}$ the most similar pairs of nests using (9). They are correspond to the minutiae m_i and m_j, which are usually identical on the genuine fingerprints (Fig. 4). We denote K_i and K_j as the number of links for nests of minutiae m_i and m_j. For evaluation, both geometric and topological characteristics are applied [22]. We use next formula for pair of nests estimation in the form of

$$Q_{ij} = \sum_{k,g=1}^{\min(K_i, K_j)} \psi\left(\omega(m_i, m_k), \omega(m_j, m_g)\right), \tag{11}$$

where $\psi\left(\omega(m_i, m_k), \omega(m_j, m_g)\right)$ is the formula (5). Dipoles are compared on the identically numbered links.

Based on the list L_1 (9) and similarity of nests according to the formula (11), we construct the list L_2 in the form of

$$L_2 = \{(m_i, m_j, Q_{ij})\}, i \in \{1, n_1\}, j \in \{1, n_2\}, \tag{12}$$

where n_1 is the number of minutiae in the request template and n_2 – in the reference template from the share of histogram $H_q(s)$. The list L_2 is ordered by matching score.

Note, that list L_2 is represented as sets of nests. Select the best pair of minutiae m_{i_1} and m_{j_1} from the list L_2, $i_1 \in \{1, n_1\}$, $j_1 \in \{1, n_2\}$ (12). Create new list L_3 and empty it. Reset values V_1 and V_2.

The algorithm of matching based on the start minutiae m_{i_1} and m_{j_1}. It consists of the following steps.

1. Push into the list L_3 pair of minutiae m_{i_1} and m_{j_1} from the list L_2 and reset the quality $Q_{i_1 j_1}$.
2. Reset the value V_1.
3. For each pair of minutiae m_i and m_j from the list L_3 estimate the quality Q_{ij} if it is zero.
4. Sort the list L_3 in ascending order.
5. Pop from the list L_3 the best pair of minutiae m_i and m_j.
6. Accumulate quality Q_{ij} for these minutiae in the variable V_1.
7. Delete pairs of minutiae with the same indexes i and j if they are duplicated in other pairs of minutiae in the list L_3.
8. Compare dipoles D_{ik} and D_{jg} in the nests according to the formulas (4)–(6) and calculate similarity of two dipoles $s_{ik,jg}$ using (7). Regardless of equality or inequality of classes, the angles φ_{ik} and φ_{jg} difference, the distances d_{ik} and d_{jg} difference, the ridge count r_{ik} and r_{jg} difference are estimated (see Fig. 4). If the values of these differences within the specified tolerances, then new pair of minutiae m_k and m_g push into the list L_3. Reset the quality Q_{kg}. Tolerances are determined at the training stage. So minutiae m_k and m_g are considered as possible candidates of identical minutiae pairs. According to formula (8) in the nest can form many pairs of minutiae (m_{k_1}, m_{g_1}), (m_{k_2}, m_{g_2})...They are all listed.
9. If the list L_3 is not empty, go to 3.
10. If $V_2 < V_1$ then $V_2 = V_1$.
11. If the number t of pairs of minutiae selected from the list L_2 is less than the specified threshold and the list L_2 is not empty, push into the list L_3 pair of minutiae $m_{i_{t+1}}$ and $m_{j_{t+1}}$ from the list L_2 and reset the quality $Q_{i_{t+1} j_{t+1}}$.
12. If the list L_3 is not empty, go to 2.
13. Yield saved to the result V_2. The end.

The best estimation of one of the paths taken as the result of matching.

On the basis of this algorithm fragments are developed along pairs of similar nests that are connected to each other. This fact is the same as in the paper [16]. Consolidation of estimations is done automatically. Finally matching score equal to

$$MS = V_2/\sqrt{n_1 n_2}, \tag{13}$$

where n_1 is the number of minutiae in the request template and n_2 – in the reference template from the share of histogram $H_q(s)$. For reference templates that are not part of $H_q(s)$, set $MS = 0$.

5 Computational Experiments

To test the proposed algorithm we selected 6,000 images from FVC 2000, 2002, 2004 and 2006 databases (only optical sensor). There are images of good, mean and bad quality. A matching was made for them. It depends on the size of the database.

The graph in Fig. 7 illustrates functional dependence of the summary error rate versus the matching rate. The solid bold line illustrates this dependence for algorithm without using of histogram, only matching stage. In this case, the speed of matching is regulated only by tolerances. The dashed bold line illustrates dependence of the same algorithm with using of histogram. In the late case, the speed of matching is regulated by both tolerances and the size of small proportion of referenced templates. The abscissa axis shows the number of comparisons, million per second. The ordinate axis represents the error in percentile@10^{-3} (see [13]). On left side both graphs correspond to the comparison of templates without any accelerators. On the right side the solid line corresponds to the comparison of templates with minimal tolerances, the dashed line corresponds to the comparison with the same tolerances combined with small proportion of referenced templates equal to 10%. As you can see, the acceleration of the algorithm is nonlinear.

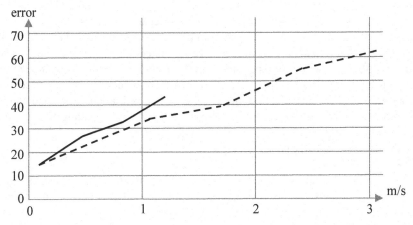

Fig. 7. Error in percentile@10^{-3} vs. number of comparisons, millions per second

To compare algorithm with other results the international testing FVC OnGoing benchmark were used [25]. Some tests results of palm identification for January 12, 2018 are presented in the Table 1. The method, described in this paper, is used in the PPM algorithm. His average match time corresponds to the beginning of the graphs. Minutiae recognition method is described in [26].

Table 1. Comparative analysis of algorithms

Algorithm	Average match time, ms	Template size, byte	FMR_{10000}
PPM	51	36922	6,179
M3gl	911	88958	9,643
MinutiaeClusterFull	86	3344160	10,214
HXKJ	705	12288	17,536
MCC-Based (Baseline)	241	214629	20,107

6 Conclusion

The paper proposes a fingerprint identification algorithm that works in two stages. At the first stage, a histogram evaluation of fingerprint similarity is performed. At the second stage, the fingerprint matching itself is performed for the database share. The use of histogram estimates further increases the performance of the algorithm due to a small loss of quality.

The method relies on topological events that are stored in a template. The size of the template increases slightly. The advantages include the fact that the comparison of nests is reduced to comparison integers and performing a logical bitwise Hamming operation (lists L_0 and L_1). This significantly reduces the identification time.

Further direction of research is seen in the development of new accelerators and adaptation of the algorithm to work with large database. Separately experiments on FVC2004 DB1 will be conducted.

References

1. Bolle, R.M., Connel, J.Y., Pankanti, S., Ratha, N.K.: Guide to Biometrics. Springer, New York (2004). https://doi.org/10.1007/978-1-4757-4036-3
2. Maltoni, D., Maio, D., Jain, A.K., Prabhakar, S.: Handbook of Fingerprint Recognition. Springer, New York (2009). https://doi.org/10.1007/978-1-84882-254-2
3. ISO/IEC 19794–2:2011. Information technology–biometric data interchange formats–Part 2: finger minutiae data
4. Bae, G., Lee, H., Hwang, S.D., Kim, J.: Secure and robust user authentication using partial fingerprint matching. In: Proceedings of 2018 IEEE International Conference on Consumer Electronics (ICCE), pp. 1–6 (2018)
5. Hidayat, R., Souvanlit, K., Bejo, A.: An improvement of minutiae-based fingerprint matching: two level of scoring system. In: Proceedings of 2016 International Symposium on Electronics and Smart Devices (ISESD), pp. 264–267 (2016)
6. Singh, P., Kaur, L.: Fingerprint feature extraction using morphological operations. In: Proceedings of 2015 International Conference on Advances in Computer Engineering and Applications, pp. 764–767 (2015)
7. Gudkov, V.J.: Methods of mathematical description and identification of fingerprints. In: Arlazarov V.L., Emeljanov, N.E. (eds.) Image Processing and Data Analysis: Proceedings of ISA RSA, vol. 38, pp. 336–356. Librokom (2008). (in Russian)

8. Liao, C.C., Chiu, C.T, Fingerprint recognition with ridge features and minutiae on distortion. In: Proceedings of 2016 IEEE International Conference on Acoustics, Speech and Signal Processing (ICASSP), pp. 2109–2113 (2016)

9. Barman, S., Chattopadhyay, S., Samanta, D., Bag, S., Show, G.: An efficient fingerprint matching approach based on minutiae to minutiae distance using indexing with effectively lower time complexity. In: Proceedings of 2014 International Conference on Information Technology, pp. 179–183 (2014)

10. Tran, M.H., Duong, T.N., Nguyen, D.M., Dang, Q.H.: A local feature vector for an adaptive hybrid fingerprint matcher. In: 2017 International Conference on Information and Communications (ICIC), pp. 249–253 (2017)

11. Gudkov, V.J.: Mathematical models of fingerprint image on the basis of lines description. Bull. Chelyabinsk State Univ. **13**, 99–108 (2011). (in Russian)

12. Jiang, X., Yau, W.Y.: Fingerprint minutiae matching based on the local and global structures. In: Proceedings of the 15th International Conference on Pattern Recognition (ICPR 2000), vol. 2, pp. 1038–1041 (2000)

13. Feng, Y., Feng, J., Chen, X., Song, V.: A novel fingerprint matching scheme based on local structure compatibility. In: Proceedings of the 18th International Conference on Pattern Recognition (ICPR 2006), pp. 374–377 (2006)

14. Cao, J., Feng, J.: A robust fingerprint matching algorithm based on compatibility of star structures. In: Proceedings of the Sixth International Symposium on Multispectral Image Processing and Pattern Recognition (MIPPR 2009), vol. 7498, Remote Sensing and GIS Data Processing and Other Applications, p. 74983X (2009)

15. Ratha, N.K., Pandit, V.D., Bolle, R.M., Vaish, V.: Robust fingerprint authentication using local structure similarity. In: Workshop on Applications of Computer Vision, pp. 29–34 (2000)

16. Chikkerur, S., Cartwright, A.N., Govindaraju, V.: K-plet and coupled BFS: a graph based fingerprint representation and matching algorithm. In: Zhang, D., Jain, A.K. (eds.) ICB 2006. LNCS, vol. 3832, pp. 309–315. Springer, Heidelberg (2005). https://doi.org/10.1007/11608288_42

17. Chen, X., Wang, L., Li, M.: An efficient graph-based algorithm for fingerprint representation and matching. In: Proceedings of the 3rd International Conference on Multimedia Technology (ICMT 2013), pp. 1019–1029 (2013)

18. Leslie, S., Sumathi, C.P.: A robust hierarchical approach to fingerprint matching based on global and local structures. Int. J. Appl. Eng. Res. **13**(7), 4730–4739 (2018)

19. Tabassi, E., Wilson, C., Watson, C.: Fingerprint image quality. NIST internal report 7151, National Institute for Standards and Technology (2004)

20. Arkabaev, D.I., Gudkov, V.J.: Method of ridge count based on topology. RU Patent 2444058, G06K 9/52, vol. 6 (2012)

21. Gudkov, V.J., Bokov, A.S.: Method for comparing finger papillary patterns. RU Patent 2185661, G06K 9/62, vol. 20 (2002)

22. Gudkov, V.J., Arkabaev, D.I.: Method for comparing fingerprints of papillary patterns. RU Patent 2331108, G06K 9/62, vol. 22 (2008)

23. Novikov, F.A.: Discrete Mathematics for Programmers: Manual, 2nd edn., St. Petersburg, Piter (2012) (in Russian)

24. Warren, H.S.: Hacker's Delight, 2nd edn., Addison-Wesley, Boston (2003)

25. Dorizzi, B., Cappelli, R., Ferrara, M., et al.: Fingerprint and on-line signature verification competitions at ICB 2009. In: Proceedings of the International Conference on Biometrics (ICB 2009), pp. 725–732 (2009)

26. Gudkov, V.J.: *Metody pervoy i vtoroy obrabotki daktiloskopicheskikh izobrazheniy* [Methods of the First and Second Processing of Fingerprint Images], p. 237. Publishing of the Geotour, Miass (2009)

Training Effective Model for Real-Time Detection of NSFW Photos and Drawings

Dmirty Zhelonkin and Nikolay Karpov$^{(\boxtimes)}$ (iD)

National Research University Higher School of Economics,
25/12 Bolshaya Pechyorskaya str., 603155 Nizhny Novgorod, Russia
dmitryzhelonkin@gmail.com, nkarpov@hse.ru
https://www.hse.ru/en/staff/nkarpov

Abstract. Convolutional Neural Networks (CNN) show state of the art results on variety of tasks. The paper presents the scheme how to prepare highly accurate (97% on the test set) and fast CNN for detection not suitable or safe for work (NSFW) images. The present research focuses on investigating questions concerning identifying NSFW pictures with nudity by neural networks. One of the main features of the present work is considering the NSFW class of images not only in terms of natural human nudity but also include cartoons and other drawn pictures containing obscene images of the primary sexual characteristics. Another important considered issue is collecting representative dataset for the problem. The research includes the review of existing nudity detection methods, which are provided by traditional machine learning techniques and quite new neural networks based approaches. In addition, several important problems in NSFW pictures filtering are considered in the study.

Keywords: Image recognition · Pattern recognition · Not suitable or safe for work · Convolutional neural networks · Pornography detection

1 Introduction

The Internet as a quite free information dissemination platform provides access for numerous texts, images, and other types of uncensored information. Freedom of information on the Internet helps us make more thoughtful decisions based on a variety of sources. But sometimes Internet resources must follow the rules and engage censorship and deny public access to some of the data. Censorship can be determined by resource subjects or audience or the company, that must follow the law. One type of information which often must be censored is not suitable or safe for work (NSFW) data. NSFW is a class of information, that contains nudity, intense sexuality, pornography, obscene texts, and other disturbing content. In our paper, we limit NSFW content to pornographic pictures. In our work such type of content is named positive class, since this class is our aim to find. Hence, normal or suitable for work images are named as a positive class. Evidently, the

© Springer Nature Switzerland AG 2020
W. M. P. van der Aalst et al. (Eds.): AIST 2019, CCIS 1086, pp. 301–312, 2020.
https://doi.org/10.1007/978-3-030-39575-9_31

NSFW filtering images problem exists for the web resources, which are opened for uploading and publishing user's content with images. If we moderate such a resource we face three main tasks. First of all, a moderation process "by hand" takes too much time to analyze all the images. Big lag between uploading picture and publishing makes web resource less attractive for users.

Secondly, it should be noted that the research is provided for the Russian market and law. The investigation of law enforcement practice leads us to the fact that there is no single-valued definition of prohibited pornographic content in different countries. To determine what is the violation of the law, and what is not, usually uses expert assessments. That is why we should take into consideration what users determine as not suitable content. As a result, the definition of such content should be wide. In this work, we use the following definition: image is not suitable for work if it contains not covered full breast's areola and genitals or part of it. These things can be drawn or photographed. Thirdly, we must minimize unsuitable pictures as much as possible. Otherwise, Internet resources can break the law or lose the audience. In other words, the aim is to minimize false positive rate, however false negative rate is less important for us.

Our experiments indicate that modern pretrained neural networks show good results in our problem on many training set configurations. As a result the most challenging task is to collect representative and diverse dataset. The dataset that most appropriate to out NSFW class definition.

The main contribution of this research is the creation of a fast automatic image pre-filtering algorithm with a minimal false positive rate. In order to archive the goal we:

- review existing nudity detection methods;
- make a procedure to collect the data for training purpose;
- train chosen neural networks;
- compare precision, recall, and speed of our approaches with others.

The remainder of this paper is organized as follows: related works overview is discussed in Sect. 2. Section 3 provides information about the dataset formation procedure, while Sect. 4 presents a proposed methodology, experimental evaluation measures, and the baselines. In the end, we conclude and discuss our results on the test dataset.

2   Background

Filtering obscene images is not a novel task. In 1996 Fleck et al. [1] made one of the first attempts to create an automatic NSFW image detector. In 2012, Marie Short, together with co-authors, [2] conducted a study of works for 2002–2012 on filtering obscene images. Further, we observe several most important works from our point of view.

2.1   Skin-Based and Bag-of-Visual-Word Approaches

Approaches which are based on the search and analysis of regions with human skin are widely presented in the literature [1, 3, 4]. For instance, Margaret Fleck and co-authors in 1996 [1] proposed a two-step algorithm. The first stage is the collection of images with a large proportion of pixels with a color close to the body. The second stage is a geometric analysis of "skin" areas.

Another widely used approach in the image classification task is the BoVW (Bag-of-Visual-Words). A dictionary of such features or words can be prepared in several ways. The most popular methods are SIFT (Scale-invariant feature transform) [5] and SURF (Speed Up Robust Features) [6].

Some of the first people who applied BoVW in order to solve the problem of filtering obscene images were Deselaers with co-authors [7]. They showed results superior to all previous algorithms based on the search and analysis of regions with human skin. Sandra Avila et al. in 2009 [8] developed the BoVW based approach and suggested to add color information to SIFT features (Hue-SIFT).

2.2   Convolutional Neural Networks

Today convolutional neural networks (CNN) show the best results in classification and other image tasks. The most well-known example of CNN application in the classification problem, which would significantly exceed all other algorithms, is the victory of AlexNet [9] in ILSVRC 2012 (ImageNet Large Scale Visual Recognition Challenge) [10].

The results shown on ImageNet not only allowed us to increase quality of the classification on a particular task but also allowed us to significantly improve the classification results for many other problems. It turned out that the model trained on ImageNet can be retrained (fine-tuned) for other tasks, even not only classification ones. If we consider the classical architecture of CNN, it turns out that the majority of layers in the neural network generate a feature vector. Then, machine learning model [9], fully connected layers [11] or convolutional layers [12] solve classification task based on the generated features. As a result, the network trained on ImageNet generates features that can effectively represent a variety of images and objects. Some of the most popular models are (ResNet [13], Inception [14], VGG [11] and their modified versions. In this paper, we will use one of the modifications of the network Inception-Xception [15], rethinking and optimization of several known CNN architectures – MobileNet [12] and one of the variants of the architecture ResNet – ResNet-50. All these types of CNN showed good results on the sample ImageNet, but at the same time, learns quickly because of compact and effective architectures.

In the problem of filtering pornographic images, neural networks are also used and show better results compared to other methods. One of the first successful applications of CNN for this task is the work of Muhammad Mustafa in 2015 [16]. He retrained two well-known architectures: AlexNet [9] and GoogleNet [17] on a sample of NPDI [18]. Sample NPDI is one of the few public samples for comparison of models filtering obscene images. Among newer works trained on the NPDI sample, one can distinguish the work of 2017 by Maurizio Perez and co-authors [19]. They experimented with filtering pornographic videos using the same network architectures and slightly improved neural network training techniques, showing better results for this dataset.

Today at least two free models are available for comparison. The first model[1] is the ResNet-50 network with a reduced number of folds on each layer, pre-trained on ImageNet and trained by specialists from Yahoo on its own proprietary sample. The second model[2] was trained to filter pornographic photos, not drawings. The model has the same architecture as the Yahoo network, also trained on a closed sample, but its main purpose is the classification of poses during sexual intercourse. Objective of the project is significantly different from ours, but the output of the network can be reduced to our formulation of the problem.

3   Dataset Formation

Generally, CNN demonstrates satisfactory results only after training on a huge dataset, but gathering sample from scratch is a challenging task since collectors should strongly identify criteria for every class, accurately choose data sources, elaborate gathering procedure and finally collect the data. All these points require time, resources and certain competencies. As a result, several carefully prepared datasets were considered.

3.1   Existing Datasets

There are several datasets for nudity detection task [8,20]. But they are not applicable to our problem as nudity and our NSFW class definition are not the same. In other words, picture with open genitals or with quite small not covered skin area is NSFW content, but it cannot be uniquely defined as nudity. Moreover usually nudity class does not contain drawn pictures. Also, there is Pornography database [18]. In contrast to the previous two, this dataset contains photos and drawn pictures and determine two classes very close to ours. The dataset is a collection of films and selected frames from the films. The benign images also divided into two subclasses: "Easy" and "Difficult". We use only the frames for our task. Close investigation ofPornography database shows us several problems. The first problem is label correctness. NSFW class consists of frames from films with NSFW content, but it seems every n-th screenshots was placed into sample without label checking. But pornographic videos from the database have benign scenes at the beginning, also pornographic scenes have frames which contain only faces, interior and etc. According to our negative class criteria, 67 images from 200 randomly taken pictures from pornographic (only 33,5%) can be named as NSFW. In the same time we did not find any wrong labeled images in the positive class as a result a benign class is correct.

---

[1] https://github.com/yahoo/open_nsfw.
[2] https://github.com/ryanjay0/miles-deep.

The second problem is frame quality. A big part of the dynamic frames are blurred and several NSFW pictures cannot be absolutely classified by human.

The third is a size of the dataset. Pornography database has 6387 and 10340 images for negative and positive classes correspondingly. Since CNN requires huge dataset for training and clear labels we suppose that such number is not enough for training effective classifier from scratch or fine-tuning. Moreover, our NSFW definition mentioned above is wider than the negative subset of this dataset. As a result, the Pornography database is not perfectly suitable for our tasks, but it was used in several experiments. The results are presented in the Experimental Results section.

3.2   Dataset Gathering Procedure

Consequently, we have to collect our own dataset. The main issue for collecting dataset is how to make it representative and able to be a good source for training classifier. In order to resolve this problem we apply an iterative procedure:

– Gather initial sample;
– Conduct a validation experiment;
– Investigate misclassified items.

If there is obvious misclassified subclass then add more images from the most problem subclass and repeat the second stage. Otherwise end procedure.

Validation experiment in the second stage is testing of fine-tuned model pretrained on ImageNet. As testing model architectures were considered some widely used neural network such as AlexNet, ResNet and Inception. Searching the best model type for validation experiment was not provided. Number of epochs determined by the moment when accuracy improvement on validation set was less than 0.005. Usually, this figure was between 3 and 10 epochs. Certainly, we can get better results if we train network longer, but we believe that such a criterion makes it possible to maintain a balance between time and result. We fine-tune only fully connected and the last convolutional layers. Intermediate dataset was split into training (80%), validation (10%) and testing (10%) sets. We explore the test set in order to observe errors and find the most problematic picture cluster. Subsets using strategy is to train network on the training data, choose the best model according to error on validation images and investigate quality on the test set.

The testing experiment strategy is not optimal, but it gives us information which helps us to improve our sample.

3.3   Application of the Iterative Procedure

At the first few iterations we exclude drawn pictures from NSFW class to simplify the searching misclassified subclasses process.

First few iterations show us that we should add more safety images with big flesh color areas and with human faces. These results are quite obvious since a huge part of NSFW images contain faces. It should be noted that every iteration network achieves more than 90% accuracy on the validation set. It can testify to the fact, that the architectures can be well trained on a specific dataset (20000 of NSFW images and 20000 of suitable data), but new examples show poor models generalizing power even with extensive use of augmentations. Since chosen architectures show great generalization on a more difficult challenge of classification of 1000 classes, we decide that the problem is in the dataset and we need to collect more data. Our aim was to collect around half a million pictures for both classes. While expanding dataset especially NSFW class we faced a problem. Whereas proven benign images are placed openly for research purposes datasets such as VOC2012, ImageNet, INRIA person and others, gathering tremendous number of pornographic images is a challenge. First of all, as it was already mentioned, there is no universal NSFW class definition. As a result, we observe that many images which are placed in pornographic resources are safe according to our definition of negative class, but at the same time, there is quite a small number of resources with similar NSFW marks as ours. As the main source of such images we used Danbooru sites and microblogs, for instance tumblr.com. Such microblogs allow users to post any images. Some users publish NSFW images in their blogs and tag images or whole blogs. In order to collect the positive class we rely on description of those images from source, because we cannot check every image manually. We did not use prepared crawlers with collected NSFW URLs pictures[3,4] since they did not exist at the time of dataset gathering formation.

From the web sites, which support picture tagging, we use images which are marked by "NSFW", "Not safe" and other similar tags or at least one tag which describes content as not suitable in our definition. From the microblogs without picture tagging we use images which are placed in microblog with the name containing NSFW keywords or phrases for example sex, naked, porno and etc.

Drawn pictures are grouped in accordance with discovered description. In other words, we firstly collect not suitable pictures according to the description. Then we gather similar in style and content drawn images, for example benign anime and hentai. And finally, we add this drawn set to the dataset and improve it by the iterative procedure. As a result, we have a structure of dataset as it is described below.

Negative class:

- 250000 – photos of the real world;
- 120000 – drawn pictures;
- 35000 – photos of people with big open skin areas;
- 15000 – face images;
- 25000 – photos of people in the crowd and alone;
- 5000 – desert images;
- 420000 – total positive sample size.

---

[3] https://github.com/GodelBose/NSFW_Detection.
[4] https://github.com/EBazarov/nsfw_data_source_urls.

Positive class:

- 300000 – images with natural human nudity;
- 120000 – drawn NSFW pictures;
- 420000 – total negative sample size.

3.4   Upgrading of Dataset A

We observe that neural networks after training on dataset A become too "strict" in terms of negative class. In other words people in bikini and other types of very open suits are constantly qualified as NSFW images by many network architectures and learning strategies. Quality metrics are presented in the Experimental Results section.

More precise investigation of dataset A shows us that there are quite a lot of examples of wrongly gathered images from negative class. It was discovered that some blogs without drawn pictures with NSFW keywords in name contain mostly erotic images with covered genitals and breast. Moreover, sometimes prohibited class definition in microblog changes over time. For instance, sometimes the author places in his blog mostly erotic images and in some time posts only images, which correspond to our NSFW class specification. Another problem appears when people mark benign pictures with few NSFW tags. We suppose that it can be explained by different understanding of what pornographic image is and a human factor. Benign images are selected on the desirable level. We haven't discovered any wrongly collected pictures for the negative class. Furthermore, networks trained on dataset A make very few mistakes in classification photos of the world. Most parts of errors are pictures that contain people in revealing clothes.

Drawn pictures have few wrongly marked instances (less than 5%), but classification quality of such images is quite poor (80% in contrast to 95% for the photos). We suppose that the reason for it is imbalanced subclasses of drawn pictures in negative class. The largest part of collected benign drawn content is anime, but positive class contains not only NSFW anime images or hentai.

As a result, we decide to upgrade dataset A by the next improvements:

- Use stronger NSFW photos gathering conditions;
- Increase benign images proportion in positive class;
- Balance drawn pictures distribution in both classes and increase its proportion in the sample.

Also, we modified conditions of NSFW content:

- From Danbooru websites, we collect images marked by "NSFW", "not safe" and other at least three similar tags which describe content as not suitable in our definition.
- From microblogs, we use image placed in microblog with a name containing NSFW keywords for example sex, porno and 98 of 100 images from this blog must be NSFW (images must be checked by the experts).

Since new conditions require manual checking we were forced to reduce new sample size. Also, as we have not saved pairs "images - tags" and "image - blog name" we had to recollect positive class. The final distribution of the new sample of dataset B described below.

Negative sample:

- 30000 – photos of the world;
- 24000 – photos of people with big open skin areas;
- 23000 – drawn pictures;
- 77000 – total positive samples size.

Positive sample:

- 44000 – images with natural human nudity;
- 33000 – drawn NSFW pictures;
- 77000 – total negative sample size.

Models trained on dataset B show more appropriate results and the last sample was chosen as the main.

4   Experimental Results

In this section we describe conducted experiments with the collected data and some other datasets.

4.1   Proposed Methodology

As it was already discussed convolutional neural networks are state of the art in various tasks. So CNN were chosen as the main classification method. For training networks fine-tuning was applied, since using pretrained networks significantly reduces training time and allows to achieve better results. Xception, MobileNet and ResNet-50 architectures were used for training, since these networks show competitive results in ILSVRC and other competitions, they are quite fast, compact and they have pretrained weights on ImageNet in Keras. We did not provide experiments with architecture modifications since vanilla networks demonstrate desirable results. Also we have limited computational budget and we cannot provide many experiments.

Most networks were trained in Keras framework with Tensorflow backend using Adam optimizer with learning rate 1e–4. We reduce learning rate by a factor 2 when validation accuracy did not improve for a three epochs. In all experiments we train networks for 100 epochs and choose the best by validation accuracy. Random flip, rotation, shear, brightness and gamma jittering augmentation techniques were used in order to improve generalization power. We convert Yahoo model in Keras for fine-tuning. Also Caffe framework was applied for testing vanilla Yahoo and Miles-Deep pretrained networks.

4.2   Evaluation Measure

In order to evaluate proposed approach we run several experiments and write the results of the experiments to the Table 1. Columns "Precision (neg/pos)" and "Recall (neg/pos)" show precision and recall of testing negative and positive subsets correspondingly. For other two columns "Recall for NSFW photos" and "Recall for NSFW drawings" we split the testing positive NSFW subset into photos and drawn pictures respectively.

Table 1. Evaluation on dataset B

Architecture	Procedure	Frame-work	Precision (neg/pos)	Recall (neg/pos)	Recall for NSFW photos	Recall for NSFW drawings
Yahoo_ResNet50_1by2	–	Caffe	0.81/0.98	0.99/0.76	0.87	0.64
Miles-Deep_ResNet50_1by2	–	Caffe	0.56/0.98	1.0/0.19	0.26	0.09
Yahoo_ResNet50_1by2	Fine-tuning	Keras	0.97/0.95	0.95/0.96	0.97	0.93
ResNet50	Fine-tuning	Keras	0.95/0.96	0.96/0.95	0.97	0.91
ResNet50 (grayscale)	Fine-tuning	Keras	0.95/0.95	0.95/0.95	0.96	0.92
Xception	Fine-tuning	Keras	0.96/0.98	0.98/0.96	0.97	0.94
Xception	Re-training	Keras	0.96/0.75	0.86/0.97	0.98	0.96
MobileNet	Fine-tuning	Keras	0.95/0.94	0.95/0.96	0.97	0.93

In Table 1 the first column notes a network structure like ResNet50, Xception or MobileNet. Such a name like Yahoo_ResNet50_1by2 means a particular implementation of the corresponding architecture. Column "Procedure" demonstrates whether we use fine-tuning procedure, pretraining on additional data or use pretrained model as-is. All models were fine-tuned only on the dataset B and pretrained on ImageNet. In retraining procedure we pretrain model using dataset A and then fine-tuned on dataset B. Next we specify a framework which was used: Caffe or Keras. As we can see, the best precisions shown by Xception network both fine-tuned and retrained. Also, we can see that fully trained Xception network works slightly better on drawn pictures than fine-tuned one. The downside is that a fully retrained network has much less precision of NSFW class. It should be noted that MobileNet architecture shows competitive precisions. One more interesting experiment with grayscale images has been done. We evaluated, how useful color information for the networks in this task. Since we have only pretrained weights for color images, standard ResNet50 architecture was fine-tuned on 3 channel gray pictures. Surprisingly results are very close to the color modification. This fact gives us possible way to speed up inference.

Also several experiments were conducted on the NPDI dataset with Xception architecture pretrained on ImageNet. Training procedure was the same as in experiments with dataset B. We can see in Table 1 that pretraining on the dataset B is better than other options. As it was already mentioned the definition of NSFW content is different from ours. Consequently using of NPDI train sample leads to metrics improvements. Also since NPDI dataset has noisy labels we did

Table 2. Evaluation on dataset NPDI

Architecture	Dataset	Precision (neg/pos)	Recall (neg/pos)
Xception	NPDI	0.93/0.96	0.94/0.95
Xception	A	0.63/0.88	0.85/0.69
Xception	B	0.92/0.76	0.50/0.97
Xception	A; NPDI	0.91/0.91	0.85/0.95
Xception	B; NPDI	0.95/0.98	0.97/0.97

not try to improve these achievements or try to use NPDI as pretraining sample and only convinced of the superiority of dataset B in close task (Table 2).

We conduct several speed tests except quality evaluation. Speed is important characteristic of algorithms especially in commercial usage. We compare several architectures on machine with NVIDIA GTX 1080, CUDA 9.0, cuDNN 7.3.1, Keras 2.2.4 and TensorFlow 1.12. It was done with 64 batch size and constant image in RAM (Random Access Memory).

As we can see in the Table 3, the slowest result shown by ResNet50. Xception architecture performs just a little bit faster. Yahoo_ResNet50_1by2 runs more than 2 times faster on Caffe than vanilla ResNet50 on Keras. The fastest architecture in this table is MobileNet.

Table 3. Images per second performance.

Architecture	Framework	Images per second
MobileNet	Keras	596
MobileNet v2 [21]	Keras	548
Yahoo_ResNet50_1by2	Keras	423
VGG 16	Keras	230
Xception	Keras	206
ResNet50	Keras	203

If we sum up our results, Tables 1 and 3 demonstrate that MobileNet shows the best tradeoff between speed and quality. That is why it is chosen to run in production system. You can check the inference of our neural network on the demonstration version, which is available at fapme.me[5] with Russian user interface.

5   Conclusion

This research is dedicated to the topic of creation of highly accurate convolutional neural network for filtering not suitable or safe for work images. As a

---

[5] https://fapme.me.

NSFW class of images we consider not only photos in terms of natural human nudity, but also include cartoons and other drawn pictures containing obscene images of the primary sexual characteristics.

In order to achieve our goal, we review existing nudity detection methods which are provided by traditional machine learning techniques and quite new neural networks based approaches. In addition, several important problems in NSFW pictures filtering are considered in the study.

We have found that existing neural networks trained on NSFW the dataset achieve more than 90% accuracy on validation. Consequently the most challenging task is to collect representative and diverse dataset.

The results of the work are:

- dataset iterative collecting procedure;
- artificial neural networks trained on the dataset;
- comparison with other existing approaches.

The best tradeoff between precision, recall and speed has been achieved using pre-trained MobileNet v1 after fine-tuning procedure on our dataset.

References

1. Fleck, M.M., Forsyth, D.A., Bregler, C.: Finding naked people. In: Buxton, B., Cipolla, R. (eds.) ECCV 1996. LNCS, vol. 1065, pp. 593–602. Springer, Heidelberg (1996). https://doi.org/10.1007/3-540-61123-1_173
2. Short, M.B., Black, L., Smith, A.H., Wetterneck, C.T., Wells, D.E.: A review of Internet pornography use research: methodology and content from the past 10 years. Cyberpsychol. Behav. Soc. Netw. **15**(1), 13–23 (2012)
3. Flores, P.I.T., Guillén, L.E.C., Prieto, O.A.N.: Approach of RSOR algorithm using HSV color model for nude detection in digital images. Comput. Inf. Sci. **4**(4), 29 (2011)
4. Platzer, C., Stuetz, M., Lindorfer, M.: Skin sheriff: a machine learning solution for detecting explicit images. In: Proceedings of the 2nd International Workshop on Security and Forensics in Communication Systems, pp. 45–56. ACM (2014)
5. Lowe, D.G.: Distinctive image features from scale-invariant keypoints. Int. J. Comput. Vision **60**(2), 91–110 (2004)
6. Bay, H., Tuytelaars, T., Van Gool, L.: SURF: speeded up robust features. In: Leonardis, A., Bischof, H., Pinz, A. (eds.) ECCV 2006. LNCS, vol. 3951, pp. 404–417. Springer, Heidelberg (2006). https://doi.org/10.1007/11744023_32
7. Deselaers, T., Pimenidis, L., Ney, H.: Bag-of-visual-words models for adult image classification and filtering. In: 19th International Conference on Pattern Recognition, ICPR 2008, pp. 1–4. IEEE (2008)
8. Lopes, A.P.B., Avila, S.E.F.d., Peixoto, A.N.A., Oliveira, R.S., Araújo, A.d.A.: A bag-of-features approach based on hue-sift descriptor for nude detection. In: Proceedings of the XVII European Signal Processing Conference (EUSIPCO), Glasgow, Scotland (2009)
9. Krizhevsky, A., Sutskever, I., Hinton, G.E.: ImageNet classification with deep convolutional neural networks. In: Advances in Neural Information Processing Systems, pp. 1097–1105 (2012)

10. Deng, J., Dong, W., Socher, R., Li, L.J., Li, K., Fei-Fei, L.: ImageNet: a large-scale hierarchical image database. In: IEEE Conference on Computer Vision and Pattern Recognition, CVPR 2009, pp. 248–255. IEEE (2009)
11. Simonyan, K., Zisserman, A.: Very deep convolutional networks for large-scale image recognition. arXiv preprint arXiv:1409.1556 (2014)
12. Howard, A.G., et al.: MobileNets: efficient convolutional neural networks for mobile vision applications. arXiv preprint arXiv:1704.04861 (2017)
13. He, K., Zhang, X., Ren, S., Sun, J.: Deep residual learning for image recognition. In: Proceedings of the IEEE Conference on Computer Vision and Pattern Recognition, pp. 770–778 (2016)
14. Szegedy, C., Vanhoucke, V., Ioffe, S., Shlens, J., Wojna, Z.: Rethinking the inception architecture for computer vision. CoRR abs/1512.00567 (2015). http://arxiv.org/abs/1512.00567
15. Chollet, F.: Xception: deep learning with depthwise separable convolutions. arXiv preprint pp. 1610–02357 (2017)
16. Moustafa, M.: Applying deep learning to classify pornographic images and videos. arXiv preprint arXiv:1511.08899 (2015)
17. Szegedy, C., et al.: Going deeper with convolutions. In: Proceedings of the IEEE Conference on Computer Vision and Pattern Recognition, pp. 1–9 (2015)
18. Avila, S., Thome, N., Cord, M., Valle, E., Araújo, A.D.A.: Pooling in image representation: the visual codeword point of view. Comput. Vis. Image Underst. $117(5)$, 453–465 (2013)
19. Perez, M., et al.: Video pornography detection through deep learning techniques and motion information. Neurocomputing 230, 279–293 (2017)
20. Jones, M.J., Rehg, J.M.: Statistical color models with application to skin detection. Int. J. Comput. Vis. $46(1)$, 81–96 (2002)
21. Sandler, M., Howard, A., Zhu, M., Zhmoginov, A., Chen, L.C.: MobileNetV2: inverted residuals and linear bottlenecks. In: 2018 IEEE/CVF Conference on Computer Vision and Pattern Recognition, pp. 4510–4520. IEEE (2018)

Optimization Problems on Graphs and Network Structures

Prediction of New Itinerary Markets for Airlines via Network Embedding

Dmitrii Kiselev[1] and Ilya Makarov[1,2]

[1] National Research University Higher School of Economics, Moscow, Russia
dakiselyov@edu.hse.ru, iamakarov@hse.ru
[2] Faculty of Computer and Information Science, University of Ljubljana, Ljubljana, Slovenia

Abstract. Network planning is one of the most significant problems that airlines solve every day. Currently, airlines utilise traveller decision choice modelling, which has certain drawbacks. It analyses each market independently, which does not consider the entire Airline network information with its dynamic structure formation based on competition factors.

In the paper, we show that Airline network structure provides an accurate prediction for the current network and for future lines. We compare several approaches for Airline network link prediction via structural network embeddings, which are interpreted as new itinerary markets estimation.

Keywords: Itinerary prediction · Network embeddings · Link prediction · Network planning · Airline market estimation

1 Introduction

The airline business is very competitive, so airlines try to act in an optimal way to gain profits. There are different ways to do it: marketing, operations or better network planning. Marketing is aimed to create brand loyalty and attract new users to a specific carrier. Operation optimisation is the cost reduction policy, such as infrastructure development or plane maintenance strategy.

Network planning deals with finding the optimal schedule and appropriate plane types while choosing proper prices for a specific route. This paper considers how an airline could efficiently enter new itinerary markets of Airline network using modern approaches in network science.

The standard way to estimate potential markets is based on a prediction of demand, market shares and corresponding profit [2,4]. Usually, one needs to

The work of I. Makarov was supported by the Russian Science Foundation under grant 17-11-01294 and performed at National Research University Higher School of Economics, Russia. The work of D. Kiselev was prepared within the framework of the HSE University Basic Research Program and funded by the Russian Academic Excellence Project '5-100'. Sections 1–3 were prepared by I. Makarov. Sections 4, 6 were prepared by D. Kiselev. Section 5 was contributed jointly and equally by both authors.

© Springer Nature Switzerland AG 2020
W. M. P. van der Aalst et al. (Eds.): AIST 2019, CCIS 1086, pp. 315–325, 2020.
https://doi.org/10.1007/978-3-030-39575-9_32

build discrete choice models or quality of service index for users who will possibly flight in such routes [2].

Discrete choice modelling is the class of methods that aimed to explain or predict individual choice over given decision space. This is a subtask of classification. Using it, airlines estimate the probability of traveller's choice across different combinations of ticket parameters. Such models vary across the industry, but the main tool is the Multinomial or (Generalized) Nested Logistic Regressions models [2]. Airlines choose such methods because it is easy to explain in comparison with other more powerful approaches like tree-based ensembles or neural networks. Recently, some works on the application of modern methods were presented in [19, 22]. It is also possible to extrapolate results from other similar markets to new interesting routes [2]. These similarities are shaped from experts' opinion, but network science methods can provide such similarities in a strictly formal way.

Quality of service index (QSI) is an empirical method assigning manually founded weights to different parameters of tickets such as departure time, route or carrier. In what follows, demand prediction is estimated as normalised QSI across different carriers [9].

Frequent flyers bring a large part of the profit to the carrier. Such travellers are sensitive to different parameters of flights: schedule, ticket type, price, airports, services and personal preferences to specific airlines. The methods described above solve the problem from travellers point of view, so it concentrates on offer-specific parameters. However, such models do not take into account information about the whole Airline network.

Another problem of these methods involves the temporal structure of new market discovering: many new routes would be closed after a small time period (about half a year). Thus, it is necessary to apply a new approach that can predict the date and duration of the new airline opening.

Taking into account that profits on a new route are of great interest as an aggregated market parameter, it is interesting to learn from how the whole network formation, rather than study behaviour of specific users. Modern methods on network science allow the accounting of all network information such as airport and origin-destination market-specific attributes.

In what follows, we describe the structure of our work. Section 2 overviews related work. Section 3 describes the experimental setup, following by Sect. 4 with results and discussion sections.

It is important to mention that there is a lack of research works in the chosen field because most data remain confidential. Lack of data also induces us to choose the United States market for research as the only available data, however, described in this paper methods are applicable for other Airline networks as well.

2   Airline Prediction via Link Prediction

In terms of network science, the new itinerary route problem can be stated as a link prediction problem (LPP). LPP is a problem of predicting new edges in the network or reconstructing missing edges in noisy graph data. It could be solved by classic machine learning methods over graph feature space [34]. The main problem in such an approach is to create feature vectors for node pairs and take into account the nodes and edges attributes together with vectorized features of the graph structure.

The basic approach suggested to find several similarity measures between vertices [20] (e.g. common neighbours, Jaccard coefficient, etc.). If nodes are similar by a certain threshold than new edge should exist. This approach can be expanded using topology and social network analysis theory bringing more complex metrics such as generalized Jaccard coefficient, hitting time between nodes or homophily and structural balance [34].

From a machine learning point of view, link prediction problem can be solved by standard classification approach using standard frameworks over feature space for pairs of nodes and predicting whether the edge between them will appear/exist in the network. The answer on how to construct such features without domain expert and independent of the task was found via the formulation of graph embedding.

Network embedding construction is a set of methods that aimed to derive meaningful features of discrete graph-structured data. These methods allow to find automatically a low-dimensional description of nodes and edges [5,6,8,11, 36,37]. It helps efficiently solve LPP because it does not require the calculation of all the similarities or manually engineering graph-related features. For dynamic networks there exist graph embedding models taking into account temporal component in structural data [10,13].

Also if we can assign a probability to each edge, then we can use probabilistic models and directly estimate edge probability in different types of networks [1,7,18]. A detailed survey of approaches on link prediction problem can be found in [34].

3   Network Embedding

There are different approaches to network embedding construction, but mainly it is based on the three following methods: matrix factorisation, random-walks and deep learning.

3.1   Matrix Factorisation

This groups of methods apply matrix factorisation to different representations of the graph structure. Laplacian Eigenmaps use graph Laplacian and aimed to preserve first-order similarities and graph community structure [3].

HOPE directly factorises proximity matrix and achieves embeddings as a concatenation of self node representation and its context [23]. Such an approach allows catching high-order proximities and asymmetric transitivity in a graph. VERSE also learn embeddings to directly reconstruct node similarities [32].

Despite this group of methods is the best theoretically learned, it is very computationally intensive.

3.2 Random Walks

Due to problems of the previous approach and good results in NLP, another method was proposed. The main idea is to sample random walks from the graph and learn node embeddings with skip-gram approach [21].

Random walk sampling varies between different methods. DeepWalk [24] use simple random walk sampling. While it is biased towards first-order proximity, LINE preserves both first and second-order proximities [29]. Node2vec adds different levels of random-walks (a mixture of breadth-first and depth-first searches), that allows the model to preserve local and global graph information [14]. Author of Diff2vec proposes the other method of random walks sampling based on diffusion on graphs [28].

Also, methods in this group can be extended to catch other graph properties. GEMSEC extends classical random-walk based method with K-means objective to save cluster information [27]. Struc2vec extends classical random-walk models with accounting for graph isomorphism and structural similarities [26]. Walklets capture the different levels of community membership [25].

3.3 Deep Learning

Graph convolution networks (GCN) provides better computational efficiency for semi-supervised tasks, like LPP or node classification [16]. One of the main advantages of GCNs is its ability to account node attributes.

Given the graph $G(V, E)$, adjacency matrix A and feature matrix X of size $(N_{nodes}, N_{features})$, where N_{nodes} refers to number of vertices and $N_{features}$ to number of node attributes.

GCN can be defined as a set of hidden layers $H^i = f(H^{i-1}, A)$, where H^0 is equal to matrix X, and each hidden layer is the tensor of node vector representations. At next hidden layer, these features are aggregated using propagation f.

It means that graph convolutions aggregate feature information of its neighbours based on the adjacency matrix.

Also one can build graph autoencoder (GAE) using graph convolutions [17]. The main idea is to learn latent representations with GCN encoder and reconstruct adjacency matrix of the given graph with an inner-product decoder. The learning of such a model is similar to classic autoencoders, so classic reconstruction loss can be used. Reconstructed adjacency matrix gives estimates for links, even it did not exist in the original graph.

Similar to other cases of autoencoders, there is a probabilistic interpretation Variational Graph Autoencoder (VGAE). It introduces conditionally independent probability distributions over latent variables and uses probabilistic inference to reconstruct adjacency matrix.

[30] introduce symmetrical graph autoencoder with weight-sharing. It is proposed to solve the memory-intensive learning of VGAE. Also, weight-sharing allows for better generalization and faster convergence.

4   Experimental Setup

4.1   Data Description

Network data is obtained from the US Department of Transportation Airline Origin and Destination Survey (DB1B) [31]. It is 10% subsample of all airline tickets reported by carriers. Minimal available time delta for DB1B data is quarter.

We analyse only data for the latter year (2018). The overall number of cities connected by graph was 414 and number of edges was 73482. Time dynamics from first to last quarter is described in the following Table 1.

Table 1. Dynamics of graph structure

	2018 Q1	2018 Q2	2018 Q3	2018 Q4	Total
Nodes	392	395	393	396	414
Edges	54212	58089	59829	58520	73482
New nodes	–	11	5	11	–
New edges	–	9966	8473	7411	–
Removed nodes	–	8	7	8	–
Removed edges	–	6089	6733	8720	–

For train period we choose the first quarter of 2018. We use the whole network for train models because our main task is to predict new routes (edges) in the future.

Test positive examples are all flights (edges) opened in the periods after the first quarter of 2018 (train period). Negative examples are generated as edges that connect nodes accessible with second-order proximity. Such edges were selected to be consistent with the shortest path length for positive test sample (11.54% for length 1, 88.37% for 2 and 0.09% for 3). In an intuitive way, such edges are the two-legs trips of the travellers.

The created test sample is imbalanced, so we undersample negative examples in the amount equal to the number of positive observations. It is necessary because evaluation metrics that we use is sensitive to imbalance classes.

4.2   Experiments

Models. We test all models described in the Network embeddings section.

Deep learning models tested in two cases: semi-supervised learning and unsupervised pretraining with supervised classifier after it.

Also, we test different edge aggregations for such embeddings: simple node vectors concatenation and Hadamard product [14].

We use gradient-boosted decision trees as supervised classifier (LightGBM implementation [15]). This method is one of the most powerful for classification. It allows catching non-linear dependencies in data.

Model Training. Firstly, all models trained on data from the first quarter of 2018.

All embeddings learned with default parameters except node vector dimensions. It is equal to 64.

After that, we use achieved node embeddings as features for GBM classifier. Edge features obtained in two ways: the concatenation of node vectors and Hadamard product of it. GBM trained on half of the test sample.

We concatenate train set with half of the test set for training end-to-end semi-supervised models. It allows comparing the results of different types of learning.

LightGBM hyperparameters set to default without fine-tuning in order to compare the quality of embeddings in a similar framework.

Evaluation. Model is evaluated using the area under the ROC curve and average precision metrics.

ROC-AUC can be interpreted as the probability of random negative example sampled from a uniform distribution is ranked lower then sampled random positive. It means, that ROC-AUC measures overall model quality independently from the chosen class threshold.

Average precision is similar to the area under the Precision-Recall curve. Higher the score, better work of classifier, because it can predict classes with high precision for a larger subsample of data independent from the threshold.

5   Results

Evaluation results are presented in Table 2.

As we can see Hadamard product shows worse performance in all cases. Concatenation provides more opportunities for feature aggregation: GBM can catch non-linear dependence between different elements of the embedding vector.

The best results show HOPE, LONGAE, node2vec and Laplacian Eigenmaps. The first three methods are similar only in accounting for high-order proximity and it can use it globally for the whole graph. But Laplacian Eigenmaps preserves only first-order. Also LINE model can catch high-order similarity, but this method works only for the local neighbourhood. LLE, DeepWalk, diff2vec, VERSE and Struc2vec accounts only first-order similarities.

As we mention above, airlines are interested in routes with high demand. Such markets usually attract travellers even if they need to perform multi-leg trips. So incorporating information of such possible travellers movement allow achieving good results.

We suppose that Laplacian Eigenmaps works good because it catches clustering. High metrics for Walklets model confirm that hypothesis. GEMSEC uses a greedy algorithm that could be insufficient to create valuable partitioning in the given graph.

LONGAE shows better results then VGAE because of weight-sharing between encoder and decoder parts. Such result is consistent to original LONGAE paper [30].

Table 2. Results of model validation (AUC)

Edge aggregation	Concatenation		Hadamard	
Model	ROC	PR	ROC	PR
VGAE semi-supervised	0.678	0.804	–	–
LONGAE semi-supervised	**0.891**	**0.864**	–	–
Node2vec	0.945	0.947	0.879	0.868
HOPE	0.951	**0.952**	0.921	0.916
Laplacian Eigenmaps	0.949	0.951	0.908	0.899
Locally Linear Embedding	0.927	0.926	0.828	0.808
LONGAE unsupervised	**0.952**	**0.952**	0.928	0.913
VGAE unsupervised	0.943	0.944	0.899	0.878
DeepWalk	0.939	0.939	0.89	0.874
LINE	0.936	0.934	0.831	0.802
VERSE	0.93	0.926	0.854	0.839
GEMSEC	0.923	0.921	0.823	0.802
diff2vec	0.931	0.928	0.849	0.837
Walklets	0.944	0.945	0.892	0.964
struc2vec	0.941	0.942	0.904	0.892

6 Conclusion

In this paper, we apply modern network science methods for the task of an airline itinerary prediction. We show that high-order similarities and graph clustering are important in that task. Standard airline approaches are based only on traveller behaviour, not taking into account network topology.

For example, our best model successfully predicts several new itineraries that connect Trenton and some other cities (Chicago, Nashville, Jacksonville, Raleigh/Durham and etc.). This flights exists nowadays and have the largest passenger flow over all routes in our test-sample. In original train graph (Q1 2018) Trenton was connected with 16 cities, 13 of it was connected with all six new itineraries from Trenton. Visualisation of subgraph for Trenton is provided on Fig. 1. The green line shows new predicted itineraries, hard dashed line shows the existing routes from Trenton and the pale dashed line shows edges to predicted cities from Trenton's neighbours. Our best model predicts 1188 false-negative edges in 2018 Q2. 672 of it was closed in 2018 Q3 and 429 in 2018 Q4 (total 1101). It means the model can help airlines to make a decision about the new itinerary market opening.

Further Research. We explore this Airline network planning from the following sides.

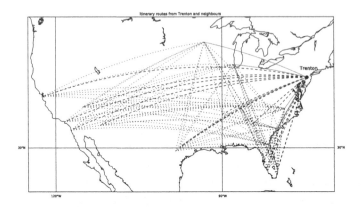

Fig. 1. Visualization of airline predictions for Trenton (dark-green) (Color figure online)

Graph Dynamics. Different attributes of graph always changing, e.g. new routes are created, price depending on booking window and planes. Such information should be included in the model.

Network Attributes. In this paper, we concentrate only on network structure, but Airline network is attributed, e.g. we can describe the city in terms of sightseeing, that is a proxy for travellers attractiveness. As shown in some papers, taking in account node attributes can give better results [16]. However, in the airline network, most data is located on edges, so we aim to study edge-based network embedding models similar to ELAINE [12], Edge2Vec [35] and GAT [33], which use incidence matrix and allows to catch edge attributes with the help of Line graph or properly chosen edge information aggregation.

Heterogeneous Network. Also, we should consider airports in one city as different nodes, but different air companies represent different edge types, and predicting of a new airline for one company is correlated with the predictions for the other air companies operating in the same market.

Domain Specific Problems. It is important not only predict whether to open a new route but also models should provide good estimates of parameters that are under airlines' interest. Such parameters can be market shares, optimal prices, profit and schedule optimizations.

References

1. Backstrom, L., Leskovec, J.: Supervised random walks: predicting and recommending links in social networks. In: Proceedings of the fourth ACM International Conference on Web Search and Data Mining, pp. 635–644. ACM (2011)
2. Barnhart, C., Smith, B.: Quantitative Problem Solving Methods in the Airline Industry. Springer, Heidelberg (2012). https://doi.org/10.1007/978-1-4614-1608-1

3. Belkin, M., Niyogi, P.: Laplacian eigenmaps for dimensionality reduction and data representation. Neural Comput. **15**(6), 1373–1396 (2003)
4. Busquets, J.G., Alonso, E., Evans, A.D.: Air itinerary shares estimation using multinomial logit models. Transp. Plann. Technol. **41**(1), 3–16 (2018). https://doi.org/10.1080/03081060.2018.1402742
5. Cai, H., Zheng, V.W., Chang, K.: A comprehensive survey of graph embedding: problems, techniques and applications. IEEE Trans. Knowl. Data Eng. **30**, 1616–1637 (2018)
6. Chen, H., Perozzi, B., Al-Rfou, R., Skiena, S.: A tutorial on network embeddings. arXiv preprint arXiv:1808.02590 (2018)
7. Chen, Z., Zhang, W.: A marginalized denoising method for link prediction in relational data, pp. 298–306. SIAM Publications Online (2014). https://doi.org/10.1137/1.9781611973440.34. https://epubs.siam.org/doi/abs/10.1137/1.9781611973440.34
8. Cui, P., Wang, X., Pei, J., Zhu, W.: A survey on network embedding. IEEE Trans. Knowl. Data Eng. **31**, 833–852 (2018)
9. Garrow, L.A.: Discrete Choice Modelling and Air Travel Demand: Theory and Applications. Routledge, London (2016)
10. Goyal, P., Chhetri, S.R., Canedo, A.: dyngraph2vec: capturing network dynamics using dynamic graph representation learning (2018)
11. Goyal, P., Ferrara, E.: Graph embedding techniques, applications, and performance: a survey. Knowl.-Based Syst. **151**, 78–94 (2018)
12. Goyal, P., Hosseinmardi, H., Ferrara, E., Galstyan, A.: Capturing edge attributes via network embedding. arXiv preprint arXiv:1805.03280 (2018)
13. Goyal, P., Kamra, N., He, X., Liu, Y.: DynGEM: deep embedding method for dynamic graphs. arXiv preprint arXiv:1805.11273 (2018)
14. Grover, A., Leskovec, J.: Node2vec: scalable feature learning for networks. In: Proceedings of the 22nd ACM SIGKDD International Conference on Knowledge Discovery and Data Mining, KDD 2016, pp. 855–864. ACM, New York (2016). https://doi.org/10.1145/2939672.2939754. http://doi.acm.org/10.1145/2939672.2939754
15. Ke, G., et al.: LightGBM: a highly efficient gradient boosting decision tree. In: Advances in Neural Information Processing Systems, pp. 3146–3154 (2017)
16. Kipf, T.N., Welling, M.: Semi-supervised classification with graph convolutional networks. arXiv preprint arXiv:1609.02907 (2016)
17. Kipf, T.N., Welling, M.: Variational graph auto-encoders. arXiv preprint arXiv:1611.07308 (2016)
18. Kuo, T.T., Yan, R., Huang, Y.Y., Kung, P.H., Lin, S.D.: Unsupervised link prediction using aggregative statistics on heterogeneous social networks. In: Proceedings of the 19th ACM SIGKDD International Conference on Knowledge Discovery and Data Mining, KDD 2013, pp. 775–783. ACM, New York (2013). https://doi.org/10.1145/2487575.2487614. http://doi.acm.org/10.1145/2487575.2487614
19. Lhéritier, A., Bocamazo, M., Delahaye, T., Acuna-Agost, R.: Airline itinerary choice modeling using machine learning. J. Choice Modell. **31**, 198–209 (2018)
20. Liben-Nowell, D., Kleinberg, J.: The link-prediction problem for social networks. J. Assoc. Inf. Sci. Technol. **58**(7), 1019–1031 (2007)
21. Mikolov, T., Sutskever, I., Chen, K., Corrado, G.S., Dean, J.: Distributed representations of words and phrases and their compositionality. In: Advances in Neural Information Processing Systems, pp. 3111–3119 (2013)

22. Mottini, A., Acuna-Agost, R.: Deep choice model using pointer networks for airline itinerary prediction. In: Proceedings of the 23rd ACM SIGKDD International Conference on Knowledge Discovery and Data Mining, pp. 1575–1583. ACM (2017)
23. Ou, M., Cui, P., Pei, J., Zhang, Z., Zhu, W.: Asymmetric transitivity preserving graph embedding. In: Proceedings of the 22nd ACM SIGKDD International Conference on Knowledge Discovery and Data Mining, pp. 1105–1114. ACM (2016)
24. Perozzi, B., Al-Rfou, R., Skiena, S.: DeepWalk: online learning of social representations. In: Proceedings of the 20th ACM SIGKDD International Conference on Knowledge Discovery and Data Mining, KDD 2014, pp. 701–710. ACM, New York (2014). https://doi.org/10.1145/2623330.2623732. http://doi.acm.org/10.1145/2623330.2623732
25. Perozzi, B., Kulkarni, V., Chen, H., Skiena, S.: Don't walk, skip!: online learning of multi-scale network embeddings. In: Proceedings of the 2017 IEEE/ACM International Conference on Advances in Social Networks Analysis and Mining, pp. 258–265. ACM (2017)
26. Ribeiro, L.F., Saverese, P.H., Figueiredo, D.R.: struc2vec: learning node representations from structural identity. In: Proceedings of the 23rd ACM SIGKDD International Conference on Knowledge Discovery and Data Mining, pp. 385–394. ACM (2017)
27. Rozemberczki, B., Davies, R., Sarkar, R., Sutton, C.: GEMSEC: graph embedding with self clustering. arXiv preprint arXiv:1802.03997 (2018)
28. Rozemberczki, B., Sarkar, R.: Fast sequence-based embedding with diffusion graphs. In: Cornelius, S., Coronges, K., Gonçalves, B., Sinatra, R., Vespignani, A. (eds.) CompleNet 2018. SPC, pp. 99–107. Springer, Cham (2018). https://doi.org/10.1007/978-3-319-73198-8_9
29. Tang, J., Qu, M., Wang, M., Zhang, M., Yan, J., Mei, Q.: LINE: large-scale information network embedding. In: Proceedings of the 24th International Conference on World Wide Web, pp. 1067–1077. International World Wide Web Conferences Steering Committee (2015)
30. Tran, P.V.: Learning to make predictions on graphs with autoencoders. In: 5th IEEE International Conference on Data Science and Advanced Analytics (2018)
31. Bureau of Transportation Statistics: Airline origin and destination survey (DB1B) (2018)
32. Tsitsulin, A., Mottin, D., Karras, P., Müller, E.: VERSE: versatile graph embeddings from similarity measures. In: Proceedings of the 2018 World Wide Web Conference, WWW 2018, pp. 539–548. International World Wide Web Conferences Steering Committee, Republic and Canton of Geneva (2018). https://doi.org/10.1145/3178876.3186120
33. Veličković, P., Cucurull, G., Casanova, A., Romero, A., Lio, P., Bengio, Y.: Graph attention networks. arXiv preprint arXiv:1710.10903 (2017)
34. Wang, P., Xu, B., Wu, Y., Zhou, X.: Link prediction in social networks: the state-of-the-art. Sci. China Inf. Sci. 58(1), 1–38 (2015). https://doi.org/10.1007/s11432-014-5237-y
35. Xiang, B., Liu, Z., Zhou, J., Li, X.: Feature propagation on graph: a new perspective to graph representation learning. CoRR abs/1804.06111 (2018). http://arxiv.org/abs/1804.06111

36. Zhang, D., Yin, J., Zhu, X., Zhang, C.: Network representation learning: a survey. IEEE Trans. Big Data (2018). https://ieeexplore.ieee.org/abstract/document/8395024
37. Zhang, Z., Cui, P., Zhu, W.: Deep learning on graphs: a survey. arXiv preprint arXiv:1812.04202 (2018)

Developing a Network-Based Method for Building Associative Series of Hashtags

Sergey Makrushin$^{(\boxtimes)}$ and Nikita Blokhin

Financial University under the Government of the Russian Federation,
49 Leningradsky Prospekt, GSP-3, 125993 Moscow, Russia
SVMakrushin@fa.ru

Abstract. Nowadays hashtags are an important mechanism of semantic navigation in social media. In the current research we considered the problem of building associative series of hashtags for Instagram. These series should meet two criteria: they should be relatively short and should not have wide semantic gaps between sequential hashtags. A generator of such series could be used for creating recommendations for increasing the quantity of hashtags in messages.

We gathered information from approximately 15 million messages from Instagram and built a co-occurrence network for hashtags from these messages. To build associative series we defined a formal definition of the semantic path building problem as a multicriteria optimization problem on the co-occurrence network.

We developed a combined optimization function for both criteria from the semantic path building problem. For measuring semantic similarity between hashtags, we use a metric based on vector embedding of hashtags by word2vec algorithm. Using empirical paths built by different algorithms, we graduated the parameters of the combined optimization function. The function with these parameters could be used in Dijkstra's algorithm and in a specialized greedy algorithm for building semantic paths which look as appropriate associative series.

Keywords: Semantic navigation · Hashtags recommendation · Co-occurrence graph · Path finding · Greedy routing

1 Introduction

Hashtags are an important mechanism of semantic navigation in social media. In the current paper we will use Instagram as a basic example of social media, but usually other contemporary social media have quite similar functionality. In Instagram a hashtag simultaneously plays the role of an ingoing and an outgoing link. A hashtag is typically first used as outgoing link from the message to the automatically generated navigation page for this hashtag. This navigation page consists of links to all messages which include this hashtag. In such a way a hashtag in a message works as ingoing link from the navigation page of the hashtag.

© Springer Nature Switzerland AG 2020
W. M. P. van der Aalst et al. (Eds.): AIST 2019, CCIS 1086, pp. 326–337, 2020.
https://doi.org/10.1007/978-3-030-39575-9_33

Every hashtag has certain semantics and its meaning is usually too narrow or too wide for the meaning of all the message. Usually a message is described by a set of hashtags, which complement and clarify each other. A large number of relevant hashtags increases the popularity of the message, because a hashtag works as an incoming link to the message. But manually finding up to 30 relevant hashtags (upper limit for Instagram) for a message is labor-consuming. Our goal is to develop an algorithm which would help to enlarge the small set of hashtags created by the message author. In particular, the algorithm will offer an associative series of hashtags which is a semantically connected sequence from the source hashtag to the target hashtag consisting of new hashtags. Any pair of hashtags from the original hashtag set could be used as a source and a target hashtag for building an associative series, and the author of the message could select appropriate hashtags from the series to enlarge the original hashtag set.

2   Related Work

Many problems which are similar to building an associative series of hashtags have been considered in the literature. In [1] the problem of building lexical chains, which are sequences of semantically related words in a text, was examined. A word is added to an existing chain only if it is related to one or more of the words already in the chain by a cohesive relation. The standard application of lexical chains is text segmentation, which was firstly suggested in [2].

The semantic distance could be used for approximation of the cohesive relation between words. Semantic distance could be computed using one of the two major classes of methods: resource-based measures, which use thesaural relations or resources alike to WordNet [3], used in [4]; or measures of distributional similarity such as word2vec [5].

Tasks similar to building an associative series are often considered in the analysis of knowledge graphs. There are usually many paths connecting a pair of entities in a knowledge graph, and the classical approach to the problem of choosing an optimal path - finding the shortest path - is extensively studied [6]. However, in real-life applications, the most-valuable paths are those that are short and meaningful, rather than just being the shortest [7].

An empirical analysis of human navigation shows that humans can easily find intuitive, but not necessarily the shortest paths in knowledge graphs. Examples presented in [8] illustrated the difference between the paths found by human and the results of a shortest-path algorithm: the shortest paths are far less semantically meaningful than the paths found by the human. It means that the distance constraint needs to be relaxed to include longer, but meaningful paths. In the paper [7], authors call the process of navigating from one entity to another, guided by entity semantics, semantic navigation. The paths which are built by the semantic navigation process can be used to connect the entities and explain their relationships. These paths could be applied in a wide range of real-world applications, such as knowledge completion [9] and entity recommendation [10].

Existence of metrics similar to semantic distance could be useful for building a suboptimal shortest path, using only local information about a network structure. It could dramatically decrease the computational complexity of building a path in big networks [11]. In [12] it was shown that the length of these suboptimal paths (authors call them semantic paths) was only slightly longer then for optimal solutions, when this approach was applied to Wikipedia network. Moreover, it was shown that semantic paths have much stronger semantic coherence in the path steps towards the destination, as compared to the shortest paths.

3 Methods

3.1 Problem Formulation

Associative series should be relatively short: it is desirable that the length of a series is less than 10 hashtags. Furthermore, it should not have wide semantic gaps: the logic of every transition in a series must be obvious to a human. It is desirable that the logic of the transition does not deviate from the direction from the source to the target: every next hashtag in the series must be semantically closer to the target hashtag and further from the source hashtag. This property could be called "semantical monotony". In Fig. 1 there are examples of associative series built for terms from Wikipedia in [12].

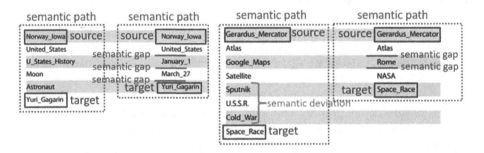

Fig. 1. Examples of associative series (from [12]) with and without semantic gaps and deviation from the direction from the source to the target term.

For the current research a corpus of 14.6 million of Instagram messages was gathered. For preventing duplicates, messages with the same hashtag sets and user ids were filtered out from the corpus. We aimed to create an algorithm for building associate series of hashtags based on using a co-occurrence network of hashtags. In the co-occurrence network, hashtags are nodes. Facts of co-occurrence of a pair of hashtags in a common message induce creation of links between nodes. We use the number of co-occurrences as the weight for a link. To exclude facts of accidental co-occurrence, we add links only between hashtags which occur in two or more messages. This way we built a network with 1.7 million nodes (hashtags) and 63.9 million of weighted links.

To create associative series of hashtags using the co-occurrence network, we need to build a path in the network which meets the associative series criteria mentioned earlier. We will denote such path as a semantic path. Let's make a formal definition for the problem of the semantic path building. Let $P = (n_0, n_1, \ldots, n_{L-1}, n_L)$ be a path of length L, from a source node a ($n_0 = a$) to a target node b ($n_L = b$), and there is a distance metric between the nodes: $d(n_i, n_{i+1}) = d_{i+1}$, which could be interpreted as a metric of semantic similarity between two consequent nodes. Then the optimal semantic path is a path which does not have wide semantic gaps, i.e. $D(P) \underset{P}{\to} \min$ where $D(P) = \underset{i \in [0, L-1]}{\max} d(n_i, n_{i+1}) = \underset{i \in [1, L]}{\max} d_i$ and also a path with shortest length, i.e. $L(P) \underset{P}{\to} \min$. Hence, we postulate the problem of the semantic path building as a multicriteria optimization problem:

$$\begin{cases} D(P) \underset{P}{\to} \min & (1) \\ L(P) \underset{P}{\to} \min & (2) \end{cases}$$

Here both criteria are important but at the same time they contradict each other. Without criterion (1) the path could be shorter, but it could include wide semantic gaps. Without criterion (2) the path could have more connected sequential nodes, but it could be too long and lose "semantical monotony" – an important implicit condition for a semantic path.

3.2   Formulation of the Combined Semantic Path Criterion

To effectively solve the semantic path building problem we need to combine the contradictory criteria (1) and (2). For this purpose, we built a combined semantic path criterion, including in it both criteria in a more adaptive form, which permits to combine them and adjust the weights for both criteria.

In the first step we consider the criterion (1) which prevents wide semantic gaps. Instead of the strict max function which is used in (1) for the combined criterion, we use a smooth maximum function S_α which tends to the strict maximum function when parameter α tends to infinity: $S_\alpha \underset{\alpha \to \infty}{\to} \max$. For our task we choose the LogSumExp (LSE) as one of smooth maximum functions: $S_\alpha = LSE_\alpha(d_1, \cdots, d_L) = \frac{1}{\alpha} \log \left(\sum_{i=1}^{L} \exp(\alpha d_i) \right)$.

By optimizing $LSE_\alpha \underset{P}{\to} \min$ instead of $D(P) \underset{P}{\to} \min$ we aim to diminish not only the semantic distance for the hop with the widest semantic gap (as it was for $D(P)$), but also other hops in the path which have wide gaps of comparative size. The monotonic log function and the multiplier $\frac{1}{\alpha}$ do not affect choosing an optimal path P. Hence for the optimization purpose, instead of LSE_α, we can use a simplified function $W_\alpha(P) = \sum_{i=1}^{L} \exp(\alpha d_i) \underset{P}{\to} \min$.

In the second step we consider the shortest path criterion (2). Optimization of function $W_\alpha(P) \underset{P}{\to} \min$ will lead to a path with a lot of semantically narrow hops, most of which are not approaching the target node. To account for the criterion (2), we add a penalty d for adding every next hop to a semantic path (for convenience we use a scaling factor γ for the exponent):

$$W_{\alpha,d,\gamma}(P) = \gamma \cdot \sum_{i=1}^{L} (\exp(\alpha d_i) + d) \underset{P}{\rightarrow} \min. \tag{3}$$

Function $W_{\alpha,d,\gamma}$ is the combined semantic path criterion with three parameters: α determines a penalty for wide semantic gap in a hop, d determines penalty for adding every next hop and γ is used to balance the two criteria.

3.3   Specification of a Semantic Distance Metrics

In the formal definition of the semantic path building problem we assume that there is a distance metric between the nodes: $d(n_i, n_{i+1}) = d_{i+1}$, which could be interpreted as a metric of semantic similarity between two consequent nodes. Instead of directly using weights of co-occurrences networks as the value of semantic similarity, we consider more advanced distance metrics based on embedding of nodes in a vector space, which has several hundred dimensions. This technique is influenced by well-known word embedding algorithms such as word2vec [5] and GloVe [13]. Vector representation of nodes (hashtags) could be interpreted as a description of their hidden semantics.

There are two different approaches to embedding network nodes in a vector space - based on network topology and based on additional features - which depend on the domain of a network. In [14] more than 10 algorithms of embedding based on network topology are described. There are several main techniques: based on matrix factorization; based on random walks; based on deep learning.

Despite embedding based on network topology been well developed, we chose to do embedding based on additional features of nodes. As the topology of the co-occurrence network of hashtags was not given initially and was derived from the corpus of Instagram messages, we chose to build hashtag embeddings directly from Instagram messages. Main advantage of this solution is the direct access to the source data instead of data retrieval from the secondary source. For example, when directly working with Instagram messages, we have a possibility to take into account the proximity of hashtags in messages in the embedding procedure. Finally, we chose to use word2vec embedding algorithm [5] with the following options. The number of dimensions of embedded vectors is 300, the size of the sliding window is 10 (the maximum distance between the current word and words in its context). We use skip-gram variant of the word2vec model, because it works better for rare words (hashtags).

3.4   Filtering the Co-occurrence Network of Hashtags

For large networks, like the co-occurrence network of hashtags, building the strict shortest path could be computationally too difficult. Vector embedding for nodes could be used as a fast working heuristic for finding the global direction to a target node. It is possible, because, unlike a metric based on network distances and weights of links, a metric based on vector embedding could be quickly calculated for any pair of nodes.

Nodes with high degree could create computational difficulties for any path building algorithm. Therefore, we decided to do a nodes degree distribution analysis for the hashtags co-occurrence network. In Fig. 2a the nodes degree distribution is shown in log-log coordinates. The tail of the distribution for the co-occurrence network is similar to scale-free networks. It means that there are hub nodes which represent popular hashtags with a very high degree ($\sim 10^5$ in our case) in the co-occurrence network. Most links from these hubs are semantically weak, but they sufficiently increase computational complexity of building a semantic path.

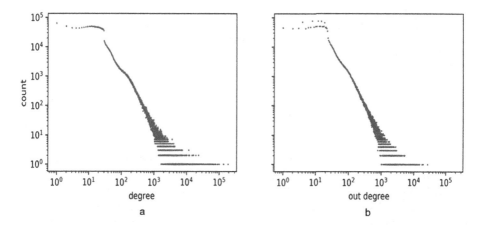

Fig. 2. Nodes degree distribution of the co-occurrence network. a. Initial network b. Network after weak links filtering

To solve this problem, we decided to filter weak links. For every node n_i we ranged all links to neighbor nodes $N(i)$ by their co-occurrence weights c_{ij}. Then we choose top k links so that the sum of their weights is 80% of the total sum of links weights from n_i: $\sum_{j \in Top(k, N(i))} c_{ij} = 0.8 \cdot \sum_{j \in N(i)} c_{ij}$. Links from n_i with lower c_{ij} which are not in the top k set are filtered out.

After filtering initial 63.9 million undirected links we got 61.9 million of directed links. Filtering procedure is not symmetrical: the link from n_i to n_j could be deleted, but the link from n_j to n_i could pass the filter. In Fig. 2b the out degree distribution of nodes for filtered network is shown in log-log coordinates. After filtering, the degree of top hub nodes dropped approximately by one order of magnitude to $\sim 10^4$ links per node.

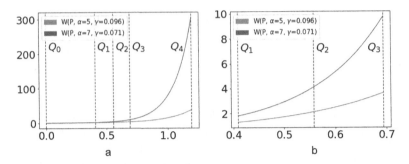

Fig. 3. Graph of the objective function $W_{\alpha,d,\gamma}$ with values of α, γ for which $W_{\alpha,d,\gamma}\left(P_{a,b}^{greedy\_H}\right) < W_{\alpha,d,\gamma}\left(P_{a,b}^{dijkstra\_hops}\right)$ for most of paths. Q_i here are right boundaries of i quartiles of the distribution of semantical distances for connected nodes a. Horizontal axis scales from the beginning of first quartile (Q_0) to the end of forth quartile (Q_4) b. Horizontal axis scales from the beginning of second quartile (Q_1) to the end of third quartile (Q_3)

3.5 Tuning Parameters of the Objective Function and Choosing Path Building Algorithms

Parameters α, d, γ of the objective function $W_{\alpha,d,\gamma}$ must be tuned before using the function for solving the optimization problem of the semantic path building. For setting the parameter d, which determines penalty for adding every next hop to a path, we use the scale from the distribution of semantical distances $d(n_i, n_j)$. We calculated distance metric values for the co-occurrence network between all connected nodes and found quartile boundaries for this distribution (see Fig. 3a). We set the parameter d equal to the median value of the distribution. Hence adding every next hop to a path would add to the value of the $W_{\alpha,d,\gamma}$ a penalty equal to median distance between connected nodes.

Algorithm 1: Greedy algorithm for building suboptimal shortest path

---

$i \longleftarrow 0$;
$path[i] \longleftarrow source\_node$;
do
$\quad$ $i \longleftarrow i+1$;
$\quad$ $path[i] \longleftarrow \arg\min_{n \in adjacent\_nodes\_of(path[i-1])} d(n, target\_node)$;
while $path[i] \neq target\_node$;

Fig. 4. The greedy algorithm for building suboptimal shortest path from [12]

For tuning α and γ we chose the following approach. We randomly chose several pairs of nodes (hashtags) from the co-occurrence network (a, b) and built two types of paths between them. Paths of the first type $P_{a,b}^{dijkstra\_hops}$ were built by Dijkstra's algorithm [15] which simply minimizes the path length (number of hops). Paths of the second type $P_{a,b}^{greedy\_H}$ were built by simple greedy algorithm from [12] (see Fig. 4 for details), which uses semantic distance to the target node in heuristics. Then we chose α and γ for which: $W_{\alpha,d,\gamma}\left(P_{a,b}^{greedy\_H}\right) < W_{\alpha,d,\gamma}\left(P_{a,b}^{dijkstra\_hops}\right)$ for most of the pairs. Using these values of parameters in object function $W_{\alpha,d,\gamma}$ in Dijkstra's algorithm will lead to preferring paths of the second type to paths of the first type.

Algorithm 2: Greedy top-k

$paths \longleftarrow \{\{nodes : [start\_node], length : 0\}\}$;
while $target\_node$ is not visited do
    $new\_paths \longleftarrow \varnothing$;
    foreach $path$ in $paths$ do
        foreach $node$ in $adjacent\_nodes\_of(path.nodes.last)$ do
            $p.nodes \longleftarrow path.nodes + [node]$;
            $p.length \longleftarrow path.length + w_{\alpha,d,\gamma}(d(path.nodes.last, node))$;
            $new\_paths \longleftarrow new\_paths + p$;
        end
    end
    $paths \longleftarrow [k$ paths from new_paths with the lowest values
    $\{path.length + w_{\alpha,d,\gamma}(h(path.nodes.last, target\_node))\}]$;
end
return $argmin_{P \in paths} D(P)$;

Fig. 5. The greedy top-k algorithm: modification of the greedy algorithm from [12] with a possibility for storing k best current paths.

Building a semantic path in a big network using Dijkstra's algorithm could be too computationally expensive because the computational complexity of Dijkstra's algorithm is $O(n \log n + m \log n)$, where n is the quantity of nodes and m is the quantity of links in a network. Instead of using algorithm which utilizes global information about a network, we could build an algorithm which utilizes only local information and heuristics about direction to target node based on the semantic distance metrics. We built such an algorithm by extending the greedy algorithm from [12] with a possibility for storing k best current paths. Details of our "top-k greedy algorithm" are described in Fig. 5. This algorithm lies between the simple greedy algorithm and the well-known A* algorithm [16], which uses heuristics for boosting Dijkstra's algorithm.

4   Results

Finally, we use the objective function $W_{\alpha,d,\gamma}$ with two chosen parameters sets with the same value of parameter d: $\alpha = 5.0$, $d = 0.55$, $\gamma = 0.096$ and $\alpha = 7.0$, $d = 0.55$, $\gamma = 0.071$. We use this function in two algorithms of building sematic path on the modified co-occurrence network which was obtained after filtering semantically weak links. The first algorithm is Dijkstra's algorithm in which function $W_{\alpha,d,\gamma}$ is used to weigh each considered link. The second algorithm is our top-k greedy algorithm, which was used for a function $W_{\alpha,d,\gamma}$ with parameters $d = 0.55$, $\gamma = 0.071$.

| | Column 1 | | Column 2 | | Column 3 | | Column 4 | | Column 5 | |
| | Greedy | | Dijkstra (min hops) | | Dijkstra (min W, α=5, γ=0.096) | | Dijkstra (min W, α=7, γ=0.071) | | Greedy top-k (min W+H, α=7, γ=0.071) | |
	Tags	$d(n_{i-1},n_i)$	Tags	$d(n_{i-1},n_i)$	Tags	$d(n_{i-1},n_i)$	Tags	$d(n_{i-1},n_i)$	Tags	$d(n_{i-1},n_i)$
1	#econometrics /#эконометрика		#econometrics /#эконометрика		#econometrics /#эконометрика		#econometrics /#эконометрика		#econometrics /#эконометрика	
2	#easymaths /#математикалегко	0,48	#Almaty /#алматы	0,55	#management /#менеджмент	0,46	#ordertest /#контрольныеназаказ	0,23	#economics /#экономика	0,40
3	#universeclue /#подсказкавселенной	0,39	#laziness /#лень	0,87	#timemanagement /#таймменеджмент	0,55	#tests /#контрольные	0,33	#entrepreneuradvices /#советыпредпринимателя	0,44
4	#numbertheory /#теориячисел	0,21			#laziness /#лень	0,57	#ordercoursework /#курсовыеработыназаказ	0,30	#timemanagement /#таймменеджмент	0,43
5	#needitforschool /#мнепошколенадо	0,07					#lazytodoit /#леньделать	0,17	#laziness /#лень	0,57
6	#wallpaperisnotanoption /#обоиневариант	0,10					#lazinessisthemother /#леньматушка	0,49		
7	#integralsv /#интегралсв	0,11					#laziness /#лень	0,36		
8	#f4f /#п4п	0,24								
9	#numbers /#числа	0,59								
10	#imtoolazy /#мнелень	0,59								
11	#newday /#новыйдень	0,62								
12	#affairs /#дела	0,60								
13	#laziness /#лень	0,53								

Fig. 6. Paths from source node #econometrics (in Russian: #эконометрика) to target node #laziness (in Russian: #лень) built by different algorithms

We get two semantically remote hashtags #econometrics (in Russian: #эконометрика) and #laziness (in Russian: #лень) and build semantic paths from #econometrics (source node) to #laziness (target node). Results of using different algorithms with different parameter sets are shown in Figs. 6 and 7. For comparison we add paths built by Dijkstra's algorithm, which simply minimizes the path length (number of hops) (see column 2) and by simple greedy algorithm (see column 1).

As one can see, semantic paths built by specialized algorithms have lengths (number of hops) between the length of the shortest path and the length of the path built by the simple greedy algorithm. It means that these sematic paths meet the criterion for associative series: they are relatively short. Also, the widest semantic gaps in paths built by specialized algorithms are sufficiently narrower than in the Dijkstra's shortest path and even narrower than in the path built by the greedy algorithm. It means that these sematic paths meet the other criterion for associative series: they do not have wide semantic gaps. In the Table 1 below there are several examples of semantic paths for different source and target nodes built by algorithms which are considered above.

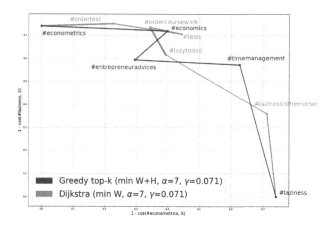

Fig. 7. Semantic paths from source node #econometrics (in Russian: #эконометрика) to target node #laziness (in Russian: #лень) built by the greedy top-k algorithm and Dijkstra's algorithm with the same parameters set for $W_{\alpha,d,\gamma}$. Coordinates of points for hashtags defined by semantic distances to source and target hashtags. A wider line means a more semantically close link.

Table 1. Examples of semantic paths built by different algorithms

Alg. types*	Semantic paths from source to target
	Value above the arrows is distance metric between the nodes
1	#**sky** $\xrightarrow{0.43}$ #*plane* $\xrightarrow{0.27}$ #*flight* $\xrightarrow{0.47}$ #*onvacation* $\xrightarrow{0.40}$ #*getinternationalpassport* $\xrightarrow{0.38}$ #**passport**
2	#**sky** $\xrightarrow{0.43}$ #*plane* $\xrightarrow{0.28}$ #*airport* $\xrightarrow{0.65}$ #**passport**
3	#**sky** $\xrightarrow{0.57}$ #*trip* $\xrightarrow{0.40}$ #*abroad* $\xrightarrow{0.50}$ #*internationalpassport* $\xrightarrow{0.35}$ #**passport**
1	#**prettyhair** $\xrightarrow{0.32}$ #*longhair* $\xrightarrow{0.38}$ #*shorthair* $\xrightarrow{0.46}$ #*facedrawing* $\xrightarrow{0.38}$ #*sketch* $\xrightarrow{0.34}$ #**art**
2	#**prettyhair** $\xrightarrow{0.32}$ #*longhair* $\xrightarrow{0.74}$ #**art**
3	#**prettyhair** $\xrightarrow{0.32}$ #*longhair* $\xrightarrow{0.50}$ #*curls* $\xrightarrow{0.49}$ #*curlygirl* $\xrightarrow{0.67}$ #*artphoto* $\xrightarrow{0.51}$ #**art**
1	#**library** $\xrightarrow{0.22}$ #*liketoread* $\xrightarrow{0.31}$ #*lovebooks* $\xrightarrow{0.29}$ #*forcedmarriage* $\xrightarrow{0.46}$ #**romance**
2	#**library** $\xrightarrow{0.44}$ #*novel* $\xrightarrow{0.59}$ #**romance**
3	#**library** $\xrightarrow{0.44}$ #*novel* $\xrightarrow{0.59}$ #**romance**
1	#**art** $\xrightarrow{0.30}$ #*imdrawing* $\xrightarrow{0.38}$ #*drawtogether* $\xrightarrow{0.47}$ #*arteducation* $\xrightarrow{0.51}$ #**forkids**
2	#**art** $\xrightarrow{0.77}$ #**forkids**
3	#**art** $\xrightarrow{0.40}$ #*drawing* $\xrightarrow{0.60}$ #*childdevelopment* $\xrightarrow{0.51}$ #**forkids**

*Types of algorithms:
1 - path built by Dijkstra algorithm (min W, $\alpha = 7$, $\gamma = 0.071$)
2 - path built by Dijkstra algorithm (min W, $\alpha = 5$, $\gamma = 0.096$)
3 - path built by Greedy top-k algorithm (min W + H, $\alpha = 7$, $\gamma = 0.071$)

Besides these formal criteria, we could make a subjective judgment about using these paths as associative series. For all the three semantic paths (see columns 3, 4, 5 in Fig. 6), logic of transitions in every step is obvious for a human. For all transitions, except steps 2, 3 in column 4, we have "semantical monotony": every next hashtag is semantically closer to the target hashtag and further from the source hashtag. And all semantic paths are short enough.

5   Discussion

We have proposed the definition for the optimization problem of the semantic path building on a co-occurrence network of hashtags, which leads to building associative series of hashtags. Moreover, our solution could be applied more widely to build associative series of any terms. For the two-criteria optimization problem of the semantic path building we propose the combined optimization function $W_{\alpha,d,\gamma}$. Three parameters of this function allow to balance the criterion of minimizing the widest semantic gap and the criterion of minimizing the length of the path. The procedure of tuning of these parameters was built by applying optimization function $W_{\alpha,d,\gamma}$ for two different path building procedures and comparing results for test examples in the real co-occurrence network of hashtags.

We gathered information about using hashtags in Instagram and built the co-occurrence network with 1.7 million nodes (hashtags) and 63.9 million of weighted links. After the analysis of nodes degree distribution in the co-occurrence network, we optimized the link structure of the network for it using in the semantic path building algorithm. We built the semantic distance metrics between nodes (hashtags) based on word2vec algorithm and used them in semantic paths building algorithms. We built examples of semantic paths using the Dijkstra's algorithm and the fast greedy algorithm and performed subjective judgment about using these paths as associative series.

References

1. Halliday, K., Hasan, R.: Cohesion in English. Longman, London (1976)
2. Morris, J., Hirst, G.: Lexical cohesion, the thesaurus, and the structure of text. Comput. Linguist. **17**(1), 21–48 (1991)
3. Fellbaum, C.: WordNet: An Electronic Lexical Database. Language, Speech, and Communication Series. MIT Press, Cambridge (1998)
4. Barzilay, R., Elhadad, M.: Using lexical chains for text summarization. In: Proceedings of the ACL Workshop on Intelligent Scalable Text Summarization, pp. 10–17, Madrid (1997)
5. Mikolov, T., Chen, K., Corrado, G.K., Dean, J.: Efficient estimation of word representations in vector space. CoRR, abs/1301.3781 (2013)
6. Sommer, C.: Shortest-path queries in static networks. ACM Comput. Surv. **46**(4), 1–31 (2014). https://doi.org/10.1145/2530531
7. He, L., et al.: Neurally-guided semantic navigation in knowledge graph. IEEE Trans. Big Data (2018). https://doi.org/10.1109/TBDATA.2018.2805363
8. West, R., Pineau, J., Precup, D.: Wikispeedia: an online game for inferring semantic distances between concepts. In: IJCAI, pp. 1598–1603. Morgan Kaufmann Publishers Inc. (2009)

9. Neelakantan, A., Roth, B., McCallum, A.: Compositional vector space models for knowledge base completion. In: Proceedings of the 53rd Annual Meeting of the Association for Computational Linguistics and the 7th International Joint Conference on Natural Language Processing, pp. 156–166, Beijing, China (2015). https://doi.org/10.3115/v1/P15-1016

10. Passant, A.: Measuring semantic distance on linking data and using it for resources recommendations. In: AAAI Spring Symposium: Linked Data Meets Artificial Intelligence, vol. 77 (2010)

11. Bringmann, K., Keusch, R., Lengler, J., Maus, Y., Molla, A.R.: Greedy routing and the algorithmic small-world phenomenon. In: Proceedings of the ACM Symposium on Principles of Distributed Computing, pp. 371–380, New York, USA (2017). https://doi.org/10.1145/3087801.3087829

12. Capitán, J.A., et al.: Local-based semantic navigation on a networked representation of information. PLoS ONE 7(8), 1–10 (2012). https://doi.org/10.1371/journal.pone.0043694

13. Pennington, J., Socher, R., Manning, C.: Glove: global vectors for word representation. In: EMNLP, pp. 1532–1543 (2014). https://doi.org/10.3115/v1/D14-1162

14. Goyal, P., Ferrara, E.: Graph embedding techniques, applications, and performance: a survey. Knowl. Based Syst. 89–94 (2018). https://doi.org/10.1016/j.knosys.2018.03.022

15. Dijkstra, E.: A note on two problems in connexion with graphs. Numer. Math. 1(1), 269–271 (1959). https://doi.org/10.1007/BF01386390

16. Hart, P., Nilsson, N.J., Raphael, B.: A formal basis for the heuristic determination of minimum cost paths. IEEE Trans. Syst. Sci. Cybern. SSC 4, 100–107 (1968). https://doi.org/10.1109/TSSC.1968.300136

Analysis of Dynamic Behavior Through Event Data

Method of Determining User Preferences for the Personalized Recommender Systems for Public Transport Passengers

Aleksandr A. Borodinov[1](✉) and Vladislav V. Myasnikov[1,2]

[1] Samara National Research University, 34, Moskovskoye Shosse, Samara 443086, Russia
aaborodinov@yandex.ru
[2] IPSI RAS – Branch of the FSRC "Crystallography and Photonics" RAS, Molodogvardeyskaya 151, 443001 Samara, Russia

Abstract. The question of creating a navigation recommender system based on user preferences arose with the development of recommender systems. The paper presents the theoretical and algorithmic aspects of making a personalized recommender system (mobile service) designed for public transport users. The main focus is to identify and formalize the concept of "user preferences", which is based on modern personalized recommender systems. Informal (verbal) and formal (mathematical) formulations of the corresponding problems of determining "user preferences" in a specific spatial-temporal context are presented: the preferred stops definition and the preferred "transport correspondences" definition. The first task can be represented as a classification problem. Thus, it represented using well-known pattern recognition and machine learning methods. In this paper, we use an approach based on the estimation algorithm proposed by Yu.I. Zhuravlev and nonparametric estimation of Parzen probability density. The second task is to find estimates for a series of conditional distributions. The experiments were conducted on data from the mobile application "Pribyvalka-63". The application is a part of the tosamara.ru service, currently used to inform Samara residents about the public transport movement.

Keywords: Recommender system · Transport correspondences · User preferences

1 Introduction

The amount of heterogeneous data characterizing the transport situation in the city has increased due to the widespread and active use of modern electronic communication systems, global navigation systems, and various active and passive sensors. Navigation and recommender systems use such information quite widely [1]. However, the expectations and demands of users and the amount of information that has to be taken into account when planning movements are growing, along with the development of services and their popularization. User demand is individualized from the classic tasks of searching for the "shortest path" [2] or getting a "forecast of arrival at a stop of public transport" [3, 4],

© Springer Nature Switzerland AG 2020
W. M. P. van der Aalst et al. (Eds.): AIST 2019, CCIS 1086, pp. 341–351, 2020.
https://doi.org/10.1007/978-3-030-39575-9_34

shifting expectations from services to Intelligent personal assistants. Although the final decision or choice in such systems remains with the person, the options of the solutions they offer significantly depend on the scenario conditions of the request as well as on previous actions and decisions of the user [5, 6]. The use for all of the indicated factors is possible in "self-tuning" systems for individual user preferences based on machine learning methods [7]. The imperfection of existing algorithms and the lack of significant experience in the use of machine learning methods in recommender systems hinders the development of such services.

Multimodal routing is the ability to use several modes of transport in one trip. The analysis of modern literature devoted to recommender systems of multimodal routing [5, 8, 9] allows us to identify some major problems:

- The cold start problem is a well-known and well-researched problem for recommender systems [8, 10]: it is essential to achieve a balance between the accuracy of the recommended routes from system initialization. Thus, the acceptable setting time for a personal preference profile should be small.
- The receiving information method from the user is not formalized [11, 12].
- Individual characteristics such as personal income, age, gender, numbers of family members, access to public transport influence the choice of the route for the same purpose of the trip [13].
- User preferences change over time. In addition, context influences user selection [14, 15].
- Typical existing solutions mainly use the Bayesian approach with a sequential parameter recalculation scheme [5, 16].
- It is possible to use transfer learning to improve recommendations [17].
- There is the problem of determining traffic flow on the vehicle route [18].

This article proposes one of the possible ways of describing and solving the problem of determining individual preferences of users of public route transport and creating a personalized recommender system. The system uses user interaction data from mobile service to solve the problem of creating a personalized recommender system. The second section of the work formalizes the basic concepts and introduces the basic notation for all objects of interaction. In the third section, we described the data received from the interaction of the public transport user interaction with the mobile application "Pribyvalka-63" and used in this research. Mobile application and service tosamara.ru are currently used to inform Samara residents about public transport movement and the estimated transport arrival time at the selected stop as shown in Fig. 1. Also in this section, the variants of non-formalized (verbally described) definitions of "user preferences", suitable for further consideration, are presented. The fourth section presents the mathematical formulations of problems, as well as methods and algorithms. We were forced to omit the description of the algorithm for determining the users-preferred "transport correspondences" in Sect. 4.2 due to the limitations of the article. However, the details of the algorithm will be presented at the conference. Finally, the fifth section presents the results of experimental studies on real data obtained using the mobile application "Pribyvalka-63".

Fig. 1. The interface of the mobile application "Pribyvalka-63".

2 Basic Definitions

Let S be the set of public transport stops. Let $\mathbf{x}_s \equiv (x_s, y_s, z_s)$ be the spatial (geographical) coordinates for each stop $s \in S$ and ID(s) is some unique stop identifier. We can assume without loss of generality that the set S is ordered (for example, by ID(s)): $S = \{s_1, s_2, \ldots, s_{|S|}\}$.

Let the value d be the calendar date, the value t is the time of day, and $w(d) \in W$ is the day of the week, taking values from the set:

$$W = W_0 \cup W_1,$$
$$W_0 \equiv \{MON, TUE, WEN, THU, FRI\}, \quad W_1 \equiv \{SAT, SUN\}. \tag{1}$$

Let V be the set of public transport vehicles, where the type characterizes each element $v \in V$ of the set:

$$type(v) \in \{BUS, TRAM, TROL, MARS\}, \tag{2}$$

and each vehicle has a unique identifier $ID(v)$. We defined the position of the vehicle v in the moment (d, t) by:

$$\mathbf{x}(v, d, t) \equiv (x(v, d, t), y(v, d, t), z(v, d, t)). \tag{3}$$

We denote the routes set of public transport objects as M. In addition, five arguments characterize each route $m \in M$:

$$m \equiv \big(ID(m), N(m), \mathbf{s}(m), N^*(m), \mathbf{x}(m)\big), \tag{4}$$

where $ID(m)$ is the route identifier, $N(m)$ is the number of stops in the route, and $\mathbf{s}(m)$ is the sequence of stops of the length $N(m)$ of the form:

$$\mathbf{s}(m) = \big(s_1^m, s_2^m, \ldots, s_{N(m)}^m\big), \tag{5}$$

where $s_n^m \in S$ $\big(m \in M, n \in \overline{1, N(m)}\big)$. Let $S(m) \equiv \{s_i^m\}_{i=\overline{1,N(m)}} \subseteq S$ be the stops set of the corresponding route, $ind(s, m)$ is the transport stop index s of the route m, such that $ind(s_n^m, m) = n$. We consider the index undefined when $s \notin S(m)$. In this case we define $ind(s, m)$ as $ind(s, m) = \Delta$. We denote the routes set passing through one or a couple stops as follows:

$$M(s) \equiv \{m \in M: s \in S(m)\}, M(s1, s2) \equiv \{m \in M: s1 \in S(m) \wedge s2 \in S(m)\}. \quad (6)$$

More detailed information about the route geometry is represented by a pair $N^*(m), \mathbf{x}(m)$, where the first value determines the number of nodes of the polyline describing the route, and the second is the vector defining the coordinates of these nodes:

$$\mathbf{x}(m) \equiv \left(\mathbf{x}_1^m, \mathbf{x}_2^m, \ldots, \mathbf{x}_{N^*(m)}^m\right). \quad (7)$$

For convenience, we will call the pair (m, k), $\left(k = \overline{1, K(d, m)}\right)$ as route implementations (RI) on the day d.

Additionally, we denote $t(d, m, k, s)$ as a vehicle arrival time assigned to RI (m, k) on day d, to stop $s \in S(m)$ (in case $s \notin S(m)$ we consider the time value as undefined).

We define the vehicle assigned to RI (m, k) on day d, as $v(d, m, k) \in V$ $\left(k = \overline{1, K(d, m)}\right)$.

In addition to the vehicles, pedestrians and passengers, considered in the paper as the users of transport services, are participants in the traffic. Let U be the set of users, and we will characterize each specific user $u \in U$ with a unique identifier $ID(u)$ (mobile device id or hash code) and spatial coordinates at a specific point in time d, t:

$$\mathbf{x}(u, d, t) \equiv (x(u, d, t), y(u, d, t), z(u, d, t)) \quad (8)$$

If there is no information about user coordinates, we consider that they have "undefined value" Δ (all three at the same time).

3 Problem Statement

We use data from the mobile application "Pribyvalka-63" in this paper for analysis. The data contains the following information:

1. public transport stop information (identifiers and coordinates);
2. public transport route information (identifiers and stop identifier list);
3. information about the vehicle (identifiers), location coordinates (the vehicle transmits its coordinates two times per minute), route identifier;
4. coordinates of users and request parameters are recorded with requests (request results are not saved since they can be restored from vehicle traffic data) in the form: $ID(s), d, t, ID(u), \mathbf{x}(u, d, t)$;
5. user response to the request is not saved.

The following two options "user preferences" seem appropriate based on the presented data:

– user-preferred public transport stops at the specific space-time coordinates of the user (Fig. 2);

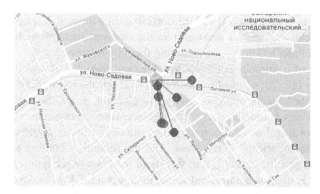

Fig. 2. Blue circle—stop location, purple circles—user location points at request time. (Color figure online)

– user-preferred "transport correspondences" in the space-time context. "Transport correspondence" refers to the actual movement from one stop to another, the route chosen and the route vehicle type. Information about the "Start" and the "End" public transport stops is additional information for a specific user (Fig. 3).

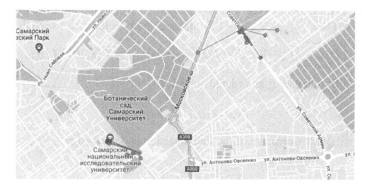

Fig. 3. "Start" and "End" stops of a specific user.

Finally, the problem statement is to get "user preferences" user-preferred public transport stops and user-preferred "transport correspondences" for some specific user $u \in U$ using information 1–5 obtained from the recommender system.

4 Methods

4.1 User-Preferred Public Transport Stops

The task of determining "user-preferred stops" for a user with specific space-time coordinates can be formalized as follows.

Let us consider the event set for a specific user. Each event in this set is a vector describing the space-time context and the corresponding answer. It is necessary to indicate the most "relevant" answer (answers) for the received vector (which may be absent in the list of examples), if necessary, ordering them by relevance.

For the case of a single issued response, the described task is a well-known classification problem [19, 20] or machine learning [7, 21] problem. In this case, the system should indicate the object/event class as a feature vector, according to the object/event description. Any recognition algorithm can be represented as two consecutive operators. The first operator translates the description of the object into a numerical value characterizing the "membership" to the class. And the second, according to the indicated value, refers to a specific class [22]. In statistical methods of recognition, the posterior probability [19] is used as a numerical value, in algebraic [23] as estimates, in a neural network as the output of the last layer of neurons, etc. We denote the specified numeric value by $\Gamma(features; class)$.

Thus, the formal description of the problem statement for a specific user $u \in U$ can be represented as follows (where $\gamma \equiv ID(s)$).

Given:

1. The event set in the form $\{\mathbf{x}_i, d_i, t_i; \gamma_i\}_{i \in \Im}$. (feature vector; answer)
2. The feature vector of the new event $\mathbf{x}, d, t$.

Wanted: The permutation of objects $\sigma : \mathbb{N}_{|S|} \to \mathbb{N}_{|S|}$ from an ordered set (public transport stops) $S = \{s_1, s_2, \ldots, s_{|S|}\}$ for the specified vector $\mathbf{x}, d, t$ such that

$$\Gamma\big(\mathbf{x}, d, t; ID\big(s_{\sigma(1)}\big)\big) \geq \Gamma\big(\mathbf{x}, d, t; ID\big(s_{\sigma(2)}\big)\big) \geq \ldots \geq \Gamma\big(\mathbf{x}, d, t; ID\big(s_{\sigma(|S|)}\big)\big). \quad (9)$$

The result for the user is an ordered stops list:

$$s_{\sigma(1)}, s_{\sigma(2)}, \ldots, s_{\sigma(|S|)}. \quad (10)$$

The formal quality measure of the final decision is described as:

$$\Gamma_\Sigma = \sum_{i=1}^{|S|} \frac{1}{i} \Gamma\big(\mathbf{x}, d, t; ID\big(s_{\sigma(i)}\big)\big). \quad (11)$$

Solution:

Theory of pattern recognition and machine learning offers a variety of methods for solving the problem. In this paper, we use an approach based on the estimation algorithm proposed by Zhuravlev [23] and nonparametric estimation of Parzen probability density

[19]. We define the value $\Gamma(features; class)$, which characterizes the belonging of the feature vector to the class, as

$$\Gamma(\mathbf{x}, d, t; \gamma) = \sum_{i \in \Im} \mu(\mathbf{x}, d, t; \mathbf{x}_i, d_i, t_i) I(\gamma_i = \gamma), \tag{12}$$

where

$$\mu(\mathbf{x}, d, t; \mathbf{x}_i, d_i, t_i) =$$

$$I\left(\begin{matrix} (w(d) \in W_0 \wedge w(d_i) \in W_0) \vee \\ (w(d) \in W_1 \wedge w(d_i) \in W_1) \end{matrix} \right) \cdot \exp(-\alpha|t - t_i|) \cdot \exp(-\beta\|\mathbf{x} - \mathbf{x}_i\|). \tag{13}$$

Event indicator:

$$I(a) = \begin{cases} 1, & a = true; \\ 0, & a = false. \end{cases} \tag{14}$$

The values $\alpha, \beta \in \mathbb{R}_+$ are coefficients, $|t - t_i|$ is the numerical value (for example, the number of seconds), which characterizes the difference between t, t_i. We denote the algorithm for solving the problem of determining "user-preferred stops" as follows:

Step 1. Calculate the values (12) for all stops from the set S:

$$\Gamma(\mathbf{x}, d, t; ID(s_i)), \quad i = \overline{1, |S|}. \tag{15}$$

Step 2. Sort by decreasing the set of values obtained in (15). Get a permutation $\sigma : \mathbb{N}_{|S|} \to \mathbb{N}_{|S|}$ (9).

The resulting permutation is the solution to the problem. An ordered list of stops provided to the user (10).

"Cold Start" problem solution:

We supplement the initially empty set of events to solve the "Cold Start" problem as follows:

$$\{(\mathbf{x}_i, d0, t0; ID(s_i))\} \cup \{(\mathbf{x}_i, d1, t0; ID(s_i))\}, \quad i = \overline{1, |S|}, \tag{16}$$

where $t0 = $ "0h00min", $d0$ and $d1$ are the dates, respectively, of the weekend and working days preceding the system launch date, and $\mathbf{x}_i$ $(i = \overline{1, |S|})$ are transport stop coordinates s_i $(i = \overline{1, |S|})$. Analysis of expressions (12)–(13) shows that with such "start" data the contribution of the time component $\exp(-\alpha|t - t_i|)$ in expression (12) will be the same, and the differences in values $\Gamma(\ldots)$ will be completely determined by differences in Euclidean distances from the point $\mathbf{x}$ to the transport stops coordinates $\mathbf{x}_i$ $(i = \overline{1, |S|})$. Thus, the value $\Gamma(\mathbf{x}, \ldots; ID(s))$ will be more significant when s is closer to $\mathbf{x}$.

4.2 User-Preferred "Transport Correspondences"

The task of determining the user-preferred "transport correspondences" can be presented as the task of estimating the probability characteristics (relative frequency) of the correspondences used by a specific user $u \in U$. That is the movements from the transport stop $s1$ to the transport stop $s2$ in the space-time context. The following values are important for specifying user correspondences:

- $p_u(t|s1, s2, m, W_a)$ $\left(m \in M(s1, s2), \ a = \overline{0,1}\right)$ is a correspondence time distribution density $s1 \rightarrow s2$ with the route choice m on the weekday W_a; the density function corresponds to the "boarding time" on the route vehicle, and is indicated to stop $s1$;
- $p_u(t|s1, s2, W_a)$ is a correspondence time distribution density $s1 \rightarrow s2$ on the weekday W_a;
- $P_u(s1, s2|W_a)$ is a correspondence probability $s1 \rightarrow s2$ on the weekday W_a;
- $P_u(m|s1, s2, W_a)$ is a probability of choosing the route m to implement the correspondence $s1 \rightarrow s2$ on the weekday W_a;
- $P_u(m|W_a)$ is a probability of choosing the route m to implement the correspondence on the weekday W_a;
- $P_u^*(s|W_a)$ is a probability that the stop s is the "end/start".

Additional information about user behavior can be obtained from additional data:

- $p_u(\rho|W_a)$ is a distance distribution that the user is able to overcome without using route vehicles;
- $p_u(\tau|\rho, W_a)$ is a time distribution that the user spends in overcoming the corresponding distance.

All specified values can be easily calculated by the user correspondences set:

$$\left\{ s_i^{start}, s_i^{end}, \ m_i^j, k_i^j, \ t\left(d, m_i^j, k_i^j, s_i^{start}\right), \tau_i^*, \sigma_i^* \right\}_{i \in I_d} \tag{17}$$

Here s_i^{start}, s_i^{end} are correspondence starting and ending stops, m_i^j, k_i^j are information about the route used by the user, $t\left(d, m_i^j, k_i^j, s_i^{start}\right)$ is the vehicle arrival time RI at the transport stop s_i^{start} and τ_i^*, σ_i^* are the mean and the standard deviation of potential boarding on the vehicle.

However, the problem is that the data set presented in (17) cannot be obtained from the mobile application "Pribyvalka-63". Therefore, the problem is to calculate the transport correspondence data (17) and calculate the statistical values from this section based on data 1–6 in Sect. 3 collected by the recommender system. The calculation algorithm (17) according to data 1–6 from Sect. 3 is not presented in the article due to limitations on the size of the article.

5   Experiments

This article presents experiments only for user-preferred public transport stops due to limitations on the size of the article.

The presented method software implementation written in Python. The results visualized based on Google Maps. We used "Pribyvalka-63" mobile application and tosamara.ru service data in the experiments.

The database obtained during the experiments contains information about requests of 57190 users. Each user is represented by a unique impersonal identifier $ID(u)$, which is defined by the device ID hash code. The database contains 4103161 user requests for an arrival forecast at a public transport stop. From 1478 stops of the tosamara.ru service, users made requests to 1417 stops.

For the experiments, we selected common user requests that represent the average user of the service. Maps with different parameters $\alpha, \beta \in \mathbb{R}_+$ used by Eq. (12) and request time were built to visualize the results of the proposed approach. The color of the area on the map corresponds to the first stop from the ordered list (10). An example of determining the preferred stop for the user, as shown in Fig. 4.

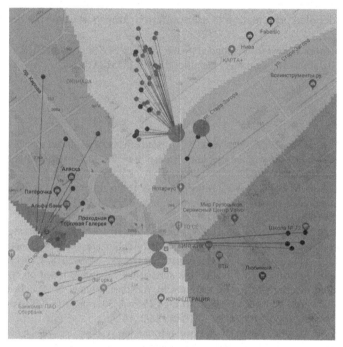

Fig. 4. Preferred stops map depending on user location.

Leave-One-Out cross-validation was applied to obtain statistical indicators characterizing the quality of the proposed algorithm. The partitions number $C_{|\Im|}^1$ of the user requests set in this case is equal to $|\Im|$, and the classification accuracy is calculated as follows:

$$Accuracy = \frac{1}{|\Im|} \sum_{i \in \Im} I\left(z_i \equiv ID(s_{\sigma_i(1)})\right) \cdot 100\% \tag{18}$$

where the indicator $I(a)$ corresponds to (14).

Also, another method was implemented for comparison with the proposed algorithm, in which the user was offered the nearest stop, without taking into account previous requests. The proposed algorithm accuracy was 93%, while the nearest stop prediction resulted in 65% of this quality measure.

6 Conclusion

In this paper, we presented the informal and mathematical problem formulations of defining user preferences. Users take public transport route in a personalized recommendation system task. We showed the results of an experimental study. We developed the algorithm using pattern recognition and machine learning methods. The determining the user preferred "transport correspondences" task was formalized, and the solution was specified. One of the directions of further work is the analysis and search for the best algorithm for predicting a user-preferred public transport stops. Another area of research is a detailed presentation of user-preferred "transport correspondences".

Acknowledgments. The work was funded by the Ministry of Science and Higher Education of the Russian Federation (unique project identifier RFMEFI57518X0177).

References

1. Chorus, C.G., Molin, E.J.E., Van Wee, B.: Use and effects of advanced traveller information services (ATIS): a review of the literature. Transp. Rev. **26**, 127–149 (2006). https://doi.org/10.1080/01441640500333677
2. Agafonov, A.A., Myasnikov, V.V.: Numerical route reservation method in the geoinformatic task of autonomous vehicle routing. Comput. Opt. **42**, 912–920 (2018). https://doi.org/10.18287/2412-6179-2018-42-5-912-920
3. Agafonov, A.A., Yumaganov, A.S., Myasnikov, V.V.: Big data analysis in a geoinformatic problem of short-term traffic flow forecasting based on a K nearest neighbors method. Comput. Opt. **42**, 1101–1111 (2018). https://doi.org/10.18287/2412-6179-2018-42-6-1101-1111
4. Agafonov, A., Myasnikov, V.: Traffic flow forecasting algorithm based on combination of adaptive elementary predictors. In: Khachay, M.Y., Konstantinova, N., Panchenko, A., Ignatov, D.I., Labunets, V.G. (eds.) AIST 2015. CCIS, vol. 542, pp. 163–174. Springer, Cham (2015). https://doi.org/10.1007/978-3-319-26123-2_16
5. Arentze, T.A.: Adaptive personalized travel information systems: a bayesian method to learn users' personal preferences in multimodal transport networks. IEEE Trans. Intell. Transp. Syst. **14**, 1957–1966 (2013). https://doi.org/10.1109/TITS.2013.2270358
6. Nuzzolo, A., Crisalli, U., Comi, A., Rosati, L.: Individual behavioural models for personal transit pre-trip planners. Presented at the Transportation Research Procedia (2015). https://doi.org/10.1016/j.trpro.2015.01.015
7. Portugal, I., Alencar, P., Cowan, D.: The use of machine learning algorithms in recommender systems: a systematic review. Expert Syst. Appl. **97**, 205–227 (2018). https://doi.org/10.1016/j.eswa.2017.12.020
8. Campigotto, P., Rudloff, C., Leodolter, M., Bauer, D.: Personalized and situation-aware multimodal route recommendations: the FAVOUR algorithm. IEEE Trans. Intell. Transp. Syst. **18**, 92–102 (2017). https://doi.org/10.1109/TITS.2016.2565643
9. Eiter, T., Krennwallner, T., Prandtstetter, M., Rudloff, C., Schneider, P., Straub, M.: Semantically enriched multi-modal routing. Int. J. Intell. Transp. Syst. Res. **14**, 20–35 (2016). https://doi.org/10.1007/s13177-014-0098-8
10. Mikic Fonte, F.A., López, M.R., Burguillo, J.C., Peleteiro, A., Barragáns Martínez, A.B.: A tagging recommender service for mobile terminals. In: Cantoni, L., Xiang, Z. (eds.) Information and Communication Technologies in Tourism 2013, pp. 424–435. Springer, Heidelberg (2013). https://doi.org/10.1007/978-3-642-36309-2_36

11. March, J.G.: Bounded rationality, ambiguity, and the engineering of choice. Bell J. Econ. **9**, 587–608 (1978). https://doi.org/10.2307/3003600

12. Campigotto, P., Passerini, A.: Adapting to a realistic decision maker: experiments towards a reactive multi-objective optimizer. In: Blum, C., Battiti, R. (eds.) LION 2010. LNCS, vol. 6073, pp. 338–341. Springer, Heidelberg (2010). https://doi.org/10.1007/978-3-642-13800-3_35

13. Zhang, J., Arentze, T.: Design and implementation of a daily activity scheduler in the context of a personal travel information system. In: Krisp, J. (ed.) Progress in Location-Based Services. LNGC, pp. 407–433. Springer, Heidelberg (2013). https://doi.org/10.1007/978-3-642-34203-5_23

14. Braunhofer, M., Ricci, F.: Selective contextual information acquisition in travel recommender systems. Inf. Technol. Tourism **17**, 5–29 (2017). https://doi.org/10.1007/s40558-017-0075-6

15. Braunhofer, M., Elahi, M., Ricci, F.: User personality and the new user problem in a context-aware point of interest recommender system. In: Tussyadiah, I., Inversini, A. (eds.) Information and Communication Technologies in Tourism 2015, pp. 537–549. Springer, Cham (2015). https://doi.org/10.1007/978-3-319-14343-9_39

16. Guo, S., Sanner, S.: Real-time multiattribute Bayesian preference elicitation with pairwise comparison queries. J. Mach. Learn. Res. **9**, 289–296 (2010)

17. Pan, S.J., Yang, Q.: A survey on transfer learning. IEEE Trans. Knowl. Data Eng. **22**, 1345–1359 (2010). https://doi.org/10.1109/TKDE.2009.191

18. Myasnikov, V.V.: Method for detection of vehicles in digital aerial and space remote sensed images. Comput. Opt. **36**, 429–438 (2012)

19. Fukunaga, K.: Introduction to Statistical Pattern Recognition. Academic Press, San Diego (1990)

20. Vorontsov, K.V.: Machine Learning: Lecture Notes. http://www.machinelearning.ru/wiki/. Date of the application 10 November 2018

21. Bishop, C.M.: Pattern Recognition and Machine Learning. Springer, New York (2006)

22. Zhuravlev, Yu.I., Nikiforov, V.V.: Recognition algorithms based on estimation calculation cybernetics. Cybernetics **7**, 387–400 (1971)

23. Zhuravlev, Yu.I., Gurevich, I.B.: Pattern recognition and image recognition, classification, forecast. Mathematical Methods and Their Application (1989)

Author Index

Printed in the United States
By Bookmasters